Data Made Flesh

Data Made Flesh

Embodying Information

Edited by **Robert Mitchell and Phillip Thurtle**

ROUTLEDGE
NEW YORK AND LONDON

Published in 2004 by
Routledge
29 West 35th Street
New York, NY 10001

Published in Great Britain by
Routledge
11 New Fetter Lane
London EC4P 4EE

Routledge is an imprint of the Taylor & Francis Group.

Grateful acknowledgment is made to the following for permission to reprint previously published material: Earlier versions of "Breeding and Training Bastards: Distinction, Information, and Inheritance in Gilded Age Trotting Horse Breeding" by Phillip Thurtle, "Flesh and Metal: Reconfiguring the Mindbody in Virtual Environments" by N. Katherine Hayles, and "The Virtual Surgeon: New Practices for an Age of Medialization" by Timothy Lenoir were originally published in *Semiotic Flesh: Information and the Human Body*, edited by Phillip Thurtle and Robert Mitchell (University of Washington Press, 2002). Copyright ©2002 by Walter Chapin Simpson Center for the Humanities. Reprinted by permission.

10 9 8 7 6 5 4 3 2 1

Library of Congress Cataloging-in-Publication Data

Data made flesh : embodying information / edited by Robert Mitchell and
Phillip Thurtle.
p. cm.
Includes bibliographical references and index.
ISBN 0-415-96904-2 (hardback : alk. paper)—ISBN 0-415-96905-0 (pbk. :)
1. Body, Human. 2. Computer science—Social aspects. 3. Information technology—Social aspects. I. Mitchell, Robert, 1969– II. Thurtle, Phillip.

HM636.D38 2003
306.4—dc21

2003008836

CONTENTS

Acknowledgments

We would like to acknowledge the following groups and individuals for the support they provided at early stages of our "Information and the Human Body" project (2000–present), from which this volume took its inspiration. We are especially grateful to the Walter B. Chapin Simpson Center for the Humanities at the University of Washington (and especially its 2000–1 interim Director, Diana Behler, and its current Director, Kathleen Woodward), the Elizabeth S. Soule Professorship in Nursing at the University of Washington, and the Department of Comparative Literature at the University of Washington for the support that these groups provided for our 2000–1 lecture series ("Information and the Human Body"), which featured superlative talks by Richard Doyle, N. Katherine Hayles, and Timothy Lenoir. We would also like to thank the Comparative History of Ideas Program at the University of Washington (and especially Tyler Fox) for sponsoring our undergraduate class entitled "Semiotic Flesh: A Genealogy of Information." We offer thanks as well to the students who took this course, as well as those who participated in the related "In Vivo" course: our Introduction owes much to you. We also extend our heartfelt thanks to the two anonymous reviewers of this volume for Routledge: we believe (and hope that they agree) that their divergent, but always insightful, comments have improved the volume significantly. We also wish to thank David Silver and Matt Byrnie, without whom this volume would not be possible. Finally, we would like to thank all of our contributors, both their intellectual contributions to this volume, but also for their unfailing patience in dealing with the many delays along the way.

Introduction
Data Made Flesh: The Material Poiesis of Informatics

PHILLIP THURTLE
ROBERT MITCHELL

Since at least the late 1940s, an apparently solid distinction between "information" and "the flesh" has enabled a wide variety of research programs, practical applications, and virtual landscapes. For example, the research program known as the Human Genome Project promises to extract an informational essence from human bodies for use by the practical applications of genetic engineering. Moreover, this distinction is essential to many elements of popular virtual landscapes from the body-teleporter technology represented in *Star Trek* to the more recent distinction between bodies and information in *The Matrix*. All of these projects and imaginings officially ground themselves on the distinction between immaterial, transcendent information and fleshy, unique bodies. Information, so this story goes, exists between elements, whereas bodies are the elements themselves. Information underwrites signs and syntax, whereas the flesh is the medium of cells and organs. Information, in short, operates through the metaphysics of absence, whereas bodies depend on the metaphysics of presence.[1]

For much of the twentieth century, this seemingly solid conceptual wall between bodies and information has allowed us to separate these two aspects of modern life, but at the same time—and especially in the last few decades—it has become increasingly clear that this conceptual wall bleeds; bodies and information continually graft themselves onto one another in a number of different cultural domains. Even as geneticists describe genetic information as an "essence" that only contingently receives "expression" in bodies, the actual goal of genetic engineering is an ever-increasing array of precisely these corporeal instantiations.[2] This confusion between "essence" and "expression" has been equally troubling to bioethicists, who have noted that the patent rights that have facilitated the massive expansion of biocommerce at the end of the twentieth and the start of the twenty-first century depend on this same distinction between "idea" and "expression," but with the valence reversed: manipulated genetic material is patentable because it is the concrete "expression" of an (otherwise unpatentable) idea, or "essence." Even in the video entertainment field, power and mastery are still imagined through the lens of the body. For instance, popular video games allow us to augment selectively body parts, skills, and physical abilities, despite programmers' commitment to the idea of an extracorporeal existence in a "matrix" of information. In the arts, this confusion has been the source of a productive poiesis, as a number of the most provocative contemporary artists have begun linking computers and biological materials to create new forms of living artistic expression. Rather than remaining on their respective sides of the conceptual wall, "information" and "bodies" seem to function almost as ripples that pass from pools of liquid across one another; it is the difference between the pools that allows the ripples to propagate themselves, and intersect, in unique and interesting ways.

The fourteen chapters in this volume, both individually and as a group, engage this rippling, seeking to illuminate what we might call the "material poiesis of informatics." Focusing on topics as diverse as horse breeding, family discipline, immortal cell line ownership, drug use, and bioart (to name just a few of the topics), each of the chapters in this collection focuses on those moments when information and flesh coconstitute one another. Understanding these moments allows us to illuminate the specific ways that we perform the transformation from the virtual world of information to the actual world of flesh and bone, as well as the implications of this moment for human biological existence (the species body), communal interactions (the social body), and the accessing and transformation of shared memory (the body of knowledge).[3]

I. Information and the New Bodies

As a group, the chapters included in *Data Made Flesh: Embodying Information* offer critical theoretical and historical studies intended to demarcate an emerging field that we might call "materialistic information studies." As such, this volume builds on the seminal work of historians such as Mark Poster, who in the late 1970s introduced the notion of the "mode of information" as a tool for historical study.[4] Sociologist Manuel Castells also contributed immeasurably to this approach through his earlier description of the *Informational City* and his later weighty explication of the dynamics of *The Networked Society.*[5] The important work of understanding the relationship of informational concepts to embodiment as developed in the field of molecular biology has progressed through the work of scholars Lily E. Kay and Richard Doyle.[6] This volume is also heavily indebted to literary critics and cultural historians such as N. Katherine Hayles, who have sought to untangle the very complicated discourses that link biology, cybernetics, chaos theory, and the fiction of authors such as William S. Burroughs and Thomas Pynchon.[7] Finally, the project of *Data Made Flesh: Embodying Emergence* would not have been possible without accounts in which information is reinterpreted as an organizing heuristic within society, such as Michael E. Hobart and Zachary S. Schiffman's *Information Ages: Literacy, Numeracy, and the Computer Revolution* and Pierre Levy's *Collective Intelligence* and *Becoming Virtual.*[8] While these latter accounts occasionally treat the concept of information as an ahistoric entity untouched by the bodies that generate it, they nevertheless have helped establish the importance of the term "information" for scholars in the arts and humanities and have offered examples of the fruitfulness of this term for understanding the dynamics of Western society and its development.[9] *Data Made Flesh: Embodying Information* seeks both to consolidate and build on the work of these researchers by locating "information" as the central term for a new field of study while at the same time emphasizing its material, embodied dimension.

As this extensive list of references suggests, there is no one theory of either "the body" or "information" that underwrites all the chapters in this collection. At the same time, however, certain commonalities in approach and conception reoccur throughout the volume. In the remaining space of this introduction, we seek to outline briefly the dominant approaches to bodies and information that emerge in this volume, as well as the theoretical framework that allows us to understand embodied information.

1. The Circuits of the Body

An emblematic—and much cited—attempt to reimagine contemporary and future relationships between bodies and information is the suggestion by Case, the protagonist of William Gibson's classic cyberpunk novel *Neuromancer,* that advanced information technologies enable the possibility of "data made flesh."[10] This provocative phrase condenses a number of hopes and fears associated with the relationship between information and living bodies. One can read this phrase as indicative of a troubling trend in the sciences, in which bodies are losing out to abstract notions of information.[11] This interpretation highlights fears that we spend more time engaged with experiences generated by informational technologies than in interpersonal interaction. Another possible interpretation of this phrase is that we are entering a period in which we have the capabilities to use informational technologies to generate new forms of fleshy experience.[12] This interpretation points to the powerful use of informational technologies in recreating current experiences of our embodied self as well as those of others. Recent developments in the cloning of mammals are only the most literal example of these new experiences. Gibson's phrase—and the wide variety of interpretations it has generated—thus highlights the importance of establishing a viable and sophisticated conception of the body itself.

For the purposes of this collection, we propose that the "the body" be understood as anything that cannot be divided without changing the fundamental pattern of its dynamics. Thus bodies are made up of organs, tissues, cells, and molecules, but a description of a body cannot be reduced to a description of the parts or their functions. This broad definition brings with it some surprising conclusions. "Machinic systems," as popularized by Gilles Deleuze and Felix Guattari, may not be "sets" but they are bodies. An autopoetic system, in the terms of Varela and Maturana, is most definitely a body. A "self-referential system," in the terms outlined by Niklas Luhmann, is a body. A "network," as described by Manuel Castells, is a body.[13] An organism bounded by flesh is a body. The time flesh continuum of my extended kinship network (so often confused by modernists with race) is a body. Finally, and most intriguingly, the database of sequenced genetic material is also a body. These systems operate around a kernel of subjectivity (in the sense outlined by Slavoj Žižek) that refuses to be explained through the determinations (either natural or social) of its parts.[14] Moreover, this understanding of the body allows us to distinguish between human and nonhuman bodies without falling into the error, diagnosed by Hans-Ulrich Gumbrecht, of employing a "concept of the human [that] exclud[es] . . . any reference to the human body."[15]

At first glance, this definition of body may seem to be overly general: for what, under this definition, *wouldn't* be a body? A quick review of the conceptual strengths and limitations of how previous authors have theorized the body, however, reveals the power of our conception as a first step in a sustained and historically informed reflection on embodiment. In order to demonstrate this, we situate our conception against four other understandings of the term: the "naturalist" body; the cultural/social determinist body; the "animal" body; and the phenomenological body.

Many authors simply take bodies for granted. There are two forms of this mistake. The first, exemplified in Armando R. Favazza's otherwise elegant book on self-mutilation, is the naturalists' error of assuming the integrity of the body outside of social and cultural frameworks.[16] The naturalist body is "just there" and, one presumes, has always been there as determined through natural process. What this overlooks, of course, are the multiple ways that bodies have been transformed through cultural practice.

The second form of taking the body for granted is the error of the social or cultural determinist. Although prevalent in contemporary scholarship, one can best locate this form of thinking in the post-Foucauldian histories of bodily practice and dispositions. Here, the body is conceived of as a blank slate where cultural practice is the sole determinant of bodily disposition. This determination can affect either the cultural meaning of embodiment (most often through a Marxist historical, semiotic, or psychoanalytic analysis)[17] or the flesh itself (as in the later work of Foucault).[18] The limitation of these analytical frameworks is that they ignore the body as a unique and interesting system in its own right. This approach fails to take into account dynamics of embodiment that cannot be reduced to cultural elements. In these interpretations, our bodies are too easily subsumed as one more cultural element and thus too easily understood as simply "reflecting" the dynamics of signs, symbols, and beliefs.

Yet even those who wish to emphasize the importance of embodiment often choose analytical frameworks that unnecessarily limit them. So, for example, in a provocative collection of essays on human embodiment entitled *ReMembering the Body,* the body is "referred to, remembered, and upheld as a residue of animalistic quality in mankind in the face of the rise of transhuman visions of the world."[19] There are a number of problems with this formulation. First, it assumes that what comprises the dynamics of the human body is its "animality." This is in part correct but begs the larger issue of what distinguishes human bodies from those of other animals. It also homogenizes the diverse category of animals (not all animals are alike), and leaves unexamined the question of what animal body one takes as a model. Moreover, why animals, rather than plants or minerals? If taken too dogmatically, this attempt to acknowledge and remember embodiment can lead to a much more simplistic framework than the "naturalistic" assumption as outlined above.[20]

The final approach to embodiment that we critique is what we will call the "phenomenological" body: that is, the body as perceived or experienced.[21] Although the benefits of phenomenological analysis are obvious (e.g., it supplies an elegant method for talking about sensate experience and understands the importance of sensing in the construction of knowledge), it too often reifies the body as the "zero point" of sense experience and does little to help us understand the integrity of mediated experience in linking and transforming bodily interaction. The cultural dimensions of mediated experience end up as a problem that begs explanation, rather than being understood as a reverberant and active part of embodied experience. For instance, many phenomenologists concentrate on describing the sensory experience of engaging with a media artifact, such as a chatroom, without adequately interrogating how this experience is constituted through the ontological extensions of technologies or the role of affect and intersubjective experience.[22]

Our view of the body builds on the strengths of the various approaches just outlined. Thus, the bodies described in the chapters in this collection acknowledge the biological aspects of the naturalists' view, the cultural reformatting of post-Foucauldian and Marxist readings, the relationship between human and nonhuman bodies emphasized by the notion of "embodiment," and sensate aspects of the phenomenological body. Yet, in addition, the approach that we adopt is intended to allow for the exploration of the materiality of embodiment in manifold forms of complexity—and in such a way that we can allow for, and emphasize, emergent organizing principles. We can thus investigate how political economic, technological, or social structures extend perceptive practices in space and time and become codified into ways of making sense of the world from a historically contingent and multifaceted perspective.[23] One way of thinking about this approach is to imagine an open ended and multidimensional "circuit" diagram of embodied experience. We find this conception useful as it delineates a larger network of potential communicative experiences on multiple conceptual levels: it allows for conception of difference beyond simple binary distinctions (circuits, for instance, may have multiple components) and it suggests the integrity of the individual component (as a body itself) while recognizing its context as a linked entity (in a body immanent on a different existential scale). Thus it recognizes that each of the approaches outlined above have developed a set of skills for understanding the specific dynamics of a component of embodied experience, and it offers a means to utilize the conceptual richness of this literature to make context specific judgments. It also explains why a collection of essays (as opposed to a monograph) is the best way for presenting material on this specific topic. In short, what we offer in the following collection is a multidimensional and conceptual varied *circuit of the body* where the irreducible element is the collection of the individual pieces.

2. The Morphogenesis of "Information"

Although the body has been long since recognized as an important term for social theory, "information" has only recently entered into the critical lexicon. This is not to deny that the word itself is old; a quick survey of the Oxford English Dictionary confirms that "information" has been in use in English since at least the fourteenth century, often used in the sense of "the giving of a form or character to something."[24] Information—understood particular items of knowledge about distant areas, topics, or groups—emerged at least as early as the newspaper revolution of the early eighteenth century, and the term became an important part of political discourse in the late eighteenth century (witnessed by, for example, the foundation of the Society for Constitutional Information in Britain in 1780, which is certainly one of the precedents for John Perry Barlow's much more recent quasi-political claim that "information wants to be free"[25]). Moreover, Geoffrey Nunberg has suggested that the modern "abstract" understanding of information—that is, information understood "as a kind of abstract stuff present in the world, disconnected from the situations that is *about*"—first emerged at least as early as the mid-nineteenth century.[26] Yet it is only in the last half century or so that information has been established as a term of central importance to fields as diverse as genetics, library sciences, and intellectual property law.

A number of different explanations have been advanced to explain why theories of information have become so central to contemporary society. So, for example, scholars have explained this rise to dominance as a function of the military's use of command-control[27]; the rise of information processing techniques to control the flow of goods in a consumer society[28]; the rise of state administered bureaucracies; the shift from orality to literacy in civil development[29]; and the rise of the service sector in a "postindustrial" economy.[30] This multitude of explanations is perhaps not surprising, for the very prevalence of the term information in contemporary society almost guarantees that it will be used in multiple senses and thus generate multiple trajectories to explain its prominence.

In our eyes, three of these stories offer special insight into the role of information in our history of human embodiment. For our purposes, we can consider them as additive stages where the concept of "information" made its presence felt in three distinct but interlinked disciplines. The first stage is the story told about communication theory and cybernetics from the viewpoint of Claude Shannon and Norbert Weiner. The second focuses on the role of the coding problem in molecular biology. The third is the emergence of information from an economic and social theory perspective and its relationship to medium theory.

The first story begins in the 1930s and 1940s with the development of mathematical probabilistic theories of information by communication scientists such as Bell Labs researchers R. V. Hartley and Claude Shannon. Bell Labs was in the business of designing and selling efficient telephone systems, and both Hartley and Shannon selected the term "information" to denote a mathematical quantity within a series of equations that described signal transmissions. Hartley's and Shannon's formulas had relatively little to do with the way information was used in colloquial speech—Shannon, for example, noted that the question of "meaning" was "irrelevant to the engineering problem" of communication system design—but they were very useful in improving telephone circuits.[31] Shannon's emphasis on information as a theoretical term that would facilitate the design of efficient human-machine systems attracted the attention of mathematicians Norbert Wiener and John von Neumann. Both Wiener and von Neumann were interested in designing and creating "cybernetic" systems that would link humans and machines in various sorts of feedback loops, and Shannon's information formulas facilitated this work. Also encouraging this process was the institutional support of the U. S. military, which (especially during the wartime years 1941–1945) was very interested in developing complex systems that integrated both humans and machines, especially in the realms of communications and gunnery.[32] The informal links between Shannon and Wiener and von Neumann were formalized, extended, and published through what have since become known as the "Macy Conferences on Cybernetics." These meetings, occurring between 1943 and 1954, were attended by researchers from a number of different fields, including such luminaries as Gregory Bateson, Margaret Mead, and Warren McCulloch, as well as von Neumann, Shannon, and Wiener, and a good deal of discussion was focused on different mathematical and non-mathematical definitions of "information."[33] As N. Katherine Hayles has noted, the technical focus of many of the key participants ensured that, in the short term, "information" was understood as an abstract mathematical quantity (akin, though not equivalent, to energy) that could be transferred from context to context. Information, in this reading, had nothing to do with meaning or interpretation, but rather concerned signals, noise, and cybernetic systems.

This understanding of information provided fertile theoretical ground for connections between the emerging science of cybernetics and the transforming field of genetics, a fact that provides a link to our second stage of the morphogenesis of "information." The history of Francis Crick and James Watson's discovery of deoxyribonucleic acid (DNA) as the substrate of chromosomes has been sufficiently recounted by other authors,[34] but vital for the history of information was physicists Erwin Schrödinger and George Gamow's repeated, and successful, attempts to position Crick and Watson's discovery as a question of "coding"; that is, as a problem of transmitting information from one system to another. This vocabulary helped facilitate the movement of the term information into genetic discourse; the problem of life, thereafter, became a problem of "information transfer" and of "cracking the genetic code."[35] Modern genetics was thus born when, in the 1950s, researchers began to describe "organisms and molecules as information transfer systems."[36] A term and approach that began as a solution to technical problems of telephone system design had become a powerful controlling metaphor for understanding the nature and generation of life itself.

Although the rise to dominance of information in the biological and computer sciences has received increasing attention in recent years, less sustained interest has been paid to a third important stage of the morphogenesis of "information." The site of emergence for this stage was in the social sciences, especially economics and sociology. Beginning in the 1950s, a number of economists and social theorists began to isolate and focus on the increasing importance to the U. S. economy of the "information sector", that is, the sector that produced and distributed knowledge, rather than, for example, agricultural or manufactured goods. Fritz Machlup first employed this terminology in the late 1950s, though it was popularized by Marshall McLuhan's *Understanding Media: The Extensions of Man* (1964) and Daniel Bell's *The Coming of the Post-Industrial Society* (1973). McLuhan proclaimed that we were living in a "new electric Age of Information," and developed several striking claims about information itself, claiming, for example, that "the electric light is pure information," because it is "a medium without a message." Although McLuhan's book was dismissed by some as not appropriately rigorous in its analysis and argumentation, his claims have proven to have continuing impact and significance (especially in the 1990s, as a generation who came to age with computers and the World Wide Web (WWW) have now found added meaning in his catch phrases, design, and aphoristic analyses). Perhaps most important was McLuhan's redefinition of media as extensions of the human body, as well as his related claim that "the effects of technology do not occur at the level of opinions or concepts, but alter sense ratios or patterns of perception steadily and without resistance."[37] McLuhan's reading of media as forms of human embodiment have been especially important for theorists interested in the communities and identities made possible through web-based interactions, such as chatrooms and multiuser games. Bell, a sociologist, sought to explain important post-World War II shifts in the economies of the United States and Western Europe, and coined the phrase "postindustrial society" to explain these changes. Although he did not mention the term "information" in the initial edition of his book, his notion of a postindustrial society, characterized by knowledge (rather than agricultural or industrial) production, helped popularize the paradigm of "information society" and the "information age."[38]

Where cybernetic and biological understandings of information stressed *transmission,* theories of the information society stress *production.* In the latter tradition, information is understood as a product of human labor, and societies are characterized as "information societies" when information production becomes a substantive and dominating component of their total economic activity. In the information age, the production of different forms of information "replaces industrial goods as the principal commodity and economic engine of the information age."[39] It is, admittedly, not always entirely clear what counts as "information" in this discourse (so, for example, how is "information" distinguished from "data" and "knowledge"?),[40] yet from the perspective of the information economy, this has been a productive ambiguity, for it has helped expand and transform the categories of both production and property.[41] Thus, "intellectual property" law has become an increasingly dominant component of the juridical field in the United States, Western Europe, and some parts of Asia, and forms of information discovered in the biological field now often enter the information economy as patented life forms and bodies.

The computer has provided an important material linkage between these three lineages of information. To begin with, the rhetoric of command and control that is central to computing was vital in establishing both the biological and cybernetic discourses of information. Thus, in biological discourse, cells and DNA often were explained, and understood, as "like" a computer and the program that runs it,[42] and many of the figures associated with the creation of the first computers (e.g., Wiener and von Neumann) were also in close contact with the first theorists of DNA as information. Yet this same rhetoric of control and command was central to the development of the "information society," as James R. Beniger has established in *The Control Revolution: Technological and Economic Origins of the Information Society*. Moreover, information management skills and technologies originally developed within corporations recently have been applied to the management of genetic databases (exemplified most spectacularly by the genome project).

The conception of "information" that underwrites this volume acknowledges these three genealogies, but also draws on a fourth, less well known, lineage, focused on the inherently contextual nature of information. From this perspective, information is understood not as the coded "content" of messages, but rather, as something that enables, and emerges through, communicative acts. In *How We Became Posthuman*, Hayles demonstrates that this understanding of information first emerged in the 1950s, when some participants of the Macy Conferences attempted to propose concepts of information that took into account structures of meaning, or focused on the effects *produced by* information. Although these alternative conceptions of information initially were rejected by many of the Macy Conference participants, they found their way into the "systems" work of Heinz von Foerster, Humberto Maturnara and Francisco Varela, Gregory Bateson, and Niklas Luhmann, and has continued to inform recent work on "self-organization," "emgergence," and virtuality.[43]

From this perspective, information only emerges as a function of choices made from a field of possibilities, and thus assumes important ontological consequences. Critical theorist William Rasch recognizes this when he encourages us to understand how "information" is not equivalent to "messages":

> The notion of information has to be distinguished from the notion of the message. Information is seen as the total field of choices from which the choice of correct message is to be made. Information is proportional to uncertainty: the greater the information, the greater the uncertainty on the part of the auditor regarding precisely which message from the manifold possible messages is the intended message.[44]

Navigating this type of world requires a great number of decisions, and all of these decisions are made from an imperfect understanding of the available choices by individuals who are immersed in the virtual landscape that they are trying to traverse. The digital media theorist Espen Aarseth calls these types of informational landscapes "cybertexts." Aarseth argues that navigating these texts requires a mode of engagement that is far more complex than the sender–receiver model: "The cybertext reader is a player, a gambler; the cybertext *is* a game-world or world-game; it is possible to explore, get lost, and discover secret paths in these texts, not metaphorically but through the topical structure of the textual machinery."[45] These decisions, or game moves, are the "work" (a physical concept as opposed to the humanistic concept of "labor") that structures a specific path through a rich informational landscape. These decisions are predicated on the embodied act of negotiating this landscape (one's perceptive experience, previous trajectories, cultural affects, forms of memory, and data processing), rather than simply understanding engagement through the categories of "command" or "message."

3. *The Methodology of Embodiment*

In addition to employing these reformatted conceptions of bodies and information, we believe that it is also useful to take advantage of an analytical device recently developed by Manuel Castells, which he calls the "informational mode of development." For Castells, the informational mode of development denotes a form of economic development in which "knowledge intervenes upon knowledge itself in order to generate productivity" as opposed to situations in which knowledge intervenes as an "intermediary element in the relationship between labor and the means of production."[46] This mode of development contains two crucial dimensions: "the technological and the organizational."[47] In his recent history of colonial history, C. A. Bayly succinctly summarizes the power of Castells' approach for humanistic study:

> What is important about his approach is that he sees the generators of knowledge, the institutions of information collection and diffusion and the discourses to which they give rise as autonomous forces for economic and social change. They are not reduced to the status of contingencies of late-industrial capitalism or the modern state. Knowledge itself is a social formation; knowledgeable people form distinct and active social segments with their own interests.[48]

Although the informational mode of development is only a frame for analysis (and should not be reified as an object in itself), its power comes from its ability to illuminate very real historical consequences of changes in the ways that information is processed.

Even more controversially, we believe that we should look to the informational mode of development itself for conceptual tools that can comprehend and manage the problems that arise from this mode of development. So, for example, as we attempt to create new collective virtual landscapes, we should think of this task in terms of qualities of modern computing technologies (and thus the informational society in general), such as the ability to store large amounts of information, and the capacity to create detailed virtual worlds that establish new configurations of self and environment. Our adoption and transformation of such terminology should not be interpreted as a blind enthusiasm for technology, but a mindful recognition of how deeply dependent contemporary society is on information technologies. Only after we come to terms with this can we pursue a fine-grained analysis that looks at how humans and technologies interact in a case- and context-specific manner.

Reformatting terminology in this manner will allow us, in turn, to make greater distinctions in practical (i.e., ethical) conduct. We must use the diverse mechanisms of our collected information storage to avoid overly confident assessments of the properties of bodies. As many of the chapters in this volume remind us, judgments concerning what is or is not a *human* body sometimes have led to large injustices. The myriad forms of social memory may not keep those in power from making similar errors; they can, however, help destabilize an arrogance that may keep them from hearing the cries of those with barely audible voices. Older conceptions of information as a message would construe these voices as the "noise" that detracts from the content of the "signal." In contrast, the conception of information we have argued for suggests that these voices are the emergent patterns that keep us listening. Our challenge is to use the electromagnetic stability of our hard drives, the emotional intractability of our ego, and the chemical stability of our nucleic acids to remember and relive the injustices we have inscribed into others.

Second, we must use the conceptual and sensory richness of novel virtual landscapes to explore more creative possibilities for the future. This requires that we realize that reducing all problems faced in an informational economy into categories constructed in response to an industrial economy is not always the best solution. Often this reduction leads to overly retrospective judgments, conservative evaluations, and entrenchment into polarized social positions. This is not a question of immanence versus transcendence; instead, it is a recognition of the importance of bodies and how registering the subtle changes in relations allows one to adjust one's own interactions in real time. Life in our informational mode of development involves a complicated coconstitution of information and bodies. This coconstitution is now understood in terms of incredible spans of distance, vast stretches of territory, and multiple levels of packaging enabled by an international consumer economy. Do we have the resources to locate that irreducible bit of integrity that comprises embodied nature in contemporary society? The answer to this question depends on how we respond to the challenge of utilizing the complex circuitries at hand for developing new "circuits" for fulfilling and engaging interactions.

II. Overview of the Volume

Although each contribution in this volume can be read in isolation, we have linked the chapters along both historical and topical axes. The first section of the volume ("Bodies Before the Information Age") highlights the historical axis by addressing pre-"informatic" notions of bodily communication and control. The chapters in this section—focusing on eighteenth-century theories of vitalism, early modern horse training, and late nineteenth-century horse breeding—provide an historical foundation that stresses the importance of earlier theories of vitality and "control" for twentieth-century relations between human bodies and information. In *On Beyond Living: Rhetorical Transformations of the Life Sciences,* Richard Doyle has argued that contemporary understandings of DNA as "information" are premised on what he calls "the postvital body"—that is, "a body without life," where "life" denotes that "unseen unity that traversed all the differences and discontinuities of living beings."[49] Yet despite the current dominance of the postvital body, conceptions of the "vitalist" body continues to inform many contemporary notions of embodiment. We thus begin this volume with a chapter that examines the vitalist position in more detail. In "Reading the 'Sensible' Body: Medicine, Philosophy, and Semiotics in Eighteenth-Century France," Anne C. Vila focuses on the version of vitalism developed by physicians associated with the Montepellier school in France in the eighteenth century, as well as Denis Diderot's "hypervitalist" response to those theories. Vila is especially attentive to the reliance of vitalist philosophy on the notion of "sensibility," and the ways in which that concept served as the foundation for eighteenth- and early nineteenth-century attempts to *read* the body's signs. At present, the biomedical semiotics of the flesh is grounded in notions of information (e.g., computer-assisted virtual imaging, genomic sequencing, etc.); in the eighteenth century, however, medical sign gathering operated through what Vila describes as an "inter-bodily communion"—mediated by sensibility—between patient and physician. Vila's chapter also provides a very useful account of the ways in which terms such as "sensibility" come to serve as what she (following Ludmilla Jordanova) calls "bridging concepts" between different discourses. Vila argues that "sensibility" served as an important bridging concept for eighteenth century medical and philosophical discourses, establishing, for example "causal connections between the physical and the moral realms" (28). We can see the same process at work in the present around the term "information," which serves as a "bridging concept" between, for example, the realms of biochemistry and computing. Vila's argument thus highlights the continued importance of the human body in serving as a platform that mediates between discursive fields.

Where Vila's chapters outlines a conception of "life" that still serves as a reference point for many contemporary conceptions of embodiment, the next two chapters focus on the notions of "command and control" that were to become so important to the cybernetic and information revolutions of the twentieth century. The implicit reference for both chapters is James R. Beniger's *The Control Revolution: Technological and Economic Origins of the Information Society.* Beniger locates the origins of the information society in what he calls a "crisis of control" that characterized the period 1840–1900 in the United States, a crisis that resulted from the fact that "innovations in information processing and communication technologies lagged behind those of energy and its application to manufacturing and transportation."[50] Thus, in Beninger's account, "cybernetics"—the science of control through communication and information processing—begins far earlier than the coinage of the term in the twentieth century would suggest. The chapters by Elisabeth LeGuin and Phillip Thurtle both build on, but also complicate, Beninger's thesis. LeGuin and Thurtle focus on a common object—the horse—and both suggest that horses provided one of the first "virtual" systems for thinking about the relationship between control and biological and social concepts of embodiment. Beninger employs a relatively broad, and ahistorical, understanding of "control,"[51] but in "Man and Horse in Harmony," LeGuin outlines the ways in which interactions between human and horse bodies informed the content of subsequent understandings of what it meant to "control" or "command." As LeGuin notes, equine metaphors of control (e.g., "to harness" the productive capacities of the gene) "are used regularly by people who have never come near a horse, to address matters of intention, control, and enactment, and they show remarkably few signs of being supplanted by automotive imagery" (47). Historians have outlined ways that horses were important to the command/control mechanisms of the nineteenth century, but LeGuin's chapter is predicated on the notion of "command as distinct from domination, command as a reciprocal condition, command as predicated upon a knowledge of when to listen as well as when to tell" (62). As LeGuin notes, her chapter also complicates Michel Foucault's account of the rise of the "disciplinary society" as outlined in *Discipline and Punish.* Foucault acknowledges the importance of the vital body to the development of techniques of discipline (*dressage*), suggesting that in the late eighteenth and early nineteenth century, "a new object was being formed; slowly, it superseded the mechanical body—the body composed of solids and assigned movements, the image of which had for so long haunted those who dreamed of disciplinary perfection. This new object is the natural body, the bearer of forces and the seat of duration."[52] LeGuin's account, however, suggests that the horse–human system in fact preceded, and continued to inform, the disciplinary body described by Foucault.

Where LeGuin focuses on horses in the early phases of the modern period, Phillip Thurtle, in "Breeding and Training Bastards: Distinction, Information, and Inheritance in Gilded Age Trotting Horse Breeding" examines the role of horses, and their breeding, in the industrial age. Thurtle notes that many of the industrialists who pioneered the new mechanisms of control were also fascinated with trotter horses. Trotter horses are connected to the emergence of new information technologies in several important ways. For the wider public, trotter horses came to serve as validation for new managerial and information techniques, for the industrialists' mastery over their horses was an expression of the same characteristics that made them successful businessmen. Industrialists also applied new managerial techniques to horse trotter breeding, creating "laboratories of speed" within which rationalized selection processes could be applied. Even more importantly, however, industrialists such as Leland Stanford (founder of Stanford University) hoped to apply these lessons and techniques to humans; these industrialists/breeders looked toward their stables for lessons on the relationship of biological beings to *social circumstances*. Thurtle demonstrates that the control of horse breeding served as a model for attempts to control human "breeding" through eugenics programs and progressive era educational reforms. In terms of Beninger's thesis, late nineteenth-century philanthropy sought to "breed" the professional class demanded by the control industries by redefining concepts of intergenerational human connectivity.

Where the chapters in the first part of *Data Made Flesh* outlined earlier paradigms of vitality, embodiment, and control, the six chapters in the second section ("Control and the New Bodies: Modes of Informational Experience"), by contrast, explore ways that contemporary discourses on informatics rely on, use, and shape the human body. The chapters here fall into two thematic groups—the first three focus on questions of control, whereas the latter three on the emergence of new forms of embodiment—but all six chapters stress the intrinsic relationship between information and embodiment. Although it may be true, as N. Katherine Hayles suggests, that "information lost its body" in the dominant cybernetic, genetic, and communication theories of the 1940s and 1950s, the chapters in this section collectively suggest that embodiment is not an "option" for information, but rather a condition for its emergence. The latter three chapters also suggest the converse: postmodern bodies do not exist outside, or beyond, information, but are rather one of the two virtual poles (along with information) between which embodiment occurs.

The first three chapters focus on the ways in which the human body mediates between information technology and questions of control. In "Desiring Information and Machines," Mark Poster investigates the relationship between cultural media and psychoanalytical conceptions of the body. In his seminal text *Foucault, Marxism, and History,* Poster asked historians to consider the ways in which new information technologies transformed traditional historical categories, such as agency, domination, and subjectivity. He continues that project here, arguing that many psychoanalytical conceptual categories (e.g., identification and object-choice) and narratives (e.g., the Oedipal drama) were dependent on the media context of *fin-de-siècle* Vienna. As Poster notes, this was a social milieu relatively poor in information technologies, and as a consequence, parents of this period were able to control their children's domestic environments in a way that has become increasingly impossible in an information society such as early twenty-first century United States. Poster considers how this psychoanalytic understanding of the body can take into account the overwhelming increase in home-centered communications media such as television and the Internet, and he concludes that categories such as "identification" may have to give way to notions of "incorporation."

The next chapter—Richard Doyle's "LSDNA: Consciousness Expansion and the Emergence of Biotechnology"—focuses on the extremely important role that deliberate attempts to "lose control" have played in the construction of contemporary bioinformatic discourse. Doyle's thesis constitutes the most recent program in the project of decoding and reformatting that he began in *On Beyond Living: Rhetorical Transformations of the Life Science*. In that text, Doyle outlined the processes by which vitalist understandings of a "deep unity" of life were overwritten by the premise of transparency inherent in modern notions of "information." But he also hinted at ways in which the notion of life was itself transforming in the wake of information and information technologies (e.g., via the notion of "artificial life"), and he has continued that exploration in his forthcoming *Wetwares! Experiments in Post-Vital Living*, as well as in his chapter for this collection. Here, Doyle outlines a peculiar, but important, connection between Albert Hoffman (inventor of LSD), Timothy Leary (populizer of LSD as a way of "reprogramming" the mind), and Kary Mullis (inventor of PCR, a DNA duplication technology). What unifies these three is a shared commitment to paradoxical forms of agency (laughter, terror, ecstasy) that put into question notions of self-control and individuality so central to the popular image of the biological scientist. The narrative Doyle constructs allows him to account for a very important, and relatively recent, change in the notion of information in biological discourse, as it has moved from a "cryptographic" paradigm (in which information was understood as the space of a revelation) to a "pragmatic" paradigm (in which one manipulates information to see what it can *do*).

One of the consequences of the emergence of this "pragmatic" paradigm has been an ever-increasing deterritorealization of elements of living bodies, as more and more hybrid bodies emerge as the result of genetic engineering (a process paralleled by a massive increase in the circulation of organs, blood, and procreative components). In "$ell: Body Wastes, Information, and Commodities," Robert Mitchell analyzes the rhetoric of contemporary attempts to control these processes—and especially their relationship to flows of money—by recourse to notions of a sacred, "dignified" human body. Mitchell outlines the legal and economic tensions that develop when traditional notions of "dignity" are applied to contemporary arguments about the ownership of body wastes and information. Focusing on the case of *Moore v. Regents of the University of California*, Mitchell notes that the California Supreme Court linked notions of waste and dignity as a means of awarding property rights in human cells to a corporate body while at the same time denying these rights to the individual who had served as the original source of the cells. Mitchell argues that the contradictions of this logic of dignity should not be seen so much as cause for despair, but rather as an opportunity to rethink what it means to "commodify" bodies and body parts.

The next three chapters of the second part of the volume investigate ways that information technologies have created new forms of embodiment. In her chapter for the first section of this volume, Anne C. Vila outlined the figure of the "sensible" physician; in "The Virtual Surgeon: New Practices for an Age of Medialization," Timothy Lenoir investigates the ways in which new imaging techniques have created "informatic surgeons" that are dispersed and extended over a wide variety of technologies and bodies. Lenoir notes that developments in surgical imaging have enabled a "fusion of digital and physical reality" (137), rather than the abstraction from the flesh that some critics have warned against. At the same time, though, Lenoir notes that the cost of this fleshy data is increasing surveillance, for these informatic tools have been developed in part to facilitate the goals of "a massive system of preventive health care from genome to lifestyle" (146).

In "The Bride Stripped Bare To Her Data: Information Flow + Digibodies," Mary Flanagan traces the emergence of a new form of information body, the gendered "digibody." Flanagan, a media theorist as well as an interactive artist and software designer, begins her narrative with a discussion of Marcel Duchamp's *The Bride Stripped Bare by her Bachelors,* arguing that it opened a "third" space between mind and body that would subsequently be occupied by what she calls "digibodies." Through a focused analysis of several of recent gendered digibodies (including Motorola's "Syndi" and UK's "Ananova"), Flanagan outlines the ways in which recent graphic design architects have sought to occupy that space by embodying information in virtual feminine forms, often employing standard stereotypes of the feminine. Yet Flanagan demonstrates that these digibodies nevertheless exceed the gendered power dynamics of consumption (in which the female form is figured as a hollow vessel of pleasure), and instead mark a form of "real semiotic" bodies. Flanagan thus demonstrates elements of what she calls "new critical practices" that will to help us "recognize this new bodily form" where the "code has become flesh, and . . . flesh, code" (177).

Lenoir's and Flanagan's chapters focus on the emergence of new informatic bodies, but Kathleen Woodward, in the final chapter of the second section—"A Feeling for the Cyborg"—focuses on the emergence of new informatic *emotions.* Through a survey of a wide range of cultural texts, Woodward suggests that, in the information age, emotions are no longer considered a form of imaginative or psychic engagement but a materially structured "prosthetic" that can "couple" technocultural systems. Woodward reminds us that it is through embodiment that we learn empathy; any effort to communicate with the unknown, therefore, will thus need to incorporate an emotional sensibility beyond conventional notions of intelligence. Woodward's chapter thus makes a nice pairing with LeGuin's piece from the first section of this volume, for both highlight the role of empathy in attempts to mediate between different biological entities.

In the third and final section of the volume ("Flesh Remembered: Art, Information, and Bodies"), we narrow our focus to investigate the recent use of bioinformatics in the visual arts. Here, we build on previous analyses to demonstrate in a context and case specific manner what happens when bodies are manipulated for aesthetic rather than practical consequences. Recent works in the arts are the most appropriate subject for this focus because these artists willingly move beyond a critique of biotechnology (which tends to remain at a purely conceptual level) and to begin using biotechnology in their creations, thus engaging bodies through practice. Because of this, art promises interesting lessons in the material poesis of informatics.

The first lesson is that information is inherently relational and does not fit easily into previous categories of essential qualities. Information emerges as a statement about a body in relation to its environment or another body. As such, relational forms of analysis are important for understanding the ways that information is constituted in specific environments. The first chapter in this section, Bernadette Wegenstein's "If you won't SHOOT me, at least DELETE me! Performance Art from 60ies Wounds to 90ies Extensions," focuses specifically on the relationship of the performer to audience in performance art from the 1960s to the present and demonstrates how this "syntax of performance" encouraged greater use of informational technologies. Wegenstein begins by outlining the role of media in early twentieth-century avant garde art and manifestos, from which she argues that 1960s performance artists inherited from their predecessors a desire to do away with mediation and representation. This desire encouraged performances that were enacted on the body itself, with the "wound" symbolizing the least mediated and most direct form of performance. Yet, perhaps paradoxically, performance artists of the 1960s and 1970s could not escape what Jay David Bolter and Richard Grusin have called the "double logic of remediation," which ensures that "[o]ur culture wants both to multiply its media and to erase all signs of mediation: ideally, it wants to erase its media in the very act of multiplying them."[53] Thus, even as performance artists sought to do away with mediation by focusing on the body, they also sought to disseminate those performances, and involve spectators, through multiple forms of media. Wegenstein argues that the result of this double logic in the 1980s and 1990s was a movement of performance *away* from the human body and toward nonorganic forms of media bodies.

In the next chapter, "Flesh and Metal: Reconfiguring the Mindbody in Virtual Environments," N. Katherine Hayles picks up where Wegenstein's chapter left off, as Hayles investigates the role of the human body in three recent virtual reality (VR) installations. As Hayles notes, "Flesh and Metal" emerged in part as a response to criticisms that the distinction between "the body" and "embodiment," articulated in her book *How We Became Posthuman,* still depended on the Cartesian dualism that she had attempted to critique. In order to counter this reading, Hayles develops Mark Hansen's notion of the "mindbody" in order to consider the following VR installations: "Traces" by Simon Penny and collaborators; "Einstein's Brain" by Alan Dunning, Paul Woodrow and collaborators; and "NØTime" by Victoria Verna and her collaborators. Hayles argues that all three works engage the non-Cartesian mindbody, but each emphasizes a different aspect: "Traces" investigates "the relation of mindbody to the immediate surroundings"; "Einstein's Brain" highlights perception; and "NØtime" "emphasizes relationality as cultural construction" (234).

One of the most pressing lessons concerning contemporary modes of coconstituting information and bodies comes from looking at the reanimation of the virtual in biological art constructions. If the medium is the message, as McLuhan contended, then what happens when the medium comes to life? The final three chapters of the volume focus on Eduardo Kac's recent bioart installation, *Genesis,* in order to articulate possible answers to this question. The first two chapters—Steve Tomasula's "Gene(sis)" and Eduardo Kac's "Transgenic Art Online"—both highlight the religious dimension of the contemporary cult of information, using Kac's installation to consider the relationship between technology, community, and the notion of a divinely sanctioned human "dominion" over earth and sea. Tomasula highlights the ways in which Kac's project seeks to create an inclusive community that will co-investigate the relationships between divine and human creation. Kac's chapter, written as a response to Tomusla's, returns us to the question of control by focusing more specifically on the notion of "dominion." Kac emphasizes the ways in which his installation forces a reflection on the complicated relationship between information technologies and a belief in divinely-guaranteed acts of possession. Although Tomasula's and Kac's chapters agree on many points, they have divergent readings of the audience participation made possible by the Genesis installation. Tomasula stresses the participatory aspect of the project (it potentially enables everyone to become coauthors of the project), whereas Kac stresses the ethical dilemma that confronts every potential user: what justifies an act (in this case, a simple click of the mouse) that will have the consequence of changing the DNA of a life form? Kac also highlights the importance of "symbolism" to his work, a discussion that hints at the fundamental changes that are underway in aesthetic theory as we enter an era of *Data Made Flesh* art, in which "symbols" biologically alter the realities they symbolize.

In "Gene(sis): Contemporary Art Explores Human Genomics," curator Robin Held situates Kac's project within a larger context of biological information art projects. Her chapter, which describes a recent exhibition, outlines some of the terrain occupied by artists interested in the intersections of art and genomics. Seconding the suggestion of geneticist Eric Lander, Held suggests that the projects included in the exhibition and her chapter are attempts to ensure that "the ultimate meaning of the human genome would be decided not by scientists alone but would be fought out in the arenas of art and culture" (263).

The works in this final section of *Data Made Flesh* provocatively sum up one of the key assumptions of the volume as a whole: we have entered an era where signifying practice and embodiment are no longer conceptually or practically separate. This suggests, in turn, that "representation" is often a problematic strategy with which to engage embodied information. The mediums for these artists are living, metabolizing bodies; they are no longer representing life, but are creating life. As a result, we can no longer be comfortable using purely literary analysis to analyze information as the structure of the message (such as seen in recent narratological analyses of virtual environments); nor can we be content using psychoanalytical crowbars to pry open the identities of biological beings to read the codes of their behaviors. Instead, we must use biology in order to understand how our communicative gestures might come alive and give birth to new possibilities. This then reveals the most urgent message of the "informational mode of development": once one recognizes the simulacrum as a productive (as opposed to a derivative) element, then we stand like gods giving life to new bodies with every communicative gesture. We thus offer this volume as collected memories of the ways that "information" emerged in the West, a demonstration of how it has changed our concept of embodiment, and a map of how widely this concept resonates in different fields. Even more importantly we provide a tour of the creative virtual landscape for future engagements with bodies of all types. We have now left the world of representations and have entered the world of *Data Made Flesh.*

Notes

1. So, for example, Norbert Wiener's postwar vision of translating the informational content of a human being into telegraph signals is still alive and well in the work of computer researcher Hans Moravec, who suggests that one might "upload" human consciousness into computers by "purée[ing] the human brain in a kind of cranial liposuction, reading the information in each molecular layer as it is stripped away and transferring the information into a computer"; see Moravec, *Mind Children: The Future of Robot and Human Intelligence* (Cambridge, Massachusetts: Harvard University Press, 1988), 109–10. For a discussion of Wiener's distinction between living bodies and information, see Lily E. Kay, *Who Wrote the Book of Life? A History of the Genetic Code* (Stanford: Stanford University Press, 2000), 78–91. For a discussion of Moravec, see N. Katherine Hayles, *How We Became Posthuman: Virtual Bodies in Cybernetics, Literature, and Informatics* (Chicago: The University of Chicago Pres, 1999), 1.
2. Which explains why one of the largest government sponsored biomedical research agendas is concerned with an informatics of the human body; in fiscal year 1999, for

instance, the federal appropriations for the Human Genome Project totaled $315.5 million. See Human Genome Program, U.S. Department of Energy, *Human Genome News* (v. 10, note 3–4).

3. We use the terms "virtual" and "actual" in the sense outlined in Gilles Deleuze, *Difference and Repetition*, trans. Paul Patton (New York: Columbia University Press, 1995), 191–214.
4. Mark Poster, *Foucault, Marxism, & History: Mode of Production versus Mode of Information* (Cambridge: Polity Press, 1984). See also Poster's *The Mode of Information: Poststructuralism and Social Context* (Chicago: The University of Chicago Press, 1990), and *The Second Media Age* (Cambridge: Polity Press, 1995).
5. See Manuel Castells, *The Informational City* and Castells, *The Information Age: Economy, Society and Culture*, 3 vols. (New York: Basil Blackwell, 1996–1998).
6. Lily E. Kay, *The Molecular Vision of Life: Caltech, the Rockefeller Foundation, and the Rise of the New Biology* (Oxford: Oxford University Press, 1993) and *Who Wrote the Book of Life?;* Richard Doyle, *On Beyond Living: Rhetorical Transformations of the Life Sciences* (Stanford: Stanford University Press, 1997). The classic discussion on this topic is found in Horace Freeland Judson's "The Number of the Beast" from *Eighth Day of Creation: Makers of the Revolution in Biology* (New York: Touchstone, 1979). More work still needs to be done on developments in enzymology that allowed for the manipulation of biological macromolecules as information entities. See Hans-Jorg Rheinberger, *Toward a History of Epistemic Things: Synthesizing Proteins in the Test Tube* (Stanford: Stanford University Press, 1997) for the beginnings of a corrective in methodology and content.
7. See Hayles, *How We Became Posthuman.*
8. Michael E. Hobart and Zachary S. Schiffman, *Information Ages: Literacy, Numeracy, and the Computer Revolution* (Baltimore: The Johns Hopkins Press, 1998); Pierre Levy, *Collective Intelligence: Mankind's Emerging World in Cyberspace*, trans. Robert Bononno (New York: Plenum Trade, 1997) and Levy, *Becoming Virtual: Reality in the Digital Age*, trans. Robert Bononno (New York: Plenum Trade, 1998). Also deserving of notice are the following: Albert Borgmann, *Holding On to Reality: The Nature of Information at the Turn of the Millennium* (Chicago: The University of Chicago Press, 1999); Richard Coyne, *Designing Information Technology in the Postmodern Age: From Method to Metaphor* (Cambridge, MA: The MIT Press, 1995) and *Technoromanticism: Digital Narrative, Holism, and the Romance of the Real* (Cambridge, MA: The MIT Press, 1999); James J. O'Donnell, *Avatars of the Word: From Papyrus to Cyberspace* (Cambridge, Massachusetts: Harvard University Press, 1998); Michael Heim, *Electric Language: A Philosophical Study of Word Processing* (New Haven: Yale University Press, 1987) and *Virtual Realism* (New York: Oxford University Press, 1998); Richard A. Lanham, *The Electronic Word: Democracy, Technology, and the Arts* (Chicago: University of Chicago Press, 1993), as well as an important early book edited by Kathleen Woodward that collected together essays by system theorists and cultural critics, *The Myths of Information: Technology and Postindustrial Culture* (London: Routledge, 1980).
9. Equally enriching for the field of information studies is the theoretical work of Hans Ulrich Gumbrecht (through his emphasis on the "materialities of communication"); Joshua Meyrowitz, Marshall McLuhan, and Harold Innis (for their explication of medium theory); Eric Havelock, Walter Ong, Elizabeth Eisenstein, Jack Goody, and Adrian Johns (for their discussions of the dynamics of orality and literacy); Elizabeth Gross (for her recognition of the importance of "embodiment" and "becoming" for feminism); Eric Michaels (through his evocative charting of informational flows in Australian Aboriginal communities); Friedrich Kittler (through a provocative mix of discourse analysis, McLuhanian medium theory, and Lacanian theories of identification); Erving Goffman (especially his argument for identity management as offered in *Stigma*); Kathleen Woodward (through her critique of information and important emphasis on the aging body in technologically centered societies); Gregory Bateson, Humberto Maturnara and Francisco Varela (for their discussion of autopoeisis); Alphonso Lingis (through his rethinking of the phenomenological tradition), and Gilles Deleuze (especially the important works *Difference and Repetition* and *The Logic of Sense*) and Brian Massumi's application of that work to question questions of embodiment and information.
10. William Gibson, *Neuromancer* (New York: Ace Books, 1984), 16.

11. In her recent history of the posthuman, N. Katherine Hayles reads Gibson's quote in this manner, suggesting that it emphasizes "cognition rather than embodiment." See *How We Became Posthuman*, 5.
12. Rich Doyle reads Gibson's quote in this manner in his forthcoming *Wetwares! Experiments in Post-Vital Living* (Minnesota: University of Minnesota Press, forthcoming 2003).
13. See Gilles Deleuze and Felix Guattari, *Anti-Oedipus: Capitalism and Schizophrenia*, trans. Robert Hurley, Mark Seem, and Helen R. Lane (Minneapolis: University of Minnesota Press, 1985); Niklas Luhmann, *Social Systems*, trans. John Bednarz Jr. (Stanford: Stanford University Press, 1995); Humberto Maturana and Fracisco Varela, *Autopoesis and Cognition: The Realization of the Living* (Dordrecht: D. Reidel Publishing Company, 1980), and Manuel Castells, *The Rise of the Network Society* (Oxford: Blackwell, 2000).
14. Slavoj Žižek, *The Ticklish Subject: The Absent Centre of Political Ontology* (New York: Verso, 1999), 1–5. See Albert-László Barabasi's *Linked: The New Science of Networks* (Cambridge, Massachusetts: Perseus Publishing, 2002) for an illuminating mathematical treatment of the notion that that which cannot be divided in the body is an element in a network exhibiting a "power law." Our conception of the body differs from traditional understandings of "systems," insofar as this latter term often depends on the notion of a collection of parts that are randomly and equally composed. In fact, certain parts provide intensive properties that other parts do not posses (an assemblage constituted thusly would be a special type of a system, a body). On the importance of intensive assemblages see Giles Deleuze and Felix Guattari, "1730: Becoming-Intense, Becoming-Animal, Becoming Imperceptible . . ." in *A Thousand Plateaus: Capitalism and Schizophrenia*, trans. Brian Massumi (Minneapolis: University of Minnesota Press, 1987).
15. Hans-Ulrich Gumbrecht, "A Farewell to Interpretation," in *Materialities of Communication*, eds. Gumbrecht and K. Ludwig Pfeiffer; trans. William Whobrey (Stanford: Stanford University Press, 1994), 391.
16. Armando R. Favazza, *Bodies under Siege: Self-mutilation and Body Modification in Culture and Psychiatry* (Baltimore: Johns Hopkins University Press, 1996).
17. See, for example, Emily Martin, *The Woman in the Body: A Cultural Analysis of Reproduction* (Boston: Beacon Press, 1992).
18. See Michel Foucault, *Power/Knowledge: Selected Interviews and Other Writings, 1972–1977*, trans. Colin Gordon, et. al. (New York: Pantheon Books, 1980), *The History of Sexuality*, trans. Robert Hurley (New York: Pantheon Books, 1978), and "The Birth of Biopolitics" from *Ethics: Subjectivity and Truth* trans. Robert Hurley, et. al. (New York: The New Press, 1994).
19. Bruce Mau, Friedrich Kittler, and Gabriele Brandstetter, *ReMembering the Body: Art and Movement in the 20th Century* (Stuttgart, Hatje Kantz, 2000), 14.
20. We note that in the biological sciences—often seen by those in the humanities as the bastion, if not source, of "naturalist" assumptions about the body–one must use specific model systems (based on a combination of biological and cultural properties) in order to understand different qualities of our bodily nature. See, for example, Robert E. Kohler, *Lords of the Fly: Drosophila Genetics and the Experimental Life* (Chicago: University of Chicago Press, 1994) and Lorraine Daston, *Biographies of Scientific Objects* (Chicago: University of Chicago Press, 2000).
21. This thematic dominates the middle work of Don Ihde. See his *Expanding Hermeneutics: Visualism in Science* (Evanston, Illinois: Northwestern University Press, 1998) for a discussion. Although the early work of Merleau-Ponty advances this perspective (especially through his use of the "zero point"), he does it much more delicately than often presumed. Compare, for instance, how he extends his theory of bodily perception by recognizing how bodies and objects are coconstituted in "The theory of the body is already a theory of perception" and "Sense Experience" from *Phenomenology of Perception* (London: Routledge, 1962). This formulation is extended in his radical reworking of these concepts as the "reversibility" of perception and the formulation of the body as the "chiasm" in *The Visible and the Invisible*, trans. Alphonso Lingis (Evanston: Northwestern University Press, 1968).
22. The recent work of Brian Massumi, especially his *Parables for the Virtual: Movement, Affect, Sensation* (Durham, N. C.: Duke University Press, 2002) goes a long way toward

correcting many of these deficiencies of the phenomenological perspective. Massumi's elaboration, in his "Introduction," of a conception of "incorporeal materiality" is especially useful, and his chapter on Stelarc ("The Evolutionary Alchemy of Reason: Stelarc") can be read productively against chapter ten of this volume. Other authors developing neo-phenomenological views consonant with the goals of this volume include M. C. Dillon, *Merleau-Ponty's Ontology* (Evanston: Northwestern University Press, 1997); Barbara M. Kennedy, *Deleuze and Cinema: The Aesthetics of Sensation* (Edinburgh: Edinburgh University Press, 2000); Manuel de Landa, *Intensive Science and Virtual Philosophy* (New York: Continuum, 2002); Dorothea Olkowski, *Merleau-Ponty, Interiority and Exteriority, Psychic Life and the World* (Albany, NY: State University of New York Press, 1999) and *Rereading Merleau-Ponty: Essays Beyond the Continental-analytic Divide* (Amherst, NY: Humanity Books, 2000); and Keith Ansell Pearson, *Viroid Life: Perspectives on Nietzsche and the Transhuman Condition* (New York: Routledge, 1997) and *Germinal Life: The Difference and Repetition of Deleuze* (New York: Routledge, 1999).

23. One of the best examples of this approach is Gilles Deleuze's analysis of the relation of the medium of film to the dynamics of image, time, and thought. See Gilles Deleuze, *Cinema 1: The Movement-Image*, trans. Hugh Tomlinson and Barbara Habberjam (Minneapolis: University of Minnesota Press, 1986) and Gilles Deleuze, *Cinema 2: The Time-Image*, trans. Hugh Tomlinson and Barbara Habberjam (Minneapolis: University of Minnesota Press, 1989). See also D. N. Rodowick's *Gilles Deleuze's Time Machine* (Durham: Duke University Press, 1997) for a sustained discussion of these two works in the context of film history.
24. Cited in Oxford English Dictionary, entry "Information."
25. See John Perry Barlow, "The Economy of Ideas," http://www.wired.com/wired/archice/2.03/economy.ideas_pr.html.
26. Geoffrey Nunberg, "A Farewell to the Information Age," in *The Future of the Book*, ed. by Geoffrey Nunberg (Berkeley: The University of California Press, 1996), 111. However, more research is required to substantiate the accuracy of Nunberg's claims about use of the term information, for something very much like the "abstract" sense of the term seems present in the early newspapers of the 1720s.
27. See Paul N. Edwards, *The Closed World: Computers and the Politics of Discourse in Cold War America* (Cambridge, MA: The MIT Press, 1996).
28. See James R. Beniger, *The Control Revolution: Technological and Economic Origins of the Information Society* (Cambridge, MA: Harvard University Press, 1986).
29. See Jack Goody, *The Logic of Writing and the Organization of Society* (Cambridge, UK: Cambridge University Press, 1986).
30. See Michael E. Hobart and Zachary S. Schiffman, *Information Ages* and Daniel Bell, *The Coming of Post-Industrial Society.*
31. Claude Shannon, "The Mathematical Theory Of Communication," *Bell System Technical Journal* 27: 3 (1948), 379. For a useful discussion of the place of Shannon in the early development of mathematical theories of information, see John R. Pierce, *Symbols, Signals, and Noise: The Nature and Process of Communication* (New York: Harper, 1961) and Kay, *Who Wrote the Book of Life?*, 91–102.
32. For an intriguing account of the military's interest in the relationships between mathematical models of information and noise and real-life battle noise, see Edwards, *The Closed World*, 209–37.
33. For accounts of the personal and institutional links between these researchers, see Steven J. Heims, *John von Neumann and Norbert Wiener: From Mathematics to the Technologies of Life and Death* (Cambridge: MIT Press, 1980) and *Constructing a Social Science for Postwar America: The Cybernetics Group, 1946–1953* (Cambridge: MIT Press, 1993), as well as N. Katherine Hayles, *How We Became Posthuman*, especially 50–112 and 131–59.
34. See, for example, James D. Watson, *The Double Helix: A Personal Account of the Discovery of the Structure of DNA* (New York: Atheneum, 1968); Robert C. Olby, *The Path to the Double Helix* (Seattle: University of Washington Press, 1974); Anne Sayre, *Rosalind Franklin and DNA* (New York: Norton, 1975); Horace Freeland Judson, *The Eighth Day of Creation: Makers of the Revolution in Biology* (New York: Simon and Schuster, 1979).
35. For a detailed historical account of the emergence of the term "information" in genetic discourse, as well as the objections of geneticists uncomfortable with that terminology,

see Lily E. Kay, *Who Wrote the Book of Life?*, 52–9 and 73–127. For a more theoretical discussion of Gamow, coding, and information, see Doyle, *On Beyond Living*, 25–64.

36. Kay, *Who Wrote the Book of Life?*, 39.
37. Marshall McLuhan, *Understanding Media: The Extensions of Man* (Cambridge, MA: The MIT Press, 1997), 8, 18.
38. For example, Manuel Castell's extraordinarily ambitious multivolume series *The Information Age: Economy, Society, and Culture* could not have been undertaken without Bell's earlier work. Moreover, it is this lineage of information, far more than the version of "information" current in the biological sciences, that encouraged Mark Poster's important attempt to establish the theoretical category of the "mode of information" in texts such as *Foucault, Marxism, and History* and *The Mode of Information*.
39. Jennifer Daryl Slack, "Introduction," in *The Ideology of the Information Age*, eds. Jennifer Daryl Slack and Fred Fejes (Norwood, NJ: Ablex Publishing Co., 1987), 4.
40. For Bell's own attempt to distinguish between these categories, see "The Axial Age of Technology Foreward: 1999" in Bell, *The Coming of Post-Industrial Society*, lxi–lxiv.
41. For discussion of the history and economics of information ownership, see Anne Wells Branscomb, *Who Owns Information? From Privacy to Public Access* (New York: Basic Books, 1994).
42. Or is the other way around? For a trenchant critique of the ways in which biologists often play something of a shell game with the correspondences between DNA/cells and program/computer, see Doyle, *On Beyond Living*, 14–7.
43. Hayles, *How We Became Posthuman*, 50–7, 6–18.
44. William Rasch, *Niklas Luhmann's Modernity: The Paradoxes of Differentiation* (Stanford: Stanford University Press, 2000), 60.
45. Espen Aarseth, *Cybertext: Perspectives on Ergodic Literature* (Baltimore, MD: Johns Hopkins University Press, 1997), 4.
46. See Manuel Castells, "The informational mode of development and the restructuring of capitalism" from *The Informational City: Information Technology, Economic Restructuring, and the Urban-regional Process* (New York: Basil Blackwell, 1992), 10. Although Castells' language often paints this change as "paradigmatic," a more than cursory reading of his texts underlines the fact that he does not see this change as revolutionary (and thus incommensurable with previous modes of development) as in Marxist conceptions of history as well as Thomas Kuhn's well-known definition of paradigmatic change in scientific knowledge.
47. Castells, *The Informational City*, 29.
48. C. A. Bayly, *Empire and Information: Intelligence Gathering and Social Communication in India, 1780–1870* (New York: Cambridge University Press, 1996), 4.
49. Richard Doyle, *On Beyond Living*, 8.
50. James R. Beniger, *The Control Revolution: Technological and Economic Origins of the Information Society* (Cambridge: MA: Harvard University Press, 1986), vii.
51. Beninger defines *control* as "any purposive influence on behavior" (p. 7).
52. Michel Foucault, *Discipline and Punish: The Birth of the Prison*, trans. Alan Sheridan, (New York: Vintage Books, 1979), 155.
53. Jay David Bolter and Richard Grusin, *Remediation: Understanding New Media* (Cambridge, MA: The MIT Press, 1999), 5.

PART I

Bodies Before the Information Age

CHAPTER 1

Reading the "Sensible" Body: Medicine, Philosophy, and Semiotics in Eighteenth-Century France

ANNE C. VILA

Over the past twenty years, through inspirations as diverse as Foucault, feminism, psychoanalysis, and cultural studies, scholars have taken a renewed interest in the body and the meanings it held during the French Enlightenment.[1] This interest has in turn brought about a rediscovery of the period's biomedical theory, a field that did much to shape knowledge during an era when the mind–body relation, the physical contours of the psyche, and the forms, limits and mechanisms of knowledge were all subjects of intense scrutiny and speculation.[2] Thanks, in part, to the secular, empiricist, monistic intellectual climate cultivated by the *philosophes*, medicine and physiology were influential not just in discussions of materialism, sensationalist philosophy, and scientific method, but also in the period's lively debates on human nature and diversity, on civilization's ambiguous effects on the mind and morality, and–last but not least–on the widespread, multifaceted culture of sensibility.

The body has, of course, served since antiquity as "man's most available metaphor."[3] However, each cultural–historical moment has its own particular ways of using the body as a metaphor or explanatory principle. In feeling-obsessed eighteenth-century France, the most productive and problematic somatic metaphor was that of the sensible body, a rich but ambiguous concept that provided physicians, moral philosophers, and literary writers with a very particular way of seeing things as they conducted their inquiries into the body, the mind, and contemporary society.[4] Several ideas and theoretical suppositions were involved in that way of seeing: a certain notion of how the internal space of the body was organized; an assumption that the physical and moral realms of human existence were closely interrelated; a conviction that all natural phenomena had a profound interconnection; and a belief that there were causal structures underlying those phenomena—structures that the philosophically minded observer could uncover, provided that he had sufficient patience, perspicacity, and instinctual feeling for the operations of nature.

With its claims to privileged knowledge of the internal operations of the body, medicine has always had a philosophical aspect. However, the vitalist physicians of the French Enlightenment used the term "philosophical" to refer specifically to their new, putatively revolutionary vision of human life—one that, in pointed opposition to the mechanistic models that had long dominated the field—saw the body as a vibrant entity driven by its own inner dynamics. They perceived those dynamics as turning almost entirely around sensibility, a property they believed was the key both to understanding the living body and to making medicine a fully philosophical endeavor. As the prominent Montpellier theorist Théophile de Bordeu put it,

> Sensibility can serve quite well as the basis for explaining all of the phenomena of life, whether it be in the state of health or in illness. . . . This is therefore the way of considering the living body that has been adopted by those who, among modern thinkers, have carried their speculations beyond practical medicine and the systems received in the schools at the beginning of the century. Such is the scope that philosophical medicine has assumed concerning the purely material functions of the body. The reign of feeling or sensibility is among the most extensive; feeling is involved in all the functions; it directs them all. It dominates over illnesses; it guides the action of remedies; it sometimes becomes so dependent upon the soul, that the soul's passions take the upper hand over all the changes of the body; it varies and modifies itself differently in almost all the [organic] parts.[5]

This account of sensibility's dominion over both somatic function and medical knowledge reflects the privileged, polysemous quality it enjoyed in Enlightenment culture at large. In the socio-moral theory of the day, sensibility was associated with notions like sympathy, virtue, pity, benevolence, tender feeling, and compassion.[6] Yet it was also central to European physiological terminology beginning in the 1740s, when it edged out "irritability" as the word most commonly used to describe the innate reactive capacity held to underlie all the phenomena of life in the human body. In fact, the various meanings attached to sensibility tended to be mutually permeable because eighteenth-century authors used the word as a bridging concept—a means of establishing causal connections between the physical and the moral realms.[7] Sensibility was, in short, fundamental to this period's effort to forge a global, unified understanding of human nature: it was seen as the root of all human perceptions and reflections, as the innate and active principle of sociability that gave rise to human society, as a kind of sixth sense whose special affective energy was essential both to virtue and to art, and finally, as the paradigmatic vital force whose actions could be detected in every bodily function, be it healthful or morbid.

Eighteenth-century French thinkers thus exalted sensibility as an essential, all-revealing quality, the mainspring of all of the relations vital to human existence, from the intraorganic to the cosmic. At the same time, however, they viewed it as a potentially dangerous quality that could lead to emotional excess, moral degeneracy, and physical debilitation. Physicians were particularly concerned about the risks involved in sensibility and frequently cited it as the cause underlying the pervasive and troublesome problem of nervous ailments in worldly women, scholars, and other oversensitive sorts. Sensibility's pathological effects seemed especially glaring in France's cities, where life seemed to be a daily battle against all kinds of potential irritants, moral as well as physical.[8] Exacerbating this threat was the putative degeneration that had occurred in the average Frenchman's constitution, as measured against the robust, relatively insensitive state of the nation's primitive ancestors. It was the darker side of sensibility that made a full-blown medical philosophy so crucially important for contemporary society, for only the physician equipped with such a philosophy could locate all of the hidden "rallying points" in the sensible body—the organs, centers, or nerve plexuses that might become trouble spots under excessive stimulation.[9]

Sensibility was, of course, strongly associated with the period's ethical codes and class distinctions: it played a pivotal role in both sentimentalism and the chesslike game of seduction, physiognomic analysis and social one-upmanship that constituted aristocratic life in eighteenth-century France.[10] However, the term's sociomoral connotations came to be subsumed within the body-based meanings that pervaded the discourse of sensibility from midcentury on. That is, after sensibility emerged as the central notion of biomedical theory in the 1740s and 1750s, the process of reading and interpreting its manifestations became a fundamentally organic endeavor: sensationalist philosophers writing in the wake of John Locke built their models of the mind on the physical operations of the senses[11]; and writers as diverse as Denis Diderot, Jean-Jacques Rousseau, Choderlos de Laclos, and the Marquis de Sade strove to incorporate sensibility's physical, material implications into their aesthetics, their social theories, and their literary works.[12] Similarly, the primary meaning of semiotics during this period was medical: although semiotics in the linguistic sense certainly existed as an idea,[13] the term itself typically denoted the signs and symptoms that physicians—along with the many nonphysicians who drew on their constructs—used to discern illness or dysfunction in the body. Among the many things that made sensibility so compelling was the fact that it seemed to provide an innovative way of connecting those signs and symptoms into a rich signifying whole.

Conditions were thus ripe in eighteenth-century France for a productive convergence between sensibility and semiotics—a convergence that created some very distinct models of intra- and interbodily communication. To uncover those models, I begin here by examining how theorists associated with the Montpellier medical school made sensibility a specifically medical concern, while also promoting medicine as the most authoritative branch of natural philosophy—and medical semiotics as the ultimate means of "observing" sensibility by tapping into the body's inner voice or voices. I then sketch how the Montpellier theory of this property was received, both in its specifics and in its larger moral–philosophical repercussions, by Denis Diderot, a celebrated *philosophe* who saw the decoding of sensibility as the cornerstone of philosophy and literature, and whose dialogue *Le Rêve de d'Alembert* (1769) stages a playful response to philosophical medicine as embodied by a fictional personification of the real-life physician, Bordeu.

I. Sensibility and Medical Semiotics in the *Encyclopédie*

Given Enlightenment medicine's overt philosophical aspirations, it is not surprising that it occupied a central place in the *Encyclopédie*, the period's most ambitious effort to coordinate all existing knowledge under the umbrella term "sciences of man."[14] Some historians have suggested that, in its medical content, the *Encyclopédie* was primarily a work of vulgarization that translated conventional medical wisdom into terms that could be grasped by a general readership.[15] Vulgarization is, however, too passive a term for describing the way in which medicine was represented in the *Encyclopédie:* as deployed by a small but energetic group of Montpellier medical vitalists, this multivolume work became a veritable *défense et illustration* of the most dynamic ideas emanating from contemporary medical theory—including some fascinating notions on how medical semiotics could be transformed into a faithful, authoritative transcription of the sensible body's true inner language.[16]

The campaign to use the *Encyclopédie* as a means of promoting sensibility and philosophical medicine was conducted by a mere handful of authors. The most prominent of the group was Bordeu: although he only contributed the article CRISE to this work, he is one of the authors most frequently cited throughout the network of Montpellier-inspired medical articles. Henri Fouquet made six contributions, most notably "SECRÉTION," an abstract of Bordeu's treatise on glands, and the long entry "SENSIBILITÉ, SENTIMENT" (*Médecine*). However, the real driving force behind this effort was Jean-Jacques Ménuret de Chambaud, who wrote at least eighty-five articles for the last ten volumes of the *Encyclopédie* (volumes 8–17, letters H to Z, all published in 1765). Ménuret was only moderately successful as a practitioner and left little else to posterity beyond his numerous *Encyclopédie* entries.[17] However, his contributions to this monumental enterprise are far more significant than has generally been acknowledged: he not only used his individual articles to maneuver vitalist medicine into a position of prominence over other, more established natural sciences, but also deftly deployed the cross-referencing system created by Diderot, the *Encyclopédie*'s energetic, influential coeditor.

In addition to underscoring the innovative physiopathology on which the medicine of sensibility was based, Ménuret and his colleagues were intent on demonstrating its profound importance for the hermeneutic art of medical semiotics. That part of their campaign took place in a series of texts that include Ménuret's articles "OBSERVATEUR, OBSERVATION, POULS," and "PROGNOSTIC," as well as the anonymous essay SEMEIOTIQUE and Bordeu's entry on CRISE. These articles served to publicize the new semiotic system put forth in Bordeu's *Recherches sur le pouls par rapport aux crises* (1757), while also portraying the *médecin philosophe* as a heroic figure uniquely qualified to read the natural signs sent out by the body.

"SEMEIOTIQUE" is relatively short, but its reformist bent is immediately apparent. The author (most probably Ménuret) begins by declaring that semiotics should be recognized as a powerful instrument for penetrating into the inner recesses of the healthy or ailing body.[18] This approach to semiotics stems directly from the new Montpellier doctrine of sensibility, which invested both the patient's body and the practitioner's with a special, dynamic mode of operation. The patient's body is thus held to be driven by inner phenomena that have a "reciprocal correspondence," a "mutual linking," and a "natural gradation" (937). The signs emitted by this body are similarly naturally interrelated, such that a physician need only follow the pattern of actions and reactions formed by these signs in order to decipher what is causing the ailment with which he is confronted. Only a certain kind of physician, however, is able to detect those patterns: "Only the enlightened observer can direct a penetrating gaze into the most hidden recesses of the body, distinguish therein the state and the disorders of the various parts, recognize through external signs the illnesses that are attacking the internal organs, and determine the particular character and seat of those illnesses." The body is a transparent machine for such an observer because he possesses a vision of its inner workings that goes beyond the usual limits of human understanding: "The mysterious veil that hides the knowledge of the future from frail mortals tears open before him; he sees with a confident eye the different changes that must occur in health or illness; he holds the chain that connects all events, and the first links that come into his hands reveal to him the nature of those that will come after, because Nature only varies in her external appearances: deep down, she is always uniform, and always follows the same course."

The science of semiotics is thus predicated on the assumption that every action, every movement, and every reaction of the body is a richly significant link in an ongoing chain of physiological or pathological events. The physician's task accordingly consists in garnering as much information as he can from the body's perceptible qualities and functions—its excretions and secretions, its color and heat, its respiratory rhythms, and above all its pulses (the reader is referred at this point to "POULS, RESPIRATION, SUEUR, URINE, etc." (938)). These signs are to be judged not by traditional symptomalogical divisions, decried here as arbitrary and metaphysical, but rather on the basis of a painstaking observation of the same body over time—one that takes into account its temperamental type, its habits of sleep, digestion, excretion and so on, and the illnesses it may have suffered in the past. Having assembled those facts and compared them with the present signs, the observing physician should be able to determine how best to maintain that patient's health, or guide him or her through an impending pathological crisis.

Ménuret's article "PROGNOSTIC" offers yet another glorifying scenario of the physician at work, this time as a prognosticator who, using the "luminous torch" of his observations and experience, can see that the triumph of nature and health is often underway even through the gravest, most frightening complications.[19] As Ménuret elaborates this tableau, his physician becomes more and more heroic in character, using his divinatory talents to instill order into both the patient's illness and the familial entourage. The physician's exceptional physical and intellectual qualities also come to the fore—most particularly his uniquely attuned sense organs, which allow him to pick up on the signs of the various disease events occurring in the patient's body, to understand their natural interrelation, and to predict the manner in which the illness will be resolved through evacuation from the body. Once he has reconstructed the script that fits the patient's specific ailment, the skilled physician acts not to interfere in the illness, but rather to guide it to its most complete denouement: "Ceaselessly devoted to following nature, and to warding off anything that might delay nature's operations or prevent its success, he will skillfully proportion his assistance to both the demands of nature and the length of the illness; he will prepare from afar a complete and healthy crisis, a prompt and speedy convalescence, and firm and constant health thereafter" (429).

Ménuret thus paints a remarkably optimistic picture of medical treatment as a mutually beneficial team effort between nature and the enlightened attending physician, who deploys his semiotic talents in order to guide the patient's impending bodily crisis to a felicitous outcome. To succeed in this endeavor, the physician must be equipped with both a reliable bank of clinical observations and a sure command of the body's most important signs. Those two related medical arts—observation and semiotic skill—have, Ménuret asserts, fallen into disuse since the days of Hippocrates, largely because so many theorists have promoted frivolous, arbitrary or nonexistent signs. Fortunately, however, a handful of contemporary physicians more philosophical in temper have revived the principles of sound prognosis: "It is only recently that prognosis has received a new luster and greater certitude, through the observations that have been made on the pulse in relation to crises" (429)—an obvious allusion to Bordeu's treatise on pulses, published eight years earlier. In Ménuret's view, one can excel in the art of prognosis only by embracing the basic tenets of Montpellier vitalism: the idea that every disease state follows a specific "critical" pattern, and the belief that, of all medical signs, pulses are the most significant.

Before ending "PROGNOSTIC," Ménuret ventures the intriguing suggestion that it might be possible to use the medically based methodology of semiotic prognosis outside the field of medicine itself, in the form of a systematic moral semiotics. This would, he argues, vastly improve our understanding of human society by revealing "the hidden forces that motivate men," which depend as much on the individual's physical disposition as on his or her present circumstances and dominant passions (430). Although Ménuret merely toys with this possibility here, he nonetheless provides a glimpse of one of his major preoccupations as an encyclopedist: to champion medicine and its divinely gifted practitioner as shining examples for all the human sciences.

This project is most explicitly carried out in the centerpiece article "OBSERVATION", an essay designed to dramatize the philosophical spirit that, in Ménuret's view, has recently reinvigorated medical thinking.[20] As he presents it, that spirit is destined to remake the medical field not just from within but also from without, by giving it a new place of prominence in the larger framework of human knowledge. Ménuret accordingly composes "OBSERVATION" to illustrate both aspects of this revolution: he continues his effort to maneuver semiotics into a privileged position in relation to the other branches of medicine, while redefining the general science of observation so as to depict medicine—the art of observing and coding the human body—as the alpha and omega of all natural–philosophical studies.

Although "OBSERVATION" is categorized as a medical article, medicine is not announced as its principal concern, at least not at the outset. Instead, Ménuret begins by denouncing experimentation, an approach to nature that he considers artificial and distorting, and promoting observation in its stead (313).[21] He compares the experimenter and the observer by way of synecdoche: each is reduced to the main sensory organ he uses for his investigations. The experimenter is thus represented by his hand—an organ that manipulates and disfigures—whereas the observer is depicted as an eye, the organ that has the most distant and unobtrusive relation to its object. Experimental methods, Ménuret insists, blind their practitioner to the virtues of simple observation and denature the object observed, producing phenomena that are artificial, arbitrary, and "ordinarily refuted by observation" (313–14). Given the multiple risks involved in experimenting, Ménuret proclaims observation to be a far surer path to knowledge. Following that principle, he presents a program for investigating nature that resembles Diderot's *De l'Interprétation de la nature* (written a decade earlier)—with the difference that Ménuret gives a distinctly organic twist to the way that the facts of natural philosophy should interact: "several facts taken separately appear dry, sterile, and fruitless; as soon as one brings them together, they acquire a certain action, and take on a life that arises from the mutual agreement, reciprocal support, and interrelation that ties all of them together" (314). By the end of this article, the homology between the facts of organic matter and the organic-like interaction of facts themselves becomes something like a founding principle for medical discourse in general.

Ménuret's discussion in "OBSERVATION" moves steadily from fields of inorganic natural observation (astronomy, geology, physics) into the organic realm of the science of man, where the medical observer reigns supreme. "From whatever angle one envisions him," he declares, "Man is the being least suited to serving as an experimental subject; he is the most appropriate, most noble, and most interesting subject of observation, and it is only through observation that one can make any progress in the human sciences" (315). The "noble" and "divine" science of medicine is, he continues, the field that most excels in observing humankind, and thus warrants a particularly close examination. Ménuret then proceeds to list every type of medical observation imaginable (anatomical, physiological, hygienic, pathological, meteorological, and therapeutic) in such a way that all appear to be united under the globalizing rubric of semiotics, which he presents as a system into which every observed bodily fact can be fit (319). If it were properly designed, such a system of facts would not be a system at all, but rather the true language of nature (320).

Sensibility, as Ménuret describes it here, provides "a general point of view that serves as a rallying point for all of the facts that observation furnishes" (318). This vitalist theory has, he adds, been endorsed by two distinguished contemporary thinkers: "A famous physician (M. de Bordeu) and an illustrious physicist (M. de Maupertuis) have agreed that Man, envisioned from this brilliant and philosophical viewpoint, can be compared to a swarm of bees that undertake to attach themselves to a tree branch; one sees them pressing together, supporting each other, and forming a kind of whole, in which each part possesses its own particular mode of life but also contributes, through the correspondence and direction of its movements, to upholding this sort of *life* of the entire body, if one can use such a term to refer to a simple connection in action" (318). The specific works to which Ménuret is alluding here are Bordeu's *Recherches anatomiques sur la position des glandes, et sur leur action* (1752) and Maupertuis's *Système de la nature, ou, Essai sur la formation des corps organisés* (1756), both of which use a version of the bee swarm metaphor that Diderot would later incorporate into *Le Rêve de d'Alembert.* What is most intriguing about Ménuret's use of this image is the way he applies it not just to the organic body, but also to the body of facts that physician–philosophers cull through their semiotic observations. A curious operational equivalence emerges in this article between the physical body being observed, and the textual body that is formed by a well-assembled group of medical facts: those facts, Ménuret declares, should "communicate" via natural connections and affinities, because they refer to a living body that is likewise driven by a set pattern of sympathetic physiological resonances—all of which links its parts into an economic whole. Unlike the plain, inert theories of old-style iatromechanism–metaphorically represented as a "flock of cranes that fly together in a certain order but without helping or depending on one another" (318)—the ideas that arise from vitalist medicine are destined to come to life. Ultimately, therefore, the force of bodily sensibility coordinates not only the physiopathological phenomena that the physician seeks to decipher, but also the way in which he obtains and disseminates his particular kind of knowledge. The bee swarm image thus figures twice in "OBSERVATION": once as a model for the animal economy, and once as a model for organizing medical discourse in the most simple, natural form possible.

Ménuret's lively resonating model of medical discourse is based, in part, on an intriguing application of analysis in the manner popularized by the sensationalist philosopher Condillac, according to which all facts are predestined to take their proper, natural place in the observer's mind—in this case, because they emanate directly from a similarly ordered animal economy. It is also a particularly dynamic realization of what Diderot envisioned an encyclopedia should be: a work that would literally arouse its readers and stimulate them to see all sorts of new connections among ideas and in nature. The textual network that Ménuret and his medical collaborators construct in the *Encyclopédie* literally embodies such connections, because it is composed of articles that not only describe but mimic, through their sympathetic interactions, the active-reactive resonating quality of the body itself. This is particularly apparent in articles on medical signs like "LANGUE," "RESPIRATION," "SUEUR," "URINE," and "POULS," each of which is written by Ménuret, and each of which is copiously cross-referenced so as to direct the reader to Ménuret's feature article, "POULS." In fact, whatever direction one takes in reading the Montpellier medical articles of the *Encyclopédie,* one almost invariably ends up at "POULS"—a subject to which Ménuret devotes an unprecedented thirty-five-page exposition in tome XIII.[22] What, one might well ask, is so special about pulses that Ménuret slips in a cross-reference to this article at every possible occasion?

At first, "POULS" appears to be a painstaking review of virtually every doctrine that had ever been proposed to explain the information transmitted by pulses: Ménuret traces the notion from ancient theorists like Galen, to the Chinese (whose poetic appellations, like the "pulse that resembles a swimming fish," seem to strike his fancy), to the seventeenth-century iatromechanists, to musically minded contemporary physicians like François Nicolas Marquet, author of the *Nouvelle méthode, facile et curieuse, pour apprendre par les notes de musique à connoître le pouls de l'homme et ses différens changemens* (Nancy, 1747). The real aim in "POULS" is not, however, to provide a comprehensive history of this medical notion. Instead, Ménuret seeks to draw attention to the new doctrine of pulses proposed by Bordeu in his *Recherches sur le pouls.* As he summarizes this treatise for the general reading public, Ménuret provides a revealing glimpse of the medical practices that were actually in use at the time—practices that, as he describes them here, culminate in the delicate art of tapping the resonances of the sensible body through its pulses.

Bordeu's treatise is, in fact, the source of Ménuret's general conviction that semiotic techniques like pulse taking are the key to understanding every aspect of human physiology. As Ménuret puts it, "the pulse is an essential object, which is causally linked to the very constitution of the machine, to the most important and most extensive of the [physiological] functions . . . When the pulse's traits are skillfully grasped and developed, it reveals the entire interior of Man."[23]

Bordeu himself provides the clearest explanation of what, precisely, is revealed through the pulse:

> The movements of the pulse depend undoubtedly on the sensibility of the nerves in the heart and arteries. . . . Every organ being sensible in its own way, and being unable to perform its functions–particularly the most forceful—without making some impression on the entire nervous system, it is evident that each organ must make a particular impression on the pulse: this impression will be almost imperceptible when, as in the natural state, an organ is no more agitated than usual; by contrast, the impression will be quite evident if, as in the state of a critical effort, the organ is impeded in its functions and tries to make some extraordinary effort.[24]

Thus, for every organ or center of sensibility, there exists a corresponding pulse–for example, a "stomach pulse," a "pectoral pulse," and a "nasal pulse." Ménuret explains to his *Encyclopédie* readers that each organ-specific pulse can be further differentiated into a "pulse of irritation," a "developed pulse," and a "critical pulse," the last of which tells the physician how a disease will be resolved, by announcing when and where its critical evacuation is going to occur in the patient's body ("POULS," 231).[25] Of course, a pulse-based prognosis can be tricky, given that simple pulses are sometimes combined with others (235). However, by consulting other available signs—and keeping in mind the resonating pattern of each individual sensible body—the medical observer can sketch a very accurate tableau of his patient's natural internal order, and thereby see very clearly what pathological imbalance is currently upsetting it (240).

The pulse, according to this doctrine, bears the signature of each organ, and thus allows the medical observer not just to translate physio-pathological phenomena into treatable terms, but to "hear" the body's inner language directly. Pulse taking is consequently a uniquely incisive and reliable tool for gauging sensibility because it retraces specific manifestations of the property back to the organ that is in distress. Bordeu's system also entails a special map of pulse points, constructed in keeping with the various structural-functional axes that, in his view, organize the body's internal space.[26] Underlying that topology is the rather peculiar gland-based model of local physiological activity that Bordeu had proposed a few years earlier, in his *Recherches sur les glandes*. As we shall see shortly when we turn to that text, Bordeu's physiological doctrine sheds a provocative new light on what is actually going on in an organ when a physician detects its specific pulsations.

First, however, let us sum up the art of medical semiotics as Bordeu himself describes it: although it requires the same set of talents as those that Ménuret ascribes to the enlightened medical practitioner—extensive clinical experience, knowledge of anatomy and physiology, a firm grasp of the science of pathological signs, and a refined faculty of sensory perception—it also demands a cerebral finesse that only true connoisseurs possess: "The most clear-seeing and confident physicians in this field of knowledge are those whose head is the most fully furnished with the images of all the various species of pulse. . . . It is for the same reason, and by virtue of the clearness of these ideas, that physicians whose tact is well trained can sometimes determine the state of a pulse through an initial sensation that is almost automatic and often precious: [this is] a fortunate sort of inspiration [*enthousiasme*] of which coolheaded, lazy minds are incapable, and which is only fully appreciated by connoisseurs."[27] In other words, to interpret pulses fully, the physician must enter into a state of fervent, inspired communication with the body before him. In practice, then, medical semiotics is not a formal theory, but a visceral understanding of another body that, as Bordeu describes it in his *Encyclopédie* article CRISE, arises from "taste, talent, and experience."[28]

Visceral understanding is, however, only the first stage in the quest for full medical knowledge: "One must return from this passive state, and paint exactly what one has perceived in this sort of ecstasy, and express it in terms that are thoughtful and arranged so as to enlighten the reader just as nature would."[29] There, in sum, is the Bordelian definition of the master physician: he is an observer who can serve as an ecstatic vessel for the truths of nature at the moment of diagnosis while also taking control of this inter-bodily communion in order to articulate it and transcribe it for posterity. He is, in other words, a philosopher figure who works in a manner very close to that of the great artist whom Diderot described in various aesthetic writings, including the *Éloge de Richardson*, the *Paradoxe sur le comédien*, and *Le Rêve de d'Alembert*.

II. Bordeu, Diderot, and the Implications of Organological Sensibility

Scholars of Diderot have generally interpreted his interest in Bordeu's ideas in a narrow, historiographic sense: they have limited their considerations of Bordeu's medical doctrine to the facts that Diderot culled from it for inclusion in the *Rêve de d'Alembert* and the *Éléments de physiologie*.[30] Such an approach, however, overlooks the dramatic form that Bordeu's ideas take in the *Rêve*, where Bordeu is cast as a fictional interlocutor along with Jean d'Alembert and his mistress, Julie de l'Espinasse. In discussing this text, many critics have treated the ideas expounded there by the character Bordeu as purely Diderotian pronouncements on selected topics in natural philosophy–specifically, materialism, sensibility, generation, and teratology. Bordeu's appearance in this philosophical dialogue has thus been characterized as poetic license, on the assumption that the Bordeu persona of the *Rêve* is as unrelated to the historical Bordeu as are the fictive d'Alembert and Mlle. de l'Espinasse to their flesh-and-blood counterparts.[31] Yet Diderot was far from arbitrary or whimsical in casting Bordeu as his "modern-day Hippocrates." Instead, he very deliberately incorporated Bordeu's doctrine of the animal economy into both the discussion and the design of the *Rêve de d'Alembert*, which one could view as an dramatization—at times parodic—of Bordeu's very particular concept of vital sensibility.

Before considering how Diderot translates Bordeu's vision of the body into the *Rêve*, let us briefly examine that vision on its own terms by considering

Bordeu's *Recherches sur les glandes*, which first established him as a major theorist on the French medical scene. It is here that Bordeu explains precisely what is going on inside the body whenever a sign of sensibility is given off, be it through a pulse variation, a humoral evacuation, or some other indication: somewhere in the body, an organ has been stimulated and roused into action.[32] Its neutral title notwithstanding, the *Recherches sur les glandes* was more than a simple anatomical treatise on glands: it was a daringly innovative effort both to dismantle the accepted mechanistic explanation of glandular function, and to take on the highly limited definition of sensibility that Albrecht von Haller had expounded in the widely influential *Dissertation on the Sensible and Irritable Parts of Animals* (1755). Whereas Haller insisted that the reactive properties of irritability and sensibility should be seen as quite distinct, Bordeu just as insistently interconnected them.

Vital reactibility is, for Bordeu as for Haller, an innately organic capacity to respond to irritations emanating from an external source; Bordeu, however, locates this elemental physiological reaction not at the level of the fiber, but at that of the organ. He insists that, contrary to mechanistic theory, glands are not passive and compressible, but rather able to engage in their own activities because they are richly innervated. In his view, the most illustrative gland of all is the male sexual organ, an obviously irritable body part whose mode of functioning cannot be attributed to mechanical compression: "one cannot help resorting to the sensibility and *vibratility* of the nerves. . . . The convulsions and tremors that precede this excretion in most animals are familiar to everyone; an anatomically informed philosopher discovers therein a good number of subjects for meditation that elude the common herd" (125). Seminal excretion is thus the first example that Bordeu invokes to dismantle the mechanistic model of glandular secretion. To reinforce his point, he proceeds to recount the steps involved in the convulsive mechanism by which semen is evacuated: the erection of the penis, which is necessary to prepare the organ for excretion; the spasm that overtakes the genital parts and everything around them when they become aroused; the *secousses* and *frottements* that "awaken" the organ into a general convulsion; and finally "the relaxation that occurs when the scene is finished."

Bordeu concludes that this observation is "already sufficient to found a theory of the mechanism of excretion" (126). In fact, however, he uses the penis only to illustrate one half of his theory of glandular function: the role of irritability, or the gland's power to perform a movement in response to a stimulus/irritant. To demonstrate the second half—the role of feeling or sensibility—Bordeu looks elsewhere in the body, to what he considers to be the feminine analogue to seminal excretion: lactation. In this observation, Bordeu describes not only the function of stimulation as carried out by the nursing infant, but also the intense, ticklish feelings that such stimulation provokes in the breast: "There is no wet-nurse who has not felt this tension, and the kind of tickling that results from it. Most of them say that they can feel the milk rising. The breast becomes rounder, harder, and swollen, and there are women who endure tugging sensations that are felt all the way to the shoulder and loins, and even to the arms. These tugging sensations are painful in some women; but usually they feel a more or less voluptuous titillation" (126). The titillation that a wet nurse experiences when breastfeeding is not incidental, but rather crucial to the process of excreting her milk; nor is the erotic language that Bordeu uses to describe this mammary "seduction" merely coincidental. Bordeu quite consciously psychologizes and eroticizes his description of sensibility in this female gland: he depicts the breast as an organ that rather capriciously decides when to yield up its milk.[33]

Once he has established his model of the autonomous action of glands, Bordeu uses it to assert that glands are active organs that exhibit individual "tastes" or sentiments of their own, thanks to the nerves that serve them and maintain their particular "tone" or "tact" (145–50). It is these local nerves—which Bordeu sees as an integral part of the gland they sensitize—that are irritated, thereby provoking the organ to manifest its "particular life" by secreting a particular humor (157). What is curious in all of this is not so much the anatomical error that, in the eyes of some of his contemporaries, Bordeu commits in constructing his glandular theory.[34] Instead, it is the fact that Bordeu vitalizes glands by sexualizing their structure and function. The innovative, provocative aspect of Bordeu's *Recherches*, therefore, resides not just in his antimechanistic model of glandular activity: it lies also in the very deliberate way that he interweaves into his physioanatomical descriptions many of the reigning cultural ideas about sensibility, like its associations with feeling, with erotic arousal, and with women. The strikingly sexy and provocative dimension of Bordeu's biomedical theory has generally been glossed over by the historians of medicine who treat him.[35] Diderot, by contrast, seemed keenly aware of it.

Late in the *Recherches*, Bordeu observes that "what we have said about the particular action of organs can shed light on what happens during an inflammation" (194). The process of inflammation, he explains, is exactly analogous to what occurs when one of the bees in a swarm becomes agitated and antagonistic—and to what happens in a gland when it enters into a secretory crisis: "An inflamed part becomes, in a way, a separate body, at least for a time. It has a kind of action that goes beyond that which creates life; it creates a separate circle . . . what happens in this part resemble what goes on in glands and in other organs toward which the blood is directed, where there occur various kinds of torrents that medical practitioners have called raptus" (194–96). A diseased state is therefore created by excessive vitality in a body part; this is the sort of crisis that medical semioticians must decipher and treat. Pulses provide a direct indication of this process: when an organ is intensely aroused, its surrounding nerves and blood vessels become so irritated that they divert the circulation of blood toward it, creating a special tension or pull on the entire vascular system. This constitutes that organ's unique pulse or signature, which the astute physician can read as a means of tracing the crisis to its specific organic locus.

Bordeu's extension of his organological model into disease theory sheds an intriguing new light on his entire system for diagnosing and transcribing the bodily language of sensibility: what that language reveals is, in essence, the occurrence of glandular orgasms. According to Diderot's free-wheeling application of Bordeu's physiology, if all organs in the body behave like glands, which in turn behave like penises, there is no reason not to assert that whole organisms—people—behave the same way. The communication of sensibility thereby becomes inherently sexual in nature, regardless of where the sensible reaction is coming from, body or brain. To reflect that idea, Diderot designs the successive dialogues of the *Rêve de d'Alembert* to take the form of an elaborate literary demonstration of sensibility and its seductive effects on the body and mind.[36] One might even call the interlocutors of the *Rêve de d'Alembert* talking glands: like the organs in Bordeu's organological confederation, they are, by turns, feverish and inflamed in their outbursts, dormant in repose, and sympathetically resonant through the course of their dialogues.[37] In the process, Diderot's fictional characters demonstrate that reading the sensible body can be an act both of physical crisis, and of "philosophical" creation.

The main dialogue of the *Rêve de d'Alembert* is set up as a physiology lesson: its ostensible purpose is to explain what organic sensibility is all about to uninformed or unconvinced characters like d'Alembert and Mlle. de l'Espinasse. Before Dr. Bordeu enters the scene, d'Alembert has spent a feverishly agitated night in the grips of sensibility, which is both the subject and the cause of his dream; he is, at the dialogue's outset, still asleep, and Mlle. de l'Espinasse is standing guard at his bedside. The presence of this sleeping interlocutor immediately undermines the possibility that there will be a clear, linear order to the conversation; for sleep, as Dr. Bordeu defines it, is "a condition during which the animal no longer exists as a coherent whole and in which any collaboration or subordination between parts is suspended. The master is given over to the good pleasure of his vassals and the uncontrolled energy of his own activity."[38] At the same time, because dreams are also etiologically linked to "some erethism or temporary disorder" (215), the setting of the *Rêve* is designed to require Dr. Bordeu's expert services: a "critical" condition like d'Alembert's passing illness calls for a physician who is able to decipher sensibility's inner workings within the body.

Dr. Bordeu is thus brought into the text not only as a medical practitioner who can treat d'Alembert's condition, but also as a *médecin philosophe* who possesses a rare capacity to interpret sensibility. His medical knowledge comes into play immediately: first, in the semiotic reading that he performs by taking d'Alembert's pulse and other vital signs ("Pulse all right . . . a bit weak . . . skin moist . . . breathing easy" (165)); and second, in his ability to make sense of d'Alembert's mumblings about "sensible molecules," bees and so on. Having pronounced that the patient has "only a bit of a temperature and there won't be any after-effects," Dr. Bordeu obligingly offers to decipher Mlle. de l'Espinasse's transcription of d'Alembert's unconscious mutterings from the night before, which she calls "some gibberish about vibrating strings and sensitive fibres" (166). The gist of the mutterings transcribed by Mlle. de l'Espinasse is, precisely, the bee swarm metaphor that Bordeu had used in his *Recherches sur les glandes* to explain the problematic passage between the local sensibility of discrete vital parts (or individual "bees"), and the general sensibility of a continuous living whole (like a swarm). Diderot's use of this organic metaphor demonstrates not just sensibility's destabilizing effects, but also its formative powers—both in the body, and in the mind of the philosophical observer. By repeating and transmuting the bee swarm analogy over the course of their dialogue, the interlocutors end up elaborating a model of sensibility that seems to explain everything, including the nature of matter, organic generation, physiological functioning, disease, and the intellectual and moral capacities of human beings.

The *Rêve de d'Alembert* is, therefore, a thought experiment on sensibility in a very literal, material sense: thinking about the property has direct consequences on both the organs of the experimenters and their mental faculties. In this ongoing observational free-for-all, each interlocutor's body is apt to become a physiological instrument for demonstrating the text's thesis that everything in nature comes down to sensibility and organic movements. Diderot's "everything," of course, covers a huge range of functions, including modes of procreation; embryonic differentiation; cognitive faculties like thinking, memory, the will, and the self; pathological conditions like vapors and erethism; the violent pleasure and pain felt by "mobile souls" during a pathetic spectacle; the natural underpinnings of vice and virtue; and the crossbreeding of animal species. The discussion of all these topics consists in a basic practice of "undressing" a body by stripping it down to its origins in a rudimentary sensitive fiber or fascicle that precedes the formation of organs.

One might call the mode of talking about sentient beings that is adopted in the *Rêve* an example of the "hypervital" style that Leo Spitzer declared to be an essential Diderotian trademark.[39] Yet it can also be read as a pointed form of mimicry: Diderot's characters imitate the discursive patterns that had been popularized through the medico-philosophical writings of the day, including those of the real Bordeu. In that sense, the "hypervital" rhythm of the *Rêve* recalls the close alliance that the Montpellier physicians strove to establish between the language of the sensible body and their own descriptive language–both of which were held to operate according to the basic principle of dynamic, organic interaction. This parallel brings to mind Jean Starobinski's classic study "Le Philosophe, le géomètre, l'hybride," where he expands upon Spitzer's analysis of Diderotian language to produce a compelling reading of the stylistic, physical, and conceptual hybridization that is at work throughout this dialogue.[40] Focusing on the sleeping d'Alembert's exalted soliloquy on life—and the wet dream with which it culminates—Starobinski interprets the *Rêve de d'Alembert* as a theoretical anticipation of modern biology: "Diderot formulates the chimera of a [field of] knowledge in which the act of knowing would be inseparable from the expansion of life itself" (Starobinski, 22). In fact, what Diderot formulates is the chimera that inspired the philosophically inclined medical theorists of his own era: namely, the dream that they might unveil the inner mysteries of the body through the pure application of their own sensibility, and thereby open up new, "audacious" perspectives on the nature of life for their readers.

In a curious way, it is precisely by reducing medicine to a series of "marvelous fact" and "historical tales" that Diderot fulfills this dream. Considered in that light, the amoral poetics that Dr. Bordeu espouses at the end of the *Rêve* in response to Mlle. de L'Espinasse's queries about mixing living species should be seen as a properly medical poetics: "'Your question involves physical science, morals and poetics.' 'Poetics!' 'Why not? The art of creating fictional beings in imitation of real ones is true poetry'" (226). For the practice of using the knowledge of sensibility to create new beings, or at least remake existing ones, was one of the principal missions of philosophical medicine during the second half of the eighteenth century. That is, the physician-philosophers of this period sought to establish a comprehensive spectrum of possible human behaviors, a spectrum grounded in a thorough knowledge of observable sensible body types and of the conditions that lead to their creation. In the hands of Diderot, that notion takes a truly fantastical turn: it creates, for example, the indefatigable "goat-men" that Dr. Bordeu envisions in the "Suite de l'Entretien" (232), and the procreative hothouse imagined earlier by the sleeping d'Alembert. What Diderot does at such moments is underscore the teratological possibilities of the contemporary medical notion that bodily sensibility could be manipulated to create certain desired combinations of physical, moral, and/or intellectual traits. The real-life physicians of the day did not, of course, go nearly so far as Diderot's Dr. Bordeu, who at the end of the text defends the idea that "Nothing that exists can be against nature or outside nature" (230). They did, however, take their own musings about sensibility's combinations and permutations among human beings to some extraordinarily imaginative length.

Notes

1. See, for example, Barbara Stafford, *Body Criticism: Imaging the Unseen in Enlightenment Art and Medicine* (Cambridge, MA: MIT Press, 1991); Marie-Hélène Huet, *Monstrous Imagination* (Cambridge, MA: Harvard University Press, 1993); Dorinda Outram, *The Body and the French Revolution: Sex, Class, and Political Culture* (New Haven: Yale University Press, 1989); Angelica Goodden, *Diderot and the Body* (Oxford: Legenda, 2001); and, looking at the body in a more metaphorical sense, Anne Deneys-Tunney, *Écritures du corps: de Descartes à Laclos* (Paris: Presses Universitaires de France, 1992).
2. As David B. Morris has aptly put it, "like theology in the Middle Ages, medicine in the Enlightenment approached the condition of a master discourse"; "The Marquis de Sade and the Discourses of Pain: Literature and Medicine at the Revolution," in *The Languages of Psyche: Mind and Body in Enlightenment Thought*, edited by G. S. Rousseau (Berkeley: University of California Press, 1990), 297.

3. Roger Cooter, "The Power of the Body: The Early Nineteenth Century" in *Natural Order: Historical Studies of Scientific Culture*, edited by Barry Barnes and Steven Shapin (Beverly Hills and London: Sage Publications, 1979), 73.
4. On the heuristic value of metaphors and other "discursive elaborations" in scientific and socio-political thought, see Judith Schlanger, *Les Métaphores de l'organisme* (Paris: Vrin, 1971), and *L'invention intellectuelle* (Paris: Fayard, 1983), esp. 182–206.
5. Théophile de Bordeu, *Recherches sur l'histoire de la médecine* in Bordeu, *Oeuvres complètes* (Paris: Caille et Ravier, 1818), t. II, 668–9; my translation.
6. Frank Baasner underscores that although "many elements of the eighteenth century definition of 'sensibilité' were inherent in the [term's] seventeenth-century connotations," its positive moral meanings were not fully developed until about 1750; in "The Changing Meaning of 'Sensibilité': 1654 till 1704," *Studies in Eighteenth century Culture*, 15 (1986): 78. See also Frank Baasner, *Der Bergriff 'sensibilité' im 18. Jahrhundert. Aufstieg und Niedergang eines Ideals* (Heidelberg: Carl Winter Universitätsverlag, 1988); John S. Spink, "'Sentiment,' 'sensible,' 'sensibilité': Les Mots, les idées, d'après les 'moralistes' français et britanniques du début du dix-huitième siècle" in *Zagadnienia Rodzajów Literackich*, vol. XX (1977): 33–47; and R. F. Brissenden, *Virtue in Distress: Studies in the Novel of Sentiment from Richardson to Sade* (London: Macmillan Press, 1974), esp. 11–55.
7. I borrow the term "bridging concept" from Ludmilla Jordanova; see her essay "Natural Facts: An Historical Perspective on Science and Sexuality" in *Sexual Visions: Images of Gender in Science and Medicine between the Eighteenth and Twentieth Centuries* (Madison, WI: University of Wisconsin Press, 1989),19–42.
8. See, for example, Bordeu's description of the excessive sensibility of city dwellers in *Recherches sur les maladies chroniques* (1775) in *Oeuvres complètes de Bordeu* (Paris: Caille et Ravier, 1818), vol. II, 806.
9. See Marin-Jacques-Clair Robert, *Traité des principaux objets de la médecine* (Paris: Lacombe, 1766), xxvii.
10. Philip Stewart uses the term "verbal chessboard" to describe the art of lovemaking in eighteenth-century worldly literature; see *Le Masque et la parole: Le langage de l'amour au XVIIIe siècle* (Paris: José Corti, 1973), esp. 100–6 and 148–84. See also Peter Brooks's analysis of the moral and communicational codes involved in worldly fiction, in *The Novel of Worldliness* (Princeton: Princeton University Press, 1969).
11. See, for example, Etienne Bonnot de Condillac, *Traité des Sensations* (1754) and Charles Bonnet, *Essai analytique sur les facultés de l'âme* (1759). For an incisive analysis of this trend in eighteenth century philosophy, see Karl Figlio, "Theories of Perception and the Physiology of Mind in the Late Eighteenth Century," *History of Science* 12 (1975): 177–212.
12. See my analyses of those authors in *Enlightenment and Pathology: Sensibility in the Literature and Medicine of Eighteenth century France* (Baltimore: Johns Hopkins University Press, 1998); an expanded version of the discussion that follows can be found in Chapter 2. On the manner in which the physical body was inscribed in eighteenth century French fiction, see also Anne Coudreuse, *Le Goût des larmes au XVIIIe siècle* (Paris: Presses Universitaires de France, 1999), esp. 195–233.
13. See on this point Sylvain Auroux, *La Sémiotique des encyclopédistes* (Payot, 1979).
14. Within the encyclopedic order described by Denis Diderot, the human being was "the sole term from which one must depart, and to which one must relate everything, if one wants to please, interest, and touch [readers] even in the most arid considerations and the driest details"; Diderot, ENCYCLOPÉDIE, in *Encyclopédie, ou Dictionnaire raisonné des sciences, des arts et des métier*, edited by Denis Diderot and Jean d'Alembert (Paris: Briasson, David, Le Breton, and Durand, 1751–65; reprint New York: Pergamon Press, 1969), volume v, 641. Daniel Brewer examines the "powerfully centralizing concept of man" which the encyclopedists used to organize their text and its representation of useful knowledge, in *The Discourse of Enlightenment in Eighteenth century France* (Cambridge, UK: Cambridge University Press, 1993), esp. 13–25.
15. See William Coleman, "Health and Hygiene in the *Encyclopédie*," *Journal of the History of Medicine* 29 (1974): 399–421.

16. On the history of Montpellier vitalism, see Elizabeth A. Williams, *The Physical and the Moral: Anthropology, Physiology, and Philosophical Medicine in France, 1750–1850* (Cambridge, UK: Cambridge University Press, 1994), 20–66; and *A Cultural History of Medical Vitalism in Enlightenment Montpellier* (Aldershot, Hampshire: Ashgate Press, forthcoming).
17. Ménuret signed seventy of his articles by name or with an "m." The rest of the articles that can be attributed to him are either referred to as his in other articles, or extremely similar in style and theme. R. N. Schwab and W. E. Rex attribute eighty-five articles to Ménuret in "Inventory of Diderot's *Encyclopédie*," *Studies on Voltaire and the Eighteenth Century* 93 (1972): 216–17. Roselyne Rey estimates that Ménuret wrote close to a hundred *Encyclopédie* entries; see Rey, "Naissance et développement du vitalisme en France de la deuxième moitié du 18e siècle à la fin du Premier Empire," *Studies on Voltaire and the Eighteenth Century* 381 (2000): 63–89. Jacques Roger was one of the first scholars to emphasize Ménuret's role in the *Encyclopédie*; see *Les Sciences de la vie dans la pensée française au XVIIIe siècle* (Paris: Albin Michel, 1963), 631–41.
18. "SEMEIOTIQUE", *ou* "SEMIOLOGIE" (*Médecin. Semeiotiq.*), in *Encyclopédie*, xiv: 937.
19. Jean-Jacques Ménuret de Chambaud, "PROGNOSTIC," in *Encyclopédie*, xiii: 429.
20. "OBSERVATION," in *Encyclopédie*, xi: 313–23.
21. This part of Ménuret's discussion reflects the methodological debate then raging over the virtues of experimental investigative methods in natural philosophy. See, by contrast, d'Alembert's article "EXPÉRIMENTAL," which was written to praise the King's recent establishment of a university chair in experimental physics; in *Encyclopédie*, vi: 298–301. See also Haller's *Supplément* article "OECONOMIE ANIMALE," where he defends vivisectionist experimentation in physiology; in *Supplément* to the *Encyclopédie*, iv: 104–5.
22. "POULS," in *Encyclopédie*, xiii: 205–40.
23. Ménuret de Chambaud, *Nouveau traité du pouls* (Paris: Vincent, 1768), ix.
24. *Recherches sur le pouls par rapport aux crises* in *Oeuvres complètes de Bordeu*, vol. 1, 421. A very similar passage appears in "POULS," 237–8. Bordeu's discovery of the natural principles of diagnostic pulse taking was prompted, as he recounts, by stumbling on a recently translated foreign treatise on pulses, M. Nihell's annotated version of the observations of Solano de Luques, an early eighteenth-century Spanish physician. Bordeu, however, recasts the practice to fit the new Montpellier doctrine of the reactive animal economy, and insists that the pulse is a phenomenon of sensibility, not motility.
25. Roselyne Rey sums up the eighteenth-century theory of the salutary crisis in these terms: "It is a question of exacerbating sensibility, of activating a vital energy which is languishing, which lacks the strength necessary to accomplish the task of fighting off the disease"; *Histoire de la douleur* (Paris: Editions de la Découverte, 1993), 153.
26. Bordeu devotes the first five chapters of his treatise to explaining how he devised a nomenclature of pulses that would take into account the anatomical architechtonics of the body—that is, the upper/lower axis at the diaphragm, and the right/left axis—as well as the natural variations caused by age, sex, and temperament. He maintains that each body type and temperament has its own natural pulse that evolves with age; hence this assertion about the female pulse: "The natural pulse of women is, in general, more lively and closer to that of children and youth, than is the pulse of men; it has its particular degrees, *its youth, its middle age, its old age*"; in *Recherches sur le pouls*, 262.
27. Bordeu, *Recherches sur le pouls*, 263.
28. This article was a reprinted version of Bordeu's *Recherches sur les crises* (1743); see Bordeu, *Oeuvres complètes*, I: 250.
29. Bordeu, *Recherches sur les crises*, 251.
30. See, for example, Herbert Dieckmann's classic essay "Théophile Bordeu und Diderots *Rêve de d'Alembert*" in *Romanische Forshungen* 52 (1938): 55–122. See also Jean Varloot's introduction to the *Rêve de d'Alembert* in Denis Diderot, *Le Neveu de Rameau /Le Rêve de d'Alembert* (Paris: Editions Sociales, 1972), 186–227.
31. Yvon Belaval reads the *Rêve* as an audaciously ironic joke on the real-life d'Alembert and Mlle. de l'Espinasse; see "Les Protagonistes du *Rêve de d'Alembert*" in *Diderot Studies* III (1961): 27–53.
32. *Recherches anatomiques sur la position des glandes, et sur leur action in* Bordeu, *Oeuvres complètes*, t. 1, 36–208. Martha Fletcher, the editor of Bordeu's *Correspondance*, describes

the *Recherches sur les glandes* as "a 'bomb' of 520 pages launched against the mechanists who formed a compact majority at the Faculty of Paris"; in Bordeu, *Correspondance* (Montpellier: Presses de l'Université Paul Valéry, 1977), vol. II, 69.

33. In the next paragraph, for example, Bordeu contends that the breasts of wet nurses only respond to the babies that excite them properly (126); he then goes on to ascribe the same sexualized capriciousness to cows (128).
34. François Duchesneau proposes the term "machina nervosa" to describe Bordeu's model of the organism, and examines the objections that were made against this model by both Haller and Paul Joseph Barthez, Bordeu's primary successor at Montpellier; see *La Physiologie des Lumières. Empirisme, Modèles, et Théories* (The Hague: Martinus Nijhoff Publishers, 1982), 361–430.
35. See, for example, Elizabeth L. Haigh, "Vitalism, the Soul, and Sensibility: The Physiology of Théophile Bordeu," in *Journal of the History of Medicine* 31 (January, 1976): 30–41; and Roselyne Rey, "La Théorie de la sécrétion chez Bordeu, modèle de la physiologie et de la pathologie vitalistes" in *Dix-huitième siècle* 23 (1991): 45–58.
36. See Wilda Anderson, "Diderot's Laboratory of Sensibility" in *Yale French Studies* 67 (1984): 72–91; and her chapter "Good, Solid Philosophy" in *Diderot's Dream* (Baltimore: Johns Hopkins University Press, 1990), 42–76. See also Rosalina de la Carrera's complementary reading of the *Rêve de d'Alembert* in *Success in Circuit Lies: Diderot's Communicational Practice* (Stanford: Stanford University Press, 1991), 127–66.
37. Aram Vartanian observes that "as occasion demands, D'Alembert becomes Diderot, Diderot becomes Bordeu, Bordeu becomes D'Alembert, and so on. One is tempted to say that each of them has surrendered a portion of his separate existence, and now functions 'organologically' in a sort of intellectual continuum"; "Diderot and the Phenomenology of the Dream," in *Diderot Studies* 8 (1966): 239. See also Michel Baridon, "L'imaginaire scientifique et la voix humaine dans *Le Rêve de D'Alembert*" in *L'Encyclopédie, Diderot, l'esthétique. Mélanges en hommage à Jacques Chouillet*, edited by Sylvain Auroux, Dominique Bourel and Charles Porset (Paris: Presses Universitaires de France, 1991), 113–21.
38. Denis Diderot, trans. Leonard Tancock, *Rameau's Nephew and D'Alembert's Dream*, (London: Penguin, 1966), 215.
39. Leo Spitzer, "The Style of Diderot" in *Linguistics and Literary History* (New York: Russell and Russell, 1962), 135–91.
40. Starobinski, "Le Philosophe, le géomètre, l'hybride," *Poétique* 21 (1975): 8–23. See also the chapter on Diderot in Jean Starobinski, *Action et réaction: vie et aventures d'un couple* (Paris: Editions du Seuil, 1999), 53–97.

CHAPTER 2
Man and Horse in Harmony

ELISABETH LEGUIN

> [A] scholar and a horse are very troublesome to one another
> —William Cavendish, Duke of Newcastle,
> *Methode et invention novvelle de dresser les chevavx*

It is only within the last century—very suddenly and very recently, given their long career—that horses are no longer intrinsic to the basic operations of Western society; few of us any longer rely on them for anything we consider indispensable. Horsemanship is still tightly woven into our speaking and our thinking, however, in countless figures that make up a kind of thoroughgoing metaphorical fabric. We speak of being back (or tall) in the saddle; being spurred to do something; reining in someone or something. We transpose equine experience into human terms with concepts like "keeping pace," "hitting one's stride," "getting off on the wrong foot," "kicking up one's heels," "feeling one's oats." Such metaphors are used regularly by people who have never come near a horse, to address matters of intention, control, and enactment, and they show remarkably few signs of being supplanted by automotive imagery. Given the readiness with which language adapts to social and technological change, it seems unlikely that the persistence of horsemanship metaphors in modern English is merely an odd pocket of resistance. I would instead suggest that it has to do with the intricate ways in which horses have represented human embodiment in Western culture.

My project here is to investigate some aspects of that representation as it evolved during the Early Modern period in Europe, most particularly in its association with another embodied practice, that of making music. This association—one that I first came to consider, not historically, but in daily practice as a professional musician and amateur horsewoman—is considerably less arbitrary than it might seem at first glance.

Recent work by Kate van Orden has followed the lead of Foucault and Mark Franko in demonstrating how both practices are linked to, indeed constitutive of, the idea of physical discipline as internalized social control, as this idea developed among the upper classes in Italy and France around the sixteenth century.[1] On the public level, music promoted bodily discipline through organizing time, and the physical traversal of time, into precise quantities whose predictable and manipulable nature was made visible through the geometrical patterning of bodies on the dance floor. On a more private level, music focused the attention of individuals discretely and discreetly onto the acquisition of socially nonthreatening bodily skills: playing the lute, singing, or improvising poetry to its accompaniment.

At the same time, Early Modern horsemen, no longer under an urgent mandate of military effectiveness, could begin to explore their work as an art—that is, in its capacity to articulate and elaborate, rather than actively defend, a *status quo*. The relationship of postmilitary horsemanship to music making is most apparent at the public (indeed, monstrously ostentatious) level: the *balletto a cavallo* used music in the same way as the geometrical dance of which it was the equestrian transposition, as temporal and gestural organizer. However, it is the private level, that of training horses so that they could be used in the *balletti* and other endeavors, that chiefly concern me here. The trainers involved in this were fortunate—notably more so than their counterparts in music—in the period determination to hearken back to Classical sources, for an excellent treatise by Xenophon, *The Art of Horsemanship*, was actually preserved entire. Six in particular among Xenophon's many followers will concern me here: Cesare Fiaschi, whose *Trattato dell'imbrigliare, atteggiare, and ferrare cavalli* was first published in 1568; Gervase Markham, prototype of the English country gentleman, whose *Cavelarice, Or The English Horseman* appeared in 1607; the very influential Antoine de Pluvinel, Master of Horses to the young Louis XIII, whose instructions to his royal charge were posthumously published in 1626 as *Le maneige Royal*; William Cavendish, Duke of Newcastle, whose 1658 *A General System of Horsemanship* is a pillar of the methods of the famous Spanish Riding School; Claude-François Menestrier, whose *Traité des tournois, ioustes, carrousels, et autres spectacles publics* appeared in 1669; and William Hope, a Scot who translated a French farriery treatise by Jacques de Solleysel, appended his own comments about riding and training, and called the whole *The compleat horseman*, publishing it in 1696.

The adoption of Xenophon's humane and commonsensical approach around 1550 marks a turning point, not only in the military purposes that had dominated horse training for millennia, but in basic European understandings of how power and command work on selfhood. Central to a Xenophonian approach is an acknowledgement of the horse as an independent intelligence, an Other, someone to be negotiated with, rather than something to be deployed. "For what the horse does under compulsion . . . is done without understanding; and there is no beauty in it either, any more than if one should whip and spur a dancer."[2] Something like this acknowledgment had been around for a long time. Metaphorically, it is of the essence that the chariot horses of the soul in Plato's *Phaedrus* are independent intelligences, such control as their charioteer has being the product of a very uncertain negotiation indeed. Historically, there is a venerable literature of equine understanding—especially military understanding—with early exemplars in the horse in the Book of Job who "saith among the trumpets, Ha, Ha; and he smelleth the battle afar off, the thunder of the captains and the shouting" (Job 39: 25), and the famous Bucephalus. Such traditions certainly inform equestrian literature in the early modern period. Menestrier tells us:

> De tous les Animaux, il n'en est aucun, qui ait plus de rapport à l'Homme, & qui semble estre plus à ses usages que le Cheval . . . Il a du cœur, il aime la gloire, & il se plait aux caresses, & aux applaudissemens.[3] (Of all animals there is not one that has more rapport with Man, and which seems better suited to his usages, than the horse . . . He has courage, he loves glory, and he is pleased by caresses and applause.)

Newcastle, in his turn, sums up the equine character warmly: "The horse [is], after man, the most noble of all animals (for he is as much superior to all other creatures as man is to him, and therefore holds a sort of middle place between man and the rest of the creation)." Greater nuance, and far greater practicality, emerges in Newcastle's characterization, however. It turns out to be as much a warning as a tribute: the horse's quasi-human, intermediate status means that "he is wise and subtile: for which reason man ought carefully to preserve his empire over him, knowing how nearly that wisdom and subtilty approaches his own."[4] Just how the horseman is to go about "preserving his empire"—that is, the practical daily science of what the horse may be said to think, why he thinks it, and how the trainer does best to respond—is the explicit concern here. Shorn of sentimentality, presented with some concern for method rather than as thrilling anecdotes, treatises like Newcastle's make a fascinating contribution to the history of discipline, in that they outline the techniques for a dialogic relationship between trainer and trained.

This being said, there is nonetheless much in these treatises that supports a reading of Early Modern horsemanship as a paradigmatic demonstration of Foucault's foundational and repellent definition of *dressage:* "uninterrupted, constant coercion . . . which assured the constant subjection of [bodily] forces and imposed upon them a relation of docility-utility."[5] This bears out in the frequent occurrence of words like "prowess" and "subjection." William Hope expresses the perfection of the horse–rider relationship as follows: "to reduce horses, and bring them under a perfect subjection and obedience, and to have no other will save that of their Riders . . . [so that] there appears such a harmony and beauty, that whatever they do contrary to them is displeasing, and cannot be endured."[6] Pluvinel's advice sounds a good deal like that kind of training more lately called "breaking" a horse:

> [I]l faut tousiours travailler le cheval à ce qui contrairie le plus sa volonté, pour le renger plus promptement à la parfaite obeyssance que nous desirons de tous les Chevaux.[7] (One must always work the horse toward what is most contrary to his will, to bring him most quickly toward that perfect obedience which we desire in all horses.)

And Newcastle takes this a step farther, requiring of the horse not just subjection, but fear:

> It is impossible to dress a horse before he obeys his rider, and by that obedience acknowledges him to be his master; that is, he must first fear him, and from this fear love must proceed, and so he must obey. For it is fear creates obedience in all creature, in man as well as in beast. Great pains then must be taken to make a horse fear his rider, that so he may obey out of self-love, that he may avoid punishment. A horse's love is not so safe to be trusted to, because it depends on his own will; whereas his fear depends on the will of the rider, and that is being a dressed horse. But when the rider depends on the will of the horse, it is the horse that manages the rider. Love then is of no use; fear does all; For which reason the rider must make himself feared, as the fundamental part of dressing a horse. Fear commands obedience, and the practice of obedience makes a horse well dressed. Believe me, for I tell it you as a friend, it is truth.[8]

This massively overdetermined passage is disturbing, coming as it does from the most urbane and humane of all equestrian writers; but I think it is a case of "the scholar and the horse being very troublesome to one another." There is a good deal that works against our reading Newcastle's advice as simplistically as Foucault's summing-up would encourage us to do. The treatises themselves frequently militate against it; and they do so most especially when they are read with some practical, embodied knowledge of the discipline in question. In the following discussion, I propose to maintain a dialogue between that knowledge and the points made in the treatises. Such dialogue is by its nature transhistorical, the same balancing act that the historian of performance negotiates everywhere: strictly documentable historicity is questionably served by reference to living praxis, yet a history of praxis, *as praxis,* is poorly served by documents alone. Nuances, sometimes essential ones, will be lost without reference to experience. As Fiaschi puts it, "No one can judge of it who is not there."[9]

Molière's Fencing Master, barking commands at the hapless Bourgeois, personifies the coercive strain in Early Modern training:

> Le Maître d'armes:
> Allons, monsieur, la révérence. Votre corps droit. Un peu penché sur la cuisse gauche. Les jambes point tant écartées. Vos pieds sur une même ligne. Votre poignet à l'opposite de votre hanche. La pointe de votre épée vis-à-vis de votre épaule. Le bras pas tout à fait si tendu. La main gauche à la hauteur de l'œil. L'épaule gauche plus quartée. La tête droite.[10] (The Fencing Master: Come sir, salute the swords. Body straight. Shift your weight to the left thigh. Legs not so far apart. Feet lined up. Wrist at hip level. Tip of your sword at shoulder level. Arm not quite so extended. Left hand at eye level. Left shoulder more in *carte* position. Head straight.)

Clearly, the obsessive level of anatomical detail, the "attentive 'malevolence' that turns everything to account,"[11] the abrasively efficient tone are not confined to the military disciplines of which Foucault writes. But of course Molière is being ironical about such "efficiency": the Bourgeois is getting tied into terrible knots as this session proceeds. Smile at him though we may, in his hopeless confusion he represents a real problem in the inculcation of discipline, and there is little indeed in the literature of his day to rescue him. Contemporary works on fencing, dance, music, or comportment tend to present a fairly rudimentary one-way flow of information between trainer and trained; even in those written as dialogues, the pupil is a kind of straight man, rarely offering any real challenge to the method being presented. Horses, however, demand another sort of dialogue altogether: even as they exhibit extraordinary ability to adapt to human purposes, they challenge human methods through their sheer otherness. By the time of Molière, horse training had developed very sophisticated means of managing that otherness. The Bourgeois, wandering down to his stables and watching his Ecuyer, or Master of Horses, would have found in use there most of the strategies of which he himself stood in such sore need: the devising of appropriate goals for practice, and their modification based on results; the incremental shaping of new neurological responses through practice and reinforcement; the recognition of and working with resistance, confusion, or tension; and a full understanding of the uses of pleasure and pain in inculcating desired habits, or systematically eradicating undesired ones.

In suggesting, as I do here, a parallel between the training of horses and the training of human bodies, it would be silly to elide the profound differences between the two species. (Indeed it would be downright dangerous, given that horse bodies weigh a thousand pounds, move with startling rapidity, and are eminently capable of injuring or killing their trainers by accident, let alone by design.) Yet it would be equally silly to disallow the parallel on that basis; if nothing else, the sheer vernacular momentum of human identification with horses constitutes a kind of obligation to pursue it. These treatises succumb neither to overidentification nor denial of commonality, but constitute a serious attempt to sort out how power and authority actually work in relation to otherness. In so doing, they move in a particularly interesting metaphorical cross-current relative to the Cartesian model, taking the separation of mind and body to a kind of logical extreme by proposing the body, not as mechanism, but as a morally independent Other. (As the Duke of Newcastle, patron and friend to Descartes, Hobbes, Jonson, and Dryden, rather irascibly puts it: "If [the horse] does not think (as the famous philosopher DES CARTES affirms of all beasts) it would be impossible to teach him what he should do.")[12]

What then are we to make of the same author's expert and strongly worded advice about maintaining fear in the horse? It is our understanding of the concept of fear, as it pertains to horses, that is key here. Fear is a central aspect of equine character, as horses are prey animals; in the wild, it is essential to their survival that they be ever ready to perceive threats and react to them. They have as a consequence a large repertory of fear-related states, from responsive alertness through suspicion through apprehension through agitation to full-blown panic.

Their reactions are similarly fine gradations of a basic response—to flee. The Ecuyer could not eradicate such a powerful instinct as this—nor indeed would he wish to, for the same swiftness that is the equine ticket to survival in the wild has recommended the horse to human purposes for millennia. But, working Aristotle fashion with nature, the Ecuyer had much the same job as the fencing or music master: one of taking an unschooled (if very well-bred) body, and training it to recognize and manage its own primary impulses. In one manner of speaking, this consisted in training the horse and (perhaps more importantly) the rider to a wider, more subtle and nuanced vocabulary of fear. But the manner of speaking is important. Within an equine vocabulary, "fear" does not necessarily mean a state of abject, quaking incompetence. Newcastle's advice above reads very differently if we consider that the "fear" of which he speaks might more closely resemble that in the old injunction, "Fear God," meaning the physical enactment of respect.[13]

It is this capacity for fear that most sharply differentiates horses from humans; we are not prey animals, and will forever lag behind equines in the precision and attunement of our responses to perceived danger. In learning to recognize and honor and anticipate his horse's fear, and in developing strategies for making use or even beauty out of it, the trainer, inevitably, trains himself: to an exigent level of physical reactivity, of constant receptivity, that is not peculiar to horsemanship, but beats at the heart of physical discipline in any performing art. In music making, this same level would be called listening.

A well-managed vocabulary of fear can be most efficiently developed through strategies of reward and punishment; as William Hope says, "hope of reward and fear of punishment governs this whole World, not only Men but Horses."[14] This apparent Lockeanism hearkens directly back to Xenophon; particularly in its emphasis on reward, it is at the heart of his radical influence. The treatises' descriptions of rewards are interestingly detailed, sometimes offering a touching glimpse of the writer's intimacy with his animals. Thus Markham:

> [T]he keepers greatest labour is but to procure love from the horse, so the onelie thing that is pleasant to the Horse, is love from the keeper; insomuch that there must be a sincere and incorporated friendshippe between them or else they cannot delight nor profit each other, of which love the keeper is to give testimonie, both by his gentle language to his horse, and by taking from him any thing which he shall beholde to annoy or hurt him, as moates, dust, superfluous hayres, flyes in Summer, or anye such like thinge, and by oft feeding him out of his hand, by which meanes the Horse will take such delight and pleasure in his keepers companye, that he shall never approach him, but the Horse shall with a kind of chearfull or inward neying, show the ioy he takes to behold him, and where this mutuall love is knit and combined, there the beast must needes prosper, and the man reap reputation and profit . . .[15]

Rewards may be elaborate, or as simple as Xenophon's suggestion that the horse appreciates a rest from time to time: "[H]e will leap and jump up and obey in all the rest if he looks forward to a season of rest on finishing what he has been directed to do."[16] This use of respite as reward results in a kind of training that proceeds through frequent alternation of work and rest. This requires that the trainer have an accurate sense of his horse's particular capacities and limitations, and that in order to accomplish his goals efficiently he respect these capacities and limitations over and above any desires of his own–essentially, that he "listen" to his horse:

> [H]orami par di dire, che al buon cavaliere fa bisogno . . . [temperarsi] secondo l'occasione, & tempi, si de batterli, come di farle carezze, ò di tenerli solamente in timore . . . hauendo l'occhio di continuo all'animo, & forze loro, & secondo quelle operare . . . [17] (Hear me then when I say that the good horseman must . . . [temper] himself as the occasion requires, to beat them, make caresses, or to keep them only in fear . . . keeping an eye on the progress of [the horses'] the spirits and strength, and taking these as his guide . . .)

Punishment is a more complex matter. The primary cautions here are nevertheless consistent among writers: it must be to the point and brief:

> A good keeper therefore must know when to correct and when to cherrish, not giving either blowe or angrie word, but in the instant of the offence, nor to punish or strike the horse any longer then whilst his present fault restes in his memorie.[18]

Here Markham is in accord with most latter-day horse trainers, who agree that the equine outer limit for accurate cause-and-effect association is about three seconds; after this the point of punishment (or, for that matter, of reward) is simply lost. Causal associations between actions and consequences ("I balk here, and I am being punished") and the decisions they make possible ("therefore I will cease to balk") are limited to this three-second envelope. Put another way: effectively admonishing or constraining an embodied intelligence requires sharp vigilance on the part of the trainer. A tardy reprimand is simply abuse, and abuse can have endless repercussions: horses may have a very limited capacity for causality, but their long-term memories are enormously retentive and detailed. It is not unusual for a horse to react with apprehension when it is brought near a place where something bad happened, not just days, but years before.

Punishing the horse is thus a delicate business, very easily made counterproductive in its activation of equine fear, which will tend to have an energy and a logic all its own, quite independent of the trainer's purposes. Should the horse be resistant to a command because it is already afraid, Xenophon reminds us that "[c]ompulsion and blows inspire only the more fear; for when horses are at all hurt at such a time, they think that what they shied at is the cause of the hurt." In other words, the fear-response has an additive tendency: "scary object" and "scary beating" will simply combine, spiraling into "doubly scary object-plus-beating." In such a situation, then, the trainer must show the horse "that there is nothing fearful in it, least of all to a courageous horse like him; but if this fails, touch the object yourself that seems so dreadful to him, and lead him up to it with gentleness."[19]

In his administration of reward and punishment the trainer must, above all, be impartial—another point on which all these writers concur, again following Xenophon: "The one great precept and practice in using a horse is this—never deal with him when you are in a fit of passion. A fit of passion is a thing that has no foresight in it, and so we often have to rue the day when we gave in to it" (37–8).

> I have seen very few passionate horsemen get the better of a horse by their anger . . . In this act there should always be a man and a beast, and not two beasts.[20]

For all the importance of vigilance and sensitivity, of "listening," of taking the horse's spirits as one's own guide, a crucial separation between horse and rider, mind and body, intention and execution, emerges here. In the Aristotelean view that informs the work of Xenophon and all his followers, training, the dressing of nature toward excellence, cannot take place without it. The moment the trainer loses his impartiality and collapses the separation, he has lost the only thing that differentiated him from what Aristotle calls the "other animals": his ability to separate himself from his own reactions and desires, by so doing becoming an independent practical reasoner. The angry trainer and his horse are become "two beasts."

A physical intelligence, proceeding without the intervention of independent practical reason, will persist in certain traits—traits which human embodiment shares quite fully with that of horses: the "vegetative principle," as Aristotle calls it, toward growth and toward reproduction; attraction to pleasure and avoidance of discomfort; conservation of energy (that is, laziness); and, most interestingly from the standpoint of issues of domination and subjection in training, a certain innate resistance to being told what to do. In the human soul "there is something contrary to the rational principle, resisting and opposing it"[21]; as far as the equine soul,

> [S]ubjection is not agreeable to a horse . . . If the wisest man in the world were put into the shape of a horse, and retained his superior understanding, he could not invent more cunning ways (I question if so many) to oppose his rider, than a horse does . . .[22]

Characterizations of horses as noble and quasi-human may have something to do with this extremely resourceful resistance to subjection, a quality long associated with moral integrity in humans. However, Early Modern horse trainers certainly do not inflate this possibility into the full-blown romanticization so prominent in current Western culture. Horses, especially wild ones, are among our most ubiquitous and imprecise symbols of freedom, galloping with manes and tails streaming across advertisements for everything from cigarettes to mortgage plans. Wildness did not have this cachet in the period in question; it was rather more likely to be seen in a Hobbesian light, as a state of violent, unregenerate war. By these lights, resistance must and should be overcome, for only through training does the good life become possible.

Newcastle's approach to entrenched resistance, resistance to training itself, is so deeply pragmatic as to be almost metaphysical. Those who have raised children may recognize his strategy for coping with an embodied intelligence in a state of existential uncooperativeness:

> If you can't gain your point therefore in one way, you must have recourse to another: I mean, that if in his extremity the horse will not agree with you, you must agree with him in the following manner. You would make your horse advance, and he to defend himself against you runs back: at that instant pull him back with all your strength. And if to oppose you he advances, immediately force him briskly forward. If you would turn to the right, and he endeavours to turn to the left, pull him round to the left as suddenly as possible: if you would turn him to the left, and he insists on the right, turn him as smartly to the right as you are able. If you would have him go sideways to one hand, and he inclines more to the other, immediately second his inclination. If he would rise, make him rise two or three times. In a word, follow his inclinations in every thing, and change as often as he. When he perceives there can be no opposition, but that you always will the same thing as he, he will be amazed, he will breathe short, snuff up his nose, and won't know what to do next . . . [23]

There is much in these extracts that applies very neatly to the process of training human bodies to an art; one can imagine the Bourgeois benefiting, for instance, from the advice about doing little but often, and about the careful use of respite. Fiaschi's treatise is one place where such parallels become explicit. He directs the student of fine riding to make reference to his own musical experience—an experience naturally presumed, since music and riding were both standard parts of a nobleman's education in Italian and French academies. Fiaschi includes short extracts of musical notation as guides for coordinating rider and horse in the execution of various figures; and he gives a beautifully concise description of equestrian "listening."

> [Gli] speroni, polpe, briglia, & uoce . . . non a tutti si dee observar un medemo [sic] modo, ma hor un poco piò, hor meno secōdo che si conosce il bisogno, il qual nō può niuno absente giudicare . . . [24] ([T]he spurs, muscles, bridle, and voice . . . must not be observed in the same way toward every [horse], but now a little more, now less, following what is known of the need; no one can judge of it who is not there . . .)

Newcastle takes the parallel even further:

> The hand, thigh, leg, and heel [of the rider] ought always to go together; for example, suppose a man playing upon the lute, and he touches the strings with

> his left hand, without touching the others with his right, he must make but a very indifferent harmony; but when both hands go together, and in right time, the musick will be good. It is the same with respect to this excellent art; what you mark with your hand should be touched at the same time with your thigh, leg, or heel, or the musick of your work will be bad. We are at present speaking of musick, and he that has not a musical head can never be a good horseman. A horse well dressed moves as true, and keeps as regular time as any musician can.[25]

In this lovely passage, "musick" means something we all recognize, riders or no, for it is fundamental in all corporeal disciplines. We have no other concise name for it. "Musick" here means the submission and organization of embodiment to a purpose.[26] I say embodiment, and not "the body," for even the solitary lute player is many in his embodiments: there is one musick of the player's hands' intention to the purpose of a melody; another to its counterpoint; another musick, a vigilant evaluative musick, takes place as the player listens to and judges the sounds he has just produced, in order to adjust the volume or the finger placement of those now forthcoming; another musick still, a demonstrative one, in how he has positioned himself to the admiring listener across the room, and in the subtle adjustment of the plucking motions of his right hand that direct his instrument's sounds toward his idea of that listener's hearing. He plays not only upon the lute, but upon his own body, the room, and the listener too; and all of them play upon his ear and his sensibility. Such musick is momentary, fluctuating, circumstantial. No one can judge of it who is not there.

In the context of riding, where two whole bodies are involved, these overlapping circles spread yet wider and more complex: there is "the musick of [the rider's] work," coordinating heels and hands and sense of balance to speak clearly to the horse in that kinetic and kinesthetic language the animal understands best—looking left, sinking weight into his left seat bone, moving his right calf back a few inches along the horse's side, the rider begins to enunciate the musick of "turn left"; meanwhile the horse's body, in readiness ("fear") toward that kinesthetic speech, feels the weight shifts of the rider's head and buttocks, anticipates the touch on his right side, and has begun the turn so that the rider need do very little more (for the horse, a reward for his readiness inheres in his rider's ceasing to insist); or, should the horse be not in readiness but in resistance (sore muscles along the right side of the neck, perhaps, due to excessive work the day before), it stiffens against the left turn; and the rider, feeling the animal shorten its stride and tense its back, hearing it swish its tail in annoyance, continues into the reinforcement of "turn left" by drawing back his left elbow, and with it the hand and rein, more or less sharply depending on the horse's degree of resistance. All of this within perhaps two seconds. Should the resistance happen a second time (the rider is vigilant here), perhaps he applies a flick of his riding whip to the horse's right shoulder at its first appearance, adding "I mean it" to "turn left." There are musicks of intention short and long: the short-term vigilance by which the rider identifies and corrects such a resistance (or, on another occasion, the horse's head has come up, its ears are tensely pricked toward the pile of wood over on the far side of the arena, which was not there yesterday, it holds its breath—now the rider's strategy will be quite different, he will relent with the left-turn music, and modulate instead to "it's really okay," enacting some version of Xenophon's "lead him up to it with gentleness" until the horse perceives that the pile of wood is actually quite boring—sighs, relaxes, resumes his pace.) The longer-term, overarching musick of the rider's intention for this time, this practice session, was that it center upon bending to the left, but it became perforce a session about refining fear: the release of inappropriate fear (fear of a woodpile) in favor of that habit of trust that must underlie any left turns, any commands at all.

Finally, in sum, "musick" means that, to the observer, the purpose of all this submission and organization emerge clearly and unambiguously: that it be, simply, beautiful, or as Pluvinel puts it, *bel,* which means also effortless, easy, unaffected. Such ease is momentary, fluctuating, circumstantial. No one can judge of it who is not there.

The necessarily unwieldy blow-by-blow description of this kinesthetic musick is itself far from *bel;* the whole process is much better served by the period metaphor of sympathetic vibration. Shared vibration in the fibrous matter of the nerves was a central image for relation among sensitive beings; and these treatises make it clear that horses were thought to have this sympathetic, musical potential quite as much as their masters. Markham describes the "Composition of Horses" in terms identical to those for the composition of human beings: horses have four humours, "choller, flegme, blood, & melancholy" (these determine the animal's color, among other things). As for their Spirit, he says, "It is the very quintessence of the blood, and being conveyed in the Arteries, gives the body a more lively and sprity Heate, and makes his feeling more quicke and tender."[27] Markham takes equine musicality as evidence of their special sensitivity:

> That a horse is a beast of a most excellent understanding, and of more rare and pure sence then anie other beast whatsoever, we have many ancient and rare records . . . we find it Written, that in the army of Sibaritanes, horsses would daunce to Musicke, and in their motions keep due time with musicke . . .[28]

As here, accounts of horses responding to music intersect substantially with equine military mythology; after all, it is not just the prospect of battle that causes the horse in Job to say, "Ha, ha," but, in specific, the sound of the trumpets. Menestrier treats of the ultimate in military–musical intersections, the great outdoor spectacles so beloved of his historical moment; he quotes accounts by Pliny, Solinus, and others to demonstrate that "Ces Animaux aiment l'harmonie, . . . qui les excite au Combat" (These animals love harmony, . . . that excites them to combat)[29] and he devotes careful consideration to the kinds of music most likely to produce such an effect.

Accounts of sixteenth- and seventeenth-century *balletti di cavallo* typically follow a certain pattern that privileges the musical responsiveness of the horse and rider by mentioning it last. Descriptions of magnificent human and equine apparel are followed by accounts of choreographed feats of virtuoso riding; the whole closes with a reference to the obedience of this stupendous assemblage to the rhythms and phrases of the music. Thus Pluvinel, recounting an equestrian ballet composed by himself and Robert Ballard, makes the following observation:

> La deuxiesme figure, les Chevaliers vis à vis l'un de l'autre Changeoient de place en faisant deux voltes à la fin, le tout à courbette en commençant, & finissant de ferme à ferme tous six en semble, tant que les cadences duroient, . . .[30] (In the second figure the horsemen, facing one another, changed places making two *voltes* at the end, the whole having begun with *courbettes*, and finishing with the *ferme à ferme*,[31] all six in unison as long as the cadences lasted.)

In Vienna at the 1667 performance of *Il contesa dell'aria e dell'acqua*, the *balletto a cavallo*, choreographed by the Roman *cavaliere* Alessandro Carducci and composed by Johann Heinrich Schmelzer, opened with a Courante, a solo for the Emperor Leopold,

> . . . il quale pel primo diè prova della sua alta perizia, facendo eseguire dal suo destriero corvette in aria, diritti e svolte, e fermate in cadenza secondo le note musicale.[32] (. . . who from the first proved his great prowess, executing upon his destrier *corvette in aria, diritte*,[33] and *svolte*[34] all held in cadence following the musical notes.)

Three hundred years later, the same privileging of certain aesthetic and philosophical values persists in the Kur, or musical freestyle, that is performed as the climax of upper-level dressage competition. It is supposed to look effortlessly "natural," as if the "quicke and tender feeling" of the horse is responding spontaneously to the rhythms and gestures of the musical selections. At its best, it is easy to believe that this is, in fact, the case; a Kur performed by a horse trained to the highest or Grand Prix level has a rhythmic alacrity, variety, and responsiveness that would be the envy of any human dancer, and there will be no sign to the casual observer that the rider is doing anything at all to influence this.[35]

The Edenic possibility of animal training as "a sincere and incorporated friendshippe" resonates pleasantly with this idea of equine musicality, the idea that things like the Terpsichorean ferocity of the Sybaritan horses, the wondrous movements of the Kur, come from the animals' own impulse to dance to music. But once we put ceremonial descriptions aside, the dynamics of this idea in practice prove intricate and equivocal. Menestrier, for instance, uses the sympathetic-vibration metaphor in order to suggest that composers and choreographers adapt their art to the animals, rather than the other way around:

> Il faut donc dire que l'harmonie de certain airs, & de certains instrumens, se trouvant sympathique avec nos corps, ou les corps des animaux, de qui les muscles peuuent estre dans vne disposition semblable à celles des cordes d'un luth bien monté, dont l'une estant touchée fait vibrer toutes les autres: il se fait de *pareils* tremoussemens dans ces corps au son son [sic] de ces instrumens . . . Cela estant ainsi, il faut estudier la nature, & le temperament des animaux, que l'on veut faire danser, & les mouvemens qui leur sont plus ordinaires, & plus naturels, pour faire choix des Instrumens, & des airs, qui sont plus propres à regler ses mouvemens.[36] (One could thus say that the harmony of certain airs and of certain instruments proves to be sympathetic with our bodies, or with the bodies of animals, in which the muscles can be in a disposition similar to that of the strings on a well-strung lute, on which the one [string] being touched, all the others vibrate: a similar excitation is made in these bodies at the sound of these instruments . . . This being so, one must study the nature and the temperament of animals if one wishes to make them dance, and the movements which are ordinary and most natural to them, in order to make that choice of instruments and airs which will be most appropriate for regulating their movements.)

The medleys to which latter-day horses and riders move in the Kur are painstakingly assembled by videoing the horse's movements in the various gaits, and then finding recorded selections in tempi that fit those movements. One can imagine the seventeenth-century composer for one of Menestrier's *Carousels*, holding in the "video" of his mind's eye the equestrian figures devised by the Ecuyer, and choosing from among the Baroque dance-types those whose character of motion most nearly fit what he had seen; the music written, the parts distributed, rehearsals would be much concerned with getting the trumpeters and timpanists to pick tempi that suited the movements of the particular horses involved. In practice, then, the music would seem to hold in cadence following the animals, quite as much as the reverse.

In "mak[ing] that choice of instruments and airs" the possibilities are manifold, since equine cadence is not a simple thing. Four legs have many more rhythmic possibilities than two. A human has only one basic gait, with various gradations from slow to fast: a two-beat affair in which diagonal pairs of limbs move together. The horse has three: the walk, a four-beat gait in which lateral pairs of limbs move hind-to-front in sequence (in so-called "gaited" horses, this can be extended into a very fast, even-tempoed "flying" motion in which only one foot touches the ground at a time); the trot, two-beat equine version of the human gait, with diagonals moving simultaneously; and the canter, which has a slightly irregular three-beat pattern, and which can lead from either the left or right side. Any of these gaits can be collected or extended, altering stride lengths and thus expressive character; in a really well-dressed horse, collection or extension will not cause the tempo of footfalls to alter unless specific additional commands are given.

The very diversity of shapes and sizes into which horses have been bred is a major preoccupation of several of the treatises under consideration, notably Menestrier, Markham and Hope. This too has impact upon matters of cadence. Their descriptions of the various breeds go avidly into subtleties like length of pasterns relative to lower leg, length of legs relative to depth of chest, amount and distribution of musculature typical to the breed, and so on—all matters with material effects upon tempo, length, and loft of stride, and verging, inescapably and almost immediately, into matters of gestural character, where things like the sex and temperament and color of the horse become factors: Markham's equine humours, musick'd.

It comes as something of a disappointment that those writers most explicitly concerned with riding to music tend to lump this wealth of physical–musical possibility into one rather dismissive category, the *terre à terre,* and then extol the "airs above the ground," the fancy hopping and rearing moves, derived from defensive and intimidating military usages, that were so beloved of this period of horsemanship. Reference to living praxis becomes impossible here: because of the strain they put on the horse's hindquarters, these moves are rarely performed anywhere today save, sparingly, at the Spanish Riding School. The "airs above the ground" include *courbettes, Corvetta* in Italian, in which the horse lifts its forequarters off the ground and hops on its hind feet—Menestrier describes this as "a [hopping] movement like a crow, which has given the name Little Crow to this air" (172); *caprioles,* in which the horse springs off the ground from a standstill, flinging out all four feet; and "the air called 'a step and a jump.'" Both the *courbette* and the *capriole* are, rhythmically speaking, single beats, adaptable to any meter. The *capriole* is obviously a massive movement, best suited to marking musical arrival points; the "step and a jump," a coupling of this move with the less-massive *courbette,* suggests a very grand triple meter articulated as LONG, short-LONG, short-LONG.

Although it is true that, as Pluvinel and others earnestly maintain, such moves occur among horses in nature, it must be mentioned that they do so only under conditions of emotional stress and agitation; a calm horse does not jump off the ground with all four feet, and the rearing posture of the *courbettes* is most likely to be seen in those highly charged instances when stallions fight one another. There will be some feedback from this fact: cause a calm horse to perform a motion that, in its native physical language, is associated with strong excitement, and it will become more or less excited. We might interpret that excitement as military enthusiasm, as Menestrier does:

> Les trompettes sont les instrumens les plus propres pour faire danser les chevaux, parce qu'ils sont loisir de reprendre haleine, quand les Trompettes la reprennent, il n'est point aussi d'instrument qui leur plaise plus, parce qu'il est martial, & que le cheval est genereux, & aime ce bruit militaire. On ne laisse pas de les dresser, & de les accoustumer à l'harmonie des violons, mais il en faut vn grand nombre, qui l'air soit de trompette, & que les basses marquent fortement les cadences

(174). (Trumpets are the most appropriate instruments for making horses dance, because they are allowed to catch their breaths when the trumpeters do so; there is also no instrument which pleases them more, because it is martial, and the horse is generous and loves such military noise. One does not let them be dressed and accustomed to the sound of violins, but a great number [of ballets] are made where the air is that of the trumpet, and where the basses mark the cadence strongly.)

Buried in this fascinating tangle of cause and effect—note the coordination of the horses' and the trumpeters' very breathing—is a simple fact: loud music, especially volume and attack, the strong suit of the trumpet-and-drum corps, agitates horses; and agitation inclines them toward certain kinds of movements. In other words, "musical" responses like this are another manifestation of equine fear.

As for other musical parameters: timbre can cause powerful responses in that is the main means of identifying a sound; along with scent, it is central to horses' capacity for recognition of and discrimination among other beings. They do indeed know their masters' voices; and they can read the moods expressed or encouraged through the timbre of those voices—or, as Newcastle goes on to attest, through the timbre of instruments; he would seem here to be referring to something softer than trumpets and drums.

You may caress him as much and as often as you please: as by patting him gently with your hand: talking kindly to him: stroking him: flattering him: or sometimes by using a certain particular tone of voice, that is common to cajole skittish and unruly horses . . . Horses take great delight in smelling to perfumed gloves, and in hearing of musick, which refreshes them very much.[37]

Timbre is inextricably linked to pitch, and while no one claims that horses find meaning in extended melody, there is good evidence for a rise of pitch being linked to excitation—trumpets are, of course, high pitched (and timbrally piercing) instruments—as well as the reverse. Thus a trainer longeing a horse (exercising it on a long line from the ground), who is not touching the animal, and who is consequently relying on voice commands, will typically signal an acceleration—say from walk to trot—with a sharply rising inflection, "Trr-OT!" and a deceleration with a falling one, "EA-sy." Fiaschi gives a particularly succinct version of this. Most of his musical examples consist of a rhythmicized single pitch, but several, including that for the step-and-a-jump, coordinates a rise in pitch (up a minor third) with a sharpening of timbre, (from "ah" to "ahi"), and the most athletic and agitated motion of this particular air (its *capriole* phase).

Horses clearly respond to many elements of music; but there is little to suggest, in these treatises or elsewhere, that they dance to it in the sense of spontaneously keeping time with a beat. Left to themselves with music, they just do not tap their feet or nod their heads. We do well to remember that keeping time, for all its feeling of atavistic inevitability, is in fact highly socialized and uniquely human—our species' culturally reinforced version of that splendid athletic regularity of motion we see, so particularly in horses of all creatures, when they are healthy, happy, and moving freely, embodying a kind of apotheosis of rhythm, rhythm as a result, not a cause, of movement. The rider to music negotiated a delicate balance between these two musics, then, the powerful silent one of the horse's kinetic bent, and the powerfully coded audible one being played as he rides.

We come here to a strange pass, for it is no longer clear who is obeying whom. Riding to music has long been understood as a demonstration of the extent to which the horse—symbolizing not only the body, but, at least in the sixteenth and seventeenth centuries, the body politic—can be brought into control and submission; but I have been at pains to demonstrate here that it has long been equally clear in practice that both musicians and riders must know very intimately how to adapt to the horse. Its bodily intelligence calls the tune, or rather the rhythm and the gestural character, and if the rider and musicians do not listen, the display will be forced, effortful: in a word, coercive. Thus Louis XIII or the Emperor Leopold, astride magnificent horses before their noble audiences, were demonstrating to their subjects nothing so crude as a capacity for domination, but rather—and this is a most crucial distinction—a capacity for command: command as distinct from domination, command as a reciprocal condition, command as predicated on a knowledge of when to listen as well as when to tell.

Through their own daily involvement in horsemanship, audiences at equestrian ballets would have had an intimate grasp of the terms, and the power, of this physically enacted metaphor for the equivocality of command. Even the most minimally effective horsemen among them, those multitudes who did not ride exquisitely or artistically, but who could consistently manage not to irritate or confuse the animal so much that it hurt or killed them—knew something, and knew it profoundly, in their bodies, about the importance of learning both to speak and to listen to corporeal intelligence. Depending on how one looks at it, such communication integrates, or hopelessly complicates, the mind–body split, and does so on a daily basis; its presence, undergirding the methodological details of numerous treatises, and unequivocally elevated to the status of art in the *balletto a cavallo*, needs a more nuanced explanation than it has yet received.

The flat reduction of discipline to coercion is symptomatic of the troubled relationship of the twenty-first century West to ideas like freedom and wildness, obedience and command. Foucault himself articulates just how poorly we deal with the necessity of discipline to the good life. We habitually conflate coercion and command; we indulge in ill-thought-out Rousseauvian fetishizations of untouched nature; we recoil from neutering our pets or training them adequately; and we worry. We worry that even those kinder, gentler visions of animal training that we find in Xenophon and his followers might, for all their reliance upon inducements, or "friendshippe," still be subtle forms of coercion. Sixteenth- and seventeenth-century horse trainers knew no such useless quandaries. By their lights, the good commander, the good listener, the horse trainer, or the emperor, does not waste time in worrying, but is exceedingly vigilant, with a vigilance that is as much moral as physical. Do I command the right, the best, the good thing? Is it the right time for this thing? Is my command being followed? Why or why not? How will that influence my next command?—precisely the musick of the lutenist's vigilance as, practicing or performing, he feels the reach or press of his hands, hears the sounds he produces, forms his judgments of what to do next. For none of these arts—horsemanship, statesmanship, music, speech—is the process of physical and ethical evaluation that I have been calling musick optional. These arts resemble one another intrinsically in that they depend, and in the end, depend absolutely for the possibility of their continuation, on the incorporation of listening into the process of command.

Notes

1. Kate van Orden, "Music and Military Virtue in Early Modern France: the Equestrian Ballet" Typescript, read at the London meeting of the International Musicological Society, August 1998.
2. Xenophon, *The Art of Horsemanship*, trans. and with notes by M. H. Morgan (London: J. A. Allen, 1962), 62.
3. Claude-François Menestrier, *Traité des tournois, ioustes, carrousels, et autres spectacles publics* (Lyon: Jacques Muguet, 1669), 182. Facsimile edition, New York, Garland, 1979. Series, *The Philosophy of Images*, edited and with introductory notes by Stephen Orgel.
4. William Cavendish, Duke of Newcastle, *Methode et invention novvelle de dresser les chevavx* (A. Anvers: Chez Iacqves van Mevrs, 1658), 122.
5. Michel Foucault, *Discipline and Punish: the Birth of the Prison*, trans. Alan Sheridan, (New York: Vintage Books, 1995), 137.
6. Jacques de Solleysel, "Some curious remarks upon Horses represented ..." (n. p.) in *Parfait mareschal ... translated from the last Paris impression, by Sir William Hope ... By whom is also added as a supplement to the first part, a most compendious and excellent collection horsemanship* (Edinburgh: George Mosman, 1696).
7. Antoine de Pluvinel, *Le Maneige Royal* (1626). Facsimile edition, with a foreword by Alois Podhajsky, (Fribourg: Office du livre, Bibliotheque des arts, 1969), 10.
8. Newcastle, *Methode et invention*, 138. This and a number of other passages in Newcastle appear, closely paraphrased but not attributed, in William Hope's treatise.
9. Cesare Fiaschi, Part Two, Cap XVI in *Trattato dell'imbrigliare, Atteggiare & Ferrare Caualli ... Opera vtilissima à Precipi à Gentil'huomini, à Soldati, & in Particolare à Manescalchi*, third edition, (Venice: 1603), 100–111.
10. Act I, Scene iii, opening in Molière (Jean-Baptiste Poquelin), *Le bourgeois gentilhomme*, trans. Stanley Applebaum (Toronto: Dover, 1998), 210.
11. Foucault, *Discipline and Punish*, 139.
12. Newcastle, *Methode et invention*, Introduction, n. p.
13. I thank Susan McClary for pointing out this small intersection between horse training and religious practice.
14. Chapter XVIII ("Of rewarding and punishing Horses, that fear doth much, but love little") in Solleysel, *Parfait mareschal*, n. p.
15. Book V, Chapter 8 ("Of the Passions which are in horses, and the love which their keepers should bear unto them") in Gervase Markham, *Cavelarice, or, The English Horseman, Contayning all the Arte of Horse-manship, as much as is necessary for any man to vnderstand* (London: Edward White, 1607), n. p.
16. Xenophon, *The Art of Horsemanship*, 49–50.
17. Fiaschi, *Trattato dell'imbrigliare*, 23–4.
18. Markham, *Cavelarice*, Book V, Chapter 8 ("Of the Passions which are in horses, and the love which their keepers should bear unto them"), n. p.
19. Xenophon, *The Art of Horsemanship*, 20.

20. Newcastle, *Methode et invention*, Introduction, n. p.
21. Aristotle, *Nichomachean Ethics*, in Aristotle, *Introduction to Aristotle*, trans. Richard McKeon, Second edition (Chicago: University of Chicago Press, 1973), Book I, Chapter 13, 1102b, 24–26.
22. Newcastle, *Methode et invention*, closing Remarks, n. p.
23. Newcastle, *Methode et invention*, closing Remarks, n. p.
24. Fiaschi, *Trattato dell'imbrigliare*, Part Two, Cap XVI, 100–111.
25. Newcastle, *Methode et invention*, 93.
26. For this concept of "musick" I am much beholden to Christopher Small, *Musicking: The Meanings of Performing and Listening* (Hanover: Wesleyan University Press, 1998), especially "Interlude I: The Language of Gesture."
27. Markham, *Cavelarice*, Book VII, Chapter 1 ("Of the Composition of Horses . . ."), n. p.
28. Markham, *Cavelarice*, Book VIII, Chapter 4 ("Of the excellency of a Horses understanding, and other qualities"), n. p.
29. Menestrier, *Traité des tournois*, 169.
30. Pluvinel, *Le Maneige Royal*, 35.
31. Any air (including those above the ground) performed in place, without traveling forward: the ultimate in control. (See description of the airs, in text.)
32. Vincenzo Forcella, *Spectacula, ossia caroselli, tornei, cavalcate e ingressi trionfali* (Bologna: A. Forni, 1975). This is Forcella's undocumented quotation of a description of the event by its librettist, Francesco Sbarra. See also Egon Wellsz, "The 'Balletto a Cavallo'," in *Essays on Opera*, trans. Patricia Kean (London: Dennis Dobson, 1950), 82–9.
33. Changes of lead at the canter (see description of this gait, below in text).
34. Tight turns made with the horse's head bent to the inside.
35. There are a number of excellent videos that demonstrate the Kur, which is performed competitively. I recommend the videos of the FEI (Fedération Equestre Internationale) World Cup finals: 1998, in Gothenborg, and 1999, in Dortmund, both by Martin Bird for Equestrian Vision Productions.
36. Menestrier, *Traité des tournois*, 179–80.
37. Newcastle, *Methode et invention*, 111.

CHAPTER **3**

Breeding and Training Bastards: Distinction, Information, and Inheritance in Gilded Age Trotting Horse Breeding

PHILLIP THURTLE

Newland Archer dined one evening in the 1870s with his mother, his sister, Miss Sophy Jackson, and the authority on New York's families, the old bachelor Sillerton Jackson. The news about the imminent financial collapse of a prominent banker had been the topic of conversations for weeks now. The participants in the "pyramid" of the New York elite always felt the banker, Julius Beaufort, to be "common." Arriving from England with letters of recommendation in his hand (but little money in his purse), the uneducated Beaufort had "speedily made himself an important position in the world of affairs; but his habits were dissipated, his tongue was bitter, his antecedents mysterious."[1]

For the guests at Mrs. Archer's table, Beaufort represented just one more crack in the edifice of the New York social elite. "Observing it from the lofty stand-point of a non-participant, [Newland's mother] was able, with the help of Mr. Sillerton Jackson and Miss Sophy, to trace each new crack in its surface, and all the strange weeds pushing up between the ordered rows of social vegetables" (256). Because of the scandal, Mrs. Archer supposed that the Beauforts would retire from polite society and live in their country house in North Carolina. "'Beaufort has always kept a racing stable, and he better breed trotting horses. I should say he had all the qualities of a successful horse dealer.' Everyone agreed with her" (279).

In this dinner table conversation from *The Age of Innocence,* Edith Wharton sketched the reaction of an old-moneyed New York elite to the brash and assuming manners of the newly wealthy. The understated confidence behind Mrs. Archer's declaration about Julius Beaufort's proper (as opposed to assumed) place in society relied on judgments shared in the patriciate regarding taste, family, and proper social form. In the Archers' social circle, one maintained social prestige by how one carried one's self in the world, and Beaufort's risky financial dealings and his predilection for fast horses distinguished him as from an inferior class. In this popular novel of late nineteenth-century manners, the character of Beaufort demonstrates how deeply intertwined personal taste was with class distinction and the conservation of familial property.

To modern readers, Wharton's reference to Beaufort's trotters may seem little more than an allusion to an obsolete form of transportation. During the Gilded Age, however, trotters were powerful symbols of industrial progress. To begin with, raising and breeding trotters was the most popular leisure activity of rich industrialists; Cornelius Vanderbilt, Leland Stanford, John D. Rockefeller, and E. H. Harriman all drove or bred trotters. Moreover, trotting horse racing was America's first "modern sport." Trotting horse owners relied on and promoted professional sports organizations, formal rules and regulations, professional specialization, the keeping of formal statistical records, and a specialized sporting press. Although the sport's origins at county agricultural fairs resonated with the industrialists' own reverence for a rural past, the sport itself would not have been possible without the groomed roads of large urban centers.[2] Because so many of the "empire builders" of the nineteenth century developed an interest in racing and breeding trotters,[3] an investigation into the sport offers an unique opportunity to study the industrialists' theories on the potential of living beings. In an age where theories on the transmutation of the species were widely debated and where all living beings were confidently assessed by a calculus of productive labor, theories for increasing the labor power of horses often shed light on ways to increase the labor power of humans.

The late nineteenth-century practice of breeding trotting horses holds at least three specific areas of interest for scholars interested in the consubstantiation of modern forms of embodiment and information flows:

1. Theories on breeding held by high profile owners were perhaps the most widely disseminated theories on heredity in the late nineteenth-century America. Written up in books, daily newspapers, and a specialized sporting press, the public eagerly awaited the application of industrial sized success to the breeding of biological beings.

2. Trotting horse farms in general were one of the first "Modern" agricultural enterprises; they were maintained with vast resources and run by extended managerial hierarchies under absentee owners, and they also sold specialized produce and utilized rationalized breeding methods. As such they provide one of the first glimpses into how the tools of managerial capital privileged certain forms of human embodiment. For many in the late nineteenth and early twentieth century, those who developed fastest were assumed to have the potential to develop furthest. Although managerial capital did not create or determine this view of human development, it actively supported it.

3. The bodies of horses supplied much more than an analogue for thinking about human potential; rather they provided a forum to think about how to select for this potential and then apply these lessons to the training and breeding of humans. Far from a blank slate, horse breeding provided an overdetermined model of breeding and training because of the physical capacities

of horses as well as its rich cultural associations. It was the uniqueness of this combination of physical and cultural associations that convinced trotting horse breeders that they had identified a powerful tool for identifying the relationship between industrial demands and biological constraints. Thus some horse breeders interpreted fraction-of-a-second improvements in horses' trotting times as a demonstration on how industrial breeding techniques opened new capabilities for living beings.

The first part of this chapter outlines how these values were formulated in opposition to an old-moneyed American elite. The second part looks at how industrial information processing tools (such as middle managers)[4] made these values seem *natural* as well as *modern* when applied to large-scale breeding projects.

This transition from an emphasis on a formal analysis of social mores of good breeding and blood inheritance to a structural functional analysis assessed by the measurement of a few well defined characters was one of the most profound changes in views of intergenerational continuity in the last few centuries. The important issue is not whether this transformation was more or less true to nature; instead, it is important to understand how we consubstantiate a vision of nature along with novel social organizations. If we want to consider how we can use our knowledge of information and bodies to create more equitable social institutions then we must also come to a greater understanding of how the models we use to understand and explore the world helps to manifest that world.

I. Middle Class Mores

1. Distinction Through Blood: Trotters and Thoroughbreds

Importantly, Edith Wharton linked Beaufort with trotting horses and not their aristocratic ancestors, the thoroughbreds. These two forms of horsemanship had different geographic distributions, appealed to different populations, and evoked different cultural associations. Trotting horses—horses that could pull carriages while at a trot—were preferred to thoroughbreds by many horse fanciers in the Northern United States at the end of the nineteenth century.[5]

Contrary to the thoroughbred's appeal to those who reenacted aristocratic turf sports, trotters were an outright celebration of utilitarianism. The whole sport—from racing to breeding—embodied the middle-class Victorian virtues of speed, democracy, and rugged masculinity. As the English tourist, John Henry Vessey, remarked on the sporting scene in 1859, "riding is not the fashionable amusement with the American people, they seem to delight in driving these fast trotting horses in light buggies."[6] By the 1850s, organized trotting horse races dominated agricultural fairs, and popular songs (such as "Old Grey Mare") and mass-produced lithographs championed the animal heroes of these races. Horse racing became the first sport "international in scope,"[7] and trotting horse racing in particular became the first "mass supported and mass endorsed spectator sport" in America.[8]

Although fast trotting horses were high-quality horses, during the middle of the nineteenth century they were distinguished more by what they accomplished than from whom they were bred. As railroads tied together urban centers, horses were increasingly relied on to bring people and goods from the surrounding countryside to the railroads. Since trotters moved quickly while in a smooth gait, they were perfectly suited for transportation with a minimum of disturbance. In the eyes of their enthusiasts, the emphasis on performance without recourse to noble ancestry allowed the fastest trotters membership in the American "aristocracy of merit."

Because trotters performed a specific function, owners did not initially emphasize the preservation of ancestral lines through pure line breeding. Instead, they selected individuals that demonstrated the ability to perform that function. In fact, the trotter's mongrel ancestry reflected the emphasis on utility at the expense of pedigree. Trotters were thought to have descended from mixes of the following four types of horses: the English thoroughbred, the Norfolk trotter, the Arab or Barb, and certain pacers of mixed breeding.[9] For the mass of Anglo middle-class Americans who defined themselves in opposition to aristocratic British culture, the mixed ancestry of the trotter qualified it as the most American of animals.[10] As Oliver Wendell Holmes stressed: "Horse-*racing* is not a republican institution; horse-*trotting* is."[11]

2. Bloodline Inheritance: The Conservation of Elite Social Privilege Through the Enactment of Proper Social Form

The appeal of European aristocratic heritage shaped many social traditions of the New York rich during the 1870s and 1880s. As America's gateway to Europe, New York's dominance as a shipping port promoted economic and social exchange with Europe. More so than their "provincial" Boston counterparts, elite New Yorkers often purchased goods from European merchants and took extended vacations at European destinations. More importantly, counting European aristocracy among one's ancestors qualified one for the very apex of the New York social triangle. As Edith Wharton reminisced, "My mother, who had a hearty contempt for the tardy discovery of aristocratic genealogies, always said that old New York was composed of Dutch and British middle-class families, and that only four or five could show a pedigree leading back to the aristocracy of their ancestral country."[12] The rest of those who had gained social privilege, had, for the most part, made their fortunes through the land speculation and mercantile trading that marked an earlier period of U.S. economic expansion. As David Hammack has argued, during the 1880s merchants and lawyers were still the most prominent of the New York economic elites.[13]

Without direct recourse to aristocratic bloodlines to support their social privileges, New York's old-moneyed elite distinguished themselves on the privileges of wealth and the observance of proper social form. Those who recently acquired their fortune, having no opportunity to acquire elite social practices or the ability to carry out these practices un-self consciously, betrayed their "common" background to members of elite society. By distinguishing who could and who could not belong to their subculture, New York elites maintained their social privilege through the *conservation* of social form.

Displaying proper social form signified that one was well bred. In use, however, the phrase "well bred" only indicated a loose aggregate of qualities. For instance, notice how easily Edith Wharton associates the nationality of ancestral heritage with the ability to use the language skills of the well bred:

> Bringing up in those days was based on what was called "good breeding." One was polite, considerate of others, carefully accepted formulas, because such were the principles of the well-bred. And probably the regard of my parents for the niceties of speech was a part of their breeding. . . . I have noticed that wherever, in old New York families, there was a strong admixture of Dutch blood, the voices were flat, the diction careless. My mother's stock was English, without Dutch blood, and this may account for the greater sensitiveness of all her people to the finer shades of English speech.[14]

Signifying a constellation of behaviors, to be "well bred" meant to observe proper social form with the assumption that that ability came from one's family origins.

Of course, this is not to say that the concepts "learning" and "inheritance" were entirely conflated. On the contrary, rough facsimiles of these two categories were continually enacted. The *Age of Innocence*, for instance, opens at the New York Academy of Music, where Gounod's *Faust* provided the counterpoint for the drama in the box seats. With opera glasses inclined toward the audience, the two guardians of New York society, Lawrence Lefferts and Stillerton Jackson, exchanged their opinions about the goings on in society. Lefferts was "the foremost authority on form in New York society," while "old Mr. Jackson was as great an authority on 'family' as Lawrence Lefferts was on form." One turned to Lefferts in order to understand "just when to wear a black tie with evening clothes," while Jackson had not only mastered "the forest of family trees" he had registered "most of the scandals and mysteries that had smouldered under the unruffled surface of New York society within the last fifty years."[15] The coherence of this tight social circle relied on the distinct judgments of Lefferts and Jackson. They simultaneously defined the boundaries of society through judgments of taste and knowledge of the intermingled kinship that wove together the social circle of the New York elite. In short, family and proper form were the two axes that structured the conservation of elite society.

Although I have read Wharton's portrayal of the guardians of elite society as representations of cultural tendencies, it is important to remember that real people assumed the responsibility for preserving proper social form in late nineteenth-century New York. Isaac Brown was one of these self-appointed guardians, and it was to Brown that many turned for advice on who to invite to social functions. Historian Frederic Jaher describes him thus: "A consummate snob, he separated society into 'old family, good stock' or 'new man' who 'had better mind his p's and q's or I will trip him up.'"[16] The most important point is that the authority for the preservation of New York society was at this time still specifically embodied; it was through the face-to-face interactions of the drawing room circle or the elaborate spectacle of the box seats at the opera that proper social form was enforced.

Alfred Chandler has argued that a similar type of face-to-face enforcement took place in the merchant businesses in which the New York elite engaged. Most commercial enterprises were still partnerships, contractual agreements between individuals: "American merchants did not yet feel the need for a legal form that could give an enterprise limited liability, the possibility of eternal life, or the ability to issue securities. Even when an enterprise was incorporated it remained a small single-unit firm run in a highly personal manner." Even the large enterprises were often partnerships built into a "chain of mutually supporting partnerships reflecting a kinship network."[17] Within these partnerships there was little delegation of tasks and the "organization and coordination of work . . . could easily be arranged in a personal and daily conversation."[18]

Edith Wharton detailed the social ostracism that followed if one did not abide by the "traditional code of family and commercial honor"[19] through which these partnerships were run:

> New York has always been a commercial community, and in my infancy the merits and defects of its citizens were those of a mercantile middle class. The first duty of such a class was to maintain a strict standard of uprightness in affairs; and the gentlemen of my father's day did maintain it, whether in the law, in banking, shipping or wholesale commercial enterprises. I well remember the horror excited by any irregularity in affairs, and the relentless social ostracism inflicted on the families of those who lapsed from professional or business integrity.[20]

In the parochial world of New York elite society, business and social interactions easily interpenetrated and both were based on a strict code of conduct enforced by personal interaction. The dinnertime discussion of Julius Beaufort's financial scandal with which I began this chapter is one dramatization of the social ostracism that could follow upon a large breach of proper business conduct. Although their distinction as an elite class relied on the conservation of social form more than on the conservation of family ties (in fact, members of the Knickerbocker elite intermarried with newcomers more frequently than their Boston counterparts) many members of a growing middle class interpreted this need to conserve in biological terms.[21] After the turn of the century, it was common to hear the interpretation that the old New York families lost their social influence because their tendencies to conserve social privilege led to physiological problems associated with inbreeding—most especially the loss of male vigor and fertility. Writing in 1916, the horse breeder and eugenicist

W. E. D. Stokes reminisced about childhood visits to old New York families:

> the ladies were all very old and distinguished looking. They dressed in black with white lace collars, and often wore lace half gloves and always talked about the dead.... There were few marriageable men, and of these many remained bachelors and some old medical records and correspondence I procured indicate that the majority were blanks. Their seed lacked fertility."[22]

II. Breeding True: Processing a New Elite

According to Alfred Chandler, the means for coordinating business operations changed in the United States at the end of the nineteenth century. With increased speeds and volumes of production and distribution, administrative coordination (as opposed to informal face-to-face interactions) proved more safe, reliable, and efficient. More recently, James Beniger reapplied Chandler's structural arguments to explain the rise of the "information society." Beniger claims that throughputs of industrial production spurred a *Crisis of Control* consisting in crises in safety, distribution, production, and marketing and advertising. The *Control Revolution* managed the Crisis of Control; throughputs and material flows were rationalized and made more efficient through the use of new communication technologies.[23]

Perhaps it is not surprising then to see that encouraging progressive evolution through competition was the driving concern for the breeding reforms instituted by National Association of Trotting Horse Breeders. In 1879, just three years after its institution, the association created a set of rules and standards designed to establish trotters as a distinct breed. Because trotters were from a mixed ancestry, the variability of "the produce" made it difficult to consistently breed a horse with the inclination to trot. In order to *fix* this inclination among the heritable constitution of trotters, breeders established a basis for a pure line. The National Association of Trotting Horse Breeders created a breeding registry "[i]n order to define what constitutes a trotting-bred horse, and establish a BREED of trotters on a more intelligent basis."[24] For a horse to qualify as a trotting horse it had to have trotted a mile in 2:30 or better, had a mare or sire who had met this performance standard, or has a grandam or grandsire that has met this performance standard. The standard was set high enough so as to assure that only the "exceptional horse should be entitled to a place in the record."[25]

Although thoroughbred owners in England had established a breeding register approximately 100 years before, the differences between the two cases were telling.[26] Whereas the original thoroughbred register traced ancestral lineages to a series of progenitor horses, trotters had to meet a single performance standard. The object was to take the handful of trotters who had had superlative racing performances and create a breed from them. Trotting horse owners were attempting to rationally establish a new breed of horse based on a utilitarian function. As a writer for *Wallace's Monthly* claimed: "These measurements have not been made to test the purity of his blood, for he is a thorough composite, but to determine its strength and power to achieve what has been claimed for it."[27]

1. "Fixing" a Function: Industrial Scales and Managerial Hierarchies

As many of the industrialists turned to racing trotters, they made use of the developments that their industries promoted: "large scale investments of money, sound promotion, the application of science and invention to the technical problems of the sport. These things the business leaders who joined up with the sport were prepared to bring it." Many at the time were excited about the progress that these individuals would bring to the sport. As one writer for the *Chicago Herald* reasoned:

> And then in the last few years men of wealth and brains have gone into the business with the ideals of producing the fastest horses in the world. . . . When such men as Governor Stanford, who built a railroad across the Rockies, and Robert Bonner, who was a successful journalist, put thousands of dollars into the training and breeding of horses, why, they are bound to have a measure of the same success they have had in other lines.[28]

Although these reports exaggerate the role of the industrialists in the changes that trotting horse breeding went through during the 1870s and 1880s, the farms that these individuals financed introduced large-scale animal husbandry practices to trotting horse breeding. These heavily capitalized "laboratories of speed" utilized vast resources, paid more attention to "organized" breeding methods, and developed intricate managerial hierarchies.[29] In contrast to previous farms that bred trotters as just one domesticated animal among many, these new farms concentrated solely on breeding trotting stock. For instance, Leland Stanford's stock farm at Palo Alto, one of the most famous of the large "laboratories of speed," had as many as 775 horses in 1889.[30]

In order to coordinate these vast enterprises, the breeders relied on one of the important social "innovations" that Beniger and Chandler have argued mark the beginnings of managerial capitalism—the creation of an extended managerial hierarchy. John Bradburn, the superintendent of a farm owned by C. J. Hamlin (who had made his money from the dry goods business and glucose manufacturing) wrote an advice book on how to set up a large-scale trotting farm. Aimed at wealthy individuals with little knowledge of horses, Bradburn detailed the necessary chain of managerial command. Beneath the proprietor of the stock farm was the superintendent who would answer directly to the owner. It was the superintendent's job to coordinate the different tasks involved in running the farm and to make sure that it operated economically. Beneath the superintendent of the farm were the trainers who were directly responsible for developing the speed of the colts. Below the trainer were the grooms who were responsible for most of the handling that the horses received. Then below the grooms were laborers who took care of many of the mundane tasks at the farm.[31] Add to this the fact that many farms also had facilities for growing and milling their own feed, a blacksmith shop, and a wheelwright shop, and it is easy to appreciate the amount of coordination it took to keep a large farm running profitably. Leland Stanford's farm alone employed up to six trainers and on the average of 150 laborers.[32]

Working on such a large scale gave these breeders a distinct advantage. Since trotters were from "mongrel" backgrounds, large-scale breeding projects gave the breeders an opportunity to locate that rare "golden cross" that would outshine other horses. The "culls," or horses that were only fit for labor, were easily distributed to nearby farms, thus not ruining the important national reputation of a stock farm. The large specialized farms had, in effect, become large arenas for screening horses with the right potential.

In their dual roles as promoters of industry and promoters of animal breeding, many of these industrialists explicitly discussed how increasing the speed of trotters would increase the productivity of the nation. Robert Bonner, for instance, argued with Mr. Garrett, president of the Baltimore and the Ohio Railroad, "that a breed of horses could be raised capable of hauling a street-car from the Astor House to the Central Park in ten minutes less time than it now takes ordinary horses to haul it, and with even more ease to the horses."[33] Leland Stanford evaluated similar arguments for increasing productivity of horses as an economic consideration:

> I have been told that there are about thirteen millions of horses in the United States. . . . It seems to me that the majority might be bred up to the standards of the best, thus increasing the average value $100 per horse. The increased value would represent a gain of thirteen hundred millions of dollars to the United States on the present number of horses . . . There is, therefore, a great economic question involved in the breeding of good horses for labor.[34]

In order to increase the speed of the trotter, Leland Stanford undertook a series of different types of experiments. Perhaps the most famous of these was the hiring of Eadward Muybridge to record his trotter "Occident" while in motion.[35] Although this experiment has shouldered a heavy interpretive burden for a constellation of fin-de-siècle discourses, one of the most interesting has been Anson Rabinbach's recent investigation into Muybridge's influence on Etienne-Jules Marey and his concerns for the conservation of labor power.[36] Marey realized, and petitioned potential sponsors, that his instantaneous photography, which broke a continuous action into a series of discrete movements, could be used to determine the "conditions the maximum speed, force, or labor which the living being can furnish may be obtained."[37]

Drawing the continuities the other way, however, back through Stanford to his other experiments for improving the productivity of trotters, emphasizes how heredity (not just motion) could be rationalized to increase productivity. Just as the disarticulation of a movement of a worker allowed for the reorganization of an action on a more efficient basis, the disarticulation of the stream of a hereditary constitution of a breed (or bloodline) could lead to the reorganization of the breed on a more efficient basis. From the perspective of the new animal breeders, the breed of the animal more than the animal itself was the object of manipulation—even superb performances of individual trotters were indicators of the "evolution" of the breed in general.[38] In the new stock farms, engineering the increase in a single trotter's speed indicated that trotters as a whole were becoming more productive.

2. Incorporating Heredity: Selecting a New Elite

In the discussion that follows we look at how theories developed on Stanford's stock farm supported suppositions about the importance of early development as a definition of excellence. We argue that the early specialization of Stanford's trotters was a consequence of the application of the values and organization promoted by industrial systems to the training of living beings. As one of the first sites to rationally produce animals for greater productivity, Stanford's stock farm is a wonderful model for investigating how presuppositions for increasing industrial productivity were identified as natural capacities.

Trotters on Stanford's farm trained at a younger stage of life than other farms. Stanford, in fact, is credited with changing the way that trotters were trained. Most trainers had followed the techniques set down in Hiram Woodruff's *The Trotting Horse of America*.[39] This text claimed that because trotters were not expected to perform at full capacity until maturity, training should not begin until the trotter's third or fourth year. Stanford, on the other hand, began training his horses during their eighth month of life in a manner similar to the races it would run in adult life. Most likely influenced by a growing interest in kindergarten education (Mrs. Stanford donated generously to kindergartens around the Bay Area), Stanford ordered a small track built for the young horses in 1879 that he named the "kindergarten track." Once broken to the halter, the colt trained on the kindergarten track.

Charles Marvin, the head trainer at the Stanford Stock Farm, recognized that trotters could be trained like humans because they both needed the direction of human guidance to fit into social circumstances:

> Every one must admit that there are a great many points of resemblance between a colt and a young child. They are both mammals, and are therefore to be conceived, begat and nourished to point of absolutely independent existence on similar lines, but the resemblance extends beyond this period for they are either as colts or children subject to the direct interference and guidance of man in the matter of education or training, each after their allotted sphere.
>
> Now, how would any sensible parent or guardian train a child? And by that word train I wish to convey the ideal of physical as well as mental education; they go hand in hand, for a healthful body makes a healthful mind.
>
> To draw the parallel closely we must bear in mind the fact that a child of seven to eight is no older in proportion than a yearling, and that a two-year-old is on a level in that respect with a healthy school-boy of from fifteen to sixteen.

> The answer to my question is conveyed by an old proverb, the truth of which has been evident from the very beginning of society: "Train up a child in the way he should go." Alter but one word, and it will be equally applicable to the little fellows we are just now most concerned about: "Train up a colt in the way he should go."[40]

Training on the kindergarten track did not mean teaching the colt new skills. The main value of the kindergarten track is that it allowed the trainer to evaluate the ability of a horse at a young age:

> They gave the promise of their future greatness there. Sunol and Palo Alto, Margeruite and Bonita, Hinda Rose and the Beautiful Bells family have been the stars of the kindergarten, just as they were afterwards stars of the sterner battle-field of the turf. And this fact proves more than any other the truth of what I have contended—that this is above all the best *natural* method of training young trotters now extant. If it were not a natural system it would not prove so true an index of the capacity which the horse is afterward destined to exhibit.[41]

Selecting the correct horse to train made good business sense. Training a world-class trotter was an expensive procedure, often requiring many hours of training time before a horse was ready to race. The kindergarten allowed the trainer to evaluate whether the horse had inherited the potential that merited the attention of the trainers. As Marvin claimed, "the miniature track enables you to select those of your colts that will best repay the labor and expense of training. Let me work allot of colts on this track for three months and I will pick out the stars" (210).

The horses that were not early bloomers were sold off as stock, saving the owner the cost of training a horse without the right potential. Or, as John Bradburn advised, these "culls can usually be disposed of to advantage among the farmers of your neighborhood, who will use them as general-purpose horses," so that the reputation of one's farm is not damaged. Training at the "large laboratories of speed" did not mean teaching the horse a skill, it meant selecting the horses with the right potential and then providing the best arena for the development of this potential.

Early training of trotting horses also made good business sense in a different way. Racing a horse brought little money directly into a stock farm; however, it did provide a forum for highlighting the major asset of a horse—its progeny. If a horse produced fast horses, then it would raise the price of the rest of its progeny. Under the old system of breeding and training, it took too long to judge the progeny of a horse.

> The business of breeding has now reached a point where few breeders have the inclination, even if they were financially able or believed it beneficial to wait six or seven years for the get of their stallions and the produce of their mares to show what their blood is worth. (185)

Under Stanford's new training system, a trainer only had to wait two or three years to display the heritable potential of a horse. In a culture where potential was marketed as a commodity, the early training of horses was reinforced by powerful financial incentives.

Early training not only selected the fastest young trotters, it also contributed to the evolution of the trotter as a distinct breed. Marvin claimed that the horses could be trained so young because the trotter had been through many rounds of selective breeding. "The older the breed grows, and hence the higher capacity in the special purpose for which it is bred, the earlier this capacity manifests itself in a high degree" (189). Early selection not only promised immediate financial incentives, but also promised to progressively evolve the trotter as a distinct breed capable of performing its function—to trot at a fast speed.

By adopting the standards of a modern industrial enterprise the new large-scale trotting farms redefined the values for training and breeding horses. Because trotters were a breed based on the performance of a function, increasing the productivity of the trotter meant increasing the ability of the trotter to perform this single function. This promoted a conception of the potential of the trotter based on a developmental hierarchy, where the horses that demonstrated this ability at an early age were thought to be those that had the natural capacity for this function.

The early specialization of trotters was a function of a conception of life promoted by the needs of the new industries and as such, Stanford later promoted a similar system for the education of humans. David Starr Jordan recognized how Stanford's experiments on the stock farm helped substantiate Stanford's theories of education for the "human colt":

> One of the interesting features of the Farm was "the kindergarten," a trotting track for the younger colts on which they were taught to maintain the proper gait from the beginning, and which thus served as a basis for an orderly progressive training. With a somewhat similar notion in regard to human education Mr. Stanford often dallied, imagining a school which should receive only a limited number of children and train them continuously from kindergarten to university. The suggestion stirred up a certain amount of ridicule, but it held more than a modicum of sound sense, although it overlooked the necessity of a broader range of environment for the human colt.[42]

When the owners of industries set up breeding farms, they applied the criteria and methods of their business to rationalize the production on these farms. This contributed to a redefinition of the quality of the product along single specialized criteria. Conscious of how their businesses demanded a new criteria of human excellence, many of the owners of these industries turned to philanthropic giving to increase the production of their industries. From the beginning, Leland Stanford intended that his university "educate boys and girls in such practical industries as will enable them to go out in the world equipped for useful labor."[43] Stanford never considered education as an end unto itself. As the head of a major industry, Stanford wearied of hiring university graduates who had no practical skill. Universities were not producing a product that fit the needs of the growing industrial sector. Stanford sought to change this with his creation of the Leland Stanford Jr. University.

After Stanford died, an anonymous author could not resist the temptation to describe Stanford's family tree as if it was the pedigree of one of his famous horses. The three paragraphs that introduce this document succinctly weave together themes argued in this chapter. More importantly, by weaving these themes together, this document demonstrates how these themes fit together to form an interlocking and mutually supportive set of beliefs:

> Stanford University, as the climax to Leland Stanford's career as a railroad builder and statesman, was a logical sequence of the heredity derived by its founder . . .
>
> If Stanford had been as familiar with his own pedigree as he was with the pedigrees of the racers which preceded the human colts on the Stanford farm, he would have known that in carrying out his desires to establish a great university he was stepping true to form. As well might some early farmer have marveled, "That Morgan mare's colt was broke to plow; where would he get the notion of trotting?" . . .
>
> . . . Before the day of specialization, that stout strain of early New England speedsters broke the ground for the oats they trained on when the leisure harvest set them free to test their inborn speed. Leland Stanford had to plow a wide field with his steel rails, and harrow it to some degree of smoothness with his administrative powers, before it grew the population which enabled his . . . blood to fulfill itself.[44]

In these paragraphs the author recognized the trotter's status as an elite horse distinguished from an utilitarian background, and how institutions could train individuals for the demands of the division of labor in an industrial society. More intriguingly, he also wonderfully mixes industrial and agricultural metaphors to suggest how the institution that Stanford created fulfilled the promise of his bloodline. The institution that Stanford funded, and not the Stanford family, became an organ for the reproduction of a social order.

For W. E. D. Stokes, this association was so strong he made no distinction between Stanford University, Stanford's stock farm, and the eugenic interests of Stanford University's first president:

> To the Trotting Horsemen, more than anyone else, is due the advancement this country is now making in eugenics. It was Governor Leland Stanford, owner of "Electioneer," and the great Palo Alto Farm, who placed David Starr Jordan at the head of Stanford University, with unlimited funds, to carry out his ideas on breeding and heredity.[45]

Stanford was only one of a group of benefactors supporting new institutions of higher education during the last half of the nineteenth century; Jonas Clark,

Johns Hopkins, Andrew Carnegie, and Ezra Cornell were only the most famous industrialists who intended to recreate higher education closer to their own image. Because these institutions were intended to address similar social problems, and since many of the philanthropists actually discussed their projects among fellow philanthropists and educators, it is not surprising that these institutions held certain ideals in common.[46] First of all, these institutions were developed in distinction to older institutions. "[A]s men of affairs" they "believed that existing institutions failed to give young people the equipment needed for successful achievement in business, agriculture, and the sciences."[47] There was a growing conviction that the older institutions were still bent upon instilling "a series of underlying responses, applicable to all future situations."[48] With an emphasis on the traditional curriculum ("Greek, Latin, mathematics, and to a lesser extent moral philosophy" (36)) older institutions had little time for the science-oriented curriculum prized by the industrialists. Even the faculty of many older institutions, with their limited range of specialties, often appeared as the product of "[i]nbreeding" (47).

"Relations with and attitudes toward family" provided another set of distinctive dispositions. In a frequently cited study, Merle Curti points out that industrialists who gave large amounts of money to "perpetuate the family name" lacked "close family connections."[49] Curti's analysis takes the first crucial step in understanding the growth of industrial philanthropy, yet he fails to adequately consider that many founders never intended to leave large inheritances to their sons, even if they had them (Ezra Cornell and Andrew Carnegie are the most well-known cases). The more general question behind the gifts is: why were institutions seen as a "logical fulfillment" of a bloodline?

3. *The Value of Blood: Trotting as a Natural Capacity*

During the last decade of the nineteenth century, owning and riding a champion trotter became a symbol of wealth and conspicuous consumption—trotting had in effect become an elite activity. Although owning a fast horse had always been expensive, the prices of champion trotters during the last part of the nineteenth century reached astonishing heights: Leland Stanford sold the three-year record holder, Sunol, to Robert Bonner for $41,000, while in 1892 he sold the promising young stud, Arion, to Malcom J. Forbes for the unheard of price of $125,000. Often even the stalls that the horses were kept in were lavish affairs, expensively decorated in order to receive the guests who came to look at the fast horses.[50]

Not all horse owners were happy with the changes that the sport had gone through; in fact, the growing commercialization of the trotter and the fantastic prices paid for the best 'horse-flesh' irritated many horsemen who idolized the rural origins of the sport. As argued earlier, one of the reasons that trotting had become popular in the first place was that the sport was open to those with hardy but not immense incomes. As the fastest trotters became more and more expensive, critics charged that rich industrialist owners were making the sport too elite. Just as the heritable constitution of Stanford's trotters came closer to approximating the hot-blooded thoroughbreds, the practices of trotting had come closer to resembling the practices of thoroughbred racing.

The growing elitism of trotting is best illustrated by the sensational sale of Stanford's trotter "Arion" to J. Malcom Forbes for the fabulous price of $125,000. Holding the record for a two-year-old trotter (2:10 3/4), Arion promised to develop into the fastest trotter ever. Forbes was not the self-made man that the earlier generation of trotting horsemen had been. A member of the Boston moneyed elite, Forbes was the son of John Murray Forbes, who had gained his money through his large railroad empire.[51] As part of the elite Boston sporting set, Forbes had been best known for known his yacht racing, where he became the first American to win the World's Cup. Forbes purchased Arion to be the progenitor sire for a breeding establishment. Although Forbes set up one of the largest breeding establishments his judgments about horses were often questioned and motives suspected:

> Did you say that you wished to see Mr. Forbes? Well you won't find him at the farm.... His horses are but a pastime, the same as a few years ago was his famous yacht, the Thistle. Is he a horseman? Only in having plenty of money to buy whatever he wishes, and not in the highest sense. His horses are to him but a pleasant way of getting rid of money and enjoying a new "fad."[52]

Forbes eventually made his name as a trotting horse breeder by applying the strategy "speed produces speed."[53] Forbes built his stock farm around a group of stallions and mares that had shown early speed and gained low records. Previously, a horse's performance was not thought to be indicative of their ability to pass these traits on to their offspring. Consequently, most breeders bred from sires and dams that were better known for their ability to produce fast horses and not for their own racing performance.

Forbes' breeding strategy demonstrates how much confidence some breeders now placed in the inheritance of the trotting gait as a distinct trait. In just a few generations of trotters, the emphasis on breeding the horse had gone from the value placed on its mongrel ancestry and the variability of its produce, to the trotter as an elite breed whose bloodline conserved qualities that once had to be rigorously selected for. But two generations of horses was not enough time to breed purebred trotters. What changes occurred in the breeding and training of trotters that convinced owners like Forbes that they could now follow pure line-breeding methods?

The new trotting farms had instituted large-scale breeding programs that selected for trotters at a young age. Breeders began rigorously selecting the fastest trotters from the rest of the stock. Thus the variability of a trotter's produce had been reduced through the "visible hand" of the trainer. A radically new type of inheritance had in effect been set up, a *corporate inheritance* that relied on early competition as an expression of natural capabilities. The horses' "natural" capabilities could only be proven, however, on the value laden system of the track, where selection occurred based on a functional standard that conformed to the needs and values of the new industries. The horse and the management of the farm had become a single reproductive entity for reproducing elite equine bloodlines.

A similar type of development had taken place under hierarchically managed business. In an almost too pat example, trotting horse fancier Andrew Carnegie chose the analogy of race horses to best describe how internal promotion was a form of selection by performance. According to business historian Harold Livesay:

> Another cornerstone of Carnegie's success was the use of systematic analysis to evaluate his men's performance. He was the first manufacturer to do so. . . . individual records were kept of "who produced the best results," "thus to compare one [man] with the other," in order to inspire what Daniel McCallum had called "an honorable spirit of emulation to excel" . . . Thus Carnegie said of a suggested promotion: "He may be just the man we need. Give him a trial. That's all we get ourselves and all we can give anyone. If he can win the race, he is our race-horse. If not, he goes to the cart." Those who turned out to be "race-horses" could aspire to promotion, even partnership. As Carnegie said, "Mr. Morgan buys his partners, I raise my own." . . . The system also fostered jealousy and, bitterness, and sometimes despair. But it produced results.[54]

Lacking the bureaucratic vocabulary of the modern personnel office, Carnegie resorted to the analogy of artificial selection. As industries vertically integrated, opportunities were created for a new class of professional workers possessing new skills. The institutions funded by the captains of industry were intended to train students in these skills.

In the industries that these individuals promoted and the institutions that their money helped to build, a new definition of human excellence emerged that relied on competition as an indicator of one's "natural" ability for leadership. This type of corporate inheritance led to a system where social privilege could be conserved through competition at the institutional level only. The selection of functional capacities through competition in hierarchically organized institutions and not the bloodline would become the major unit for the reproduction of social privilege.

4. A "Backward Glance": Corporate Inheritance as Bastard Birth

Sitting alone in his study on a day near the end of the first decade of the twentieth century, Newland Archer contemplated the changes that had occurred over the last twenty-six years. The material changes in his world were immense: automobiles transformed personal transportation (making trotting horses even less useful than their thoroughbred counterparts), and telephones enabled communication at a distance. Professionally, young men of means now chose occupations other than law or business. Archer's eldest son, for instance, studied architecture. Other young men indulged their interests in "municipal reform," "archeology," or even "landscape engineering."[55]

The changes that most preoccupied Newland, however, were the changes to his family. His oldest son had recently announced his engagement to Fanny Beaufort, the daughter of Julius Beaufort and his mistress, Fanny Ring. After the death of his wife, Beaufort quietly married his mistress, moved to South America, and recovered his fortune through international business ventures. His daughter appeared in New York after Beaufort and his wife died in Buenos Aires. Archer remembered a comment that Lawrence Lefferts, the guardian of proper social form, dropped decades ago:

> What was left of the little world he had grown up in, and whose standards had bent and bound him? He remembered a sneering prophecy of poor Lawrence Leffert's, uttered years ago in that very room: "If things go on at this rate, our children will be marrying Beaufort's bastards."
>
> It was just what Archer's eldest son, the pride of his life, was doing; and nobody wondered or reproved. (352)

Put into the context of the ideas explored in this essay, Wharton's use of the analogy of bastard birth suggests more than a relaxation of social mores. New York society embraced Beaufort's bastard with little fear because the enforcement of proper social form no longer depended on face-to-face interactions. As the loci of political, social, and cultural power became more fragmented, greater emphasis was placed on the specialization of knowledge and opinions of experts. The institutions that housed these experts, however, were often supported by those who had gained their fortunes through the expansion of industries. Their values were built into the tools used to coordinate and process the data that enabled production on industrial scales. New forms for conceiving of inheritance became possible when these tools were applied to looking at the transmission of hereditary character.

For those of Wharton's generation and class, reproducing social privilege on the institutional scale changed the perceptions of human relationships. The world had in effect grown larger and more impersonal:

> Nothing could more clearly give the measure of the distance that the world had traveled. People nowadays were too busy—busy with reforms and "movements," with fads, and fetishes and frivolities—to bother much about their neighbours. And of what account was anybody's past, in the huge kaleidoscope where all the social atoms spun around the same plane? (353)

From the expanded perspective of the institution, people were only statistical entities—kinetic particles sorted by managerial hierarchies. With the realization that "systematic management was built on the assumption that individuals were less important than the systems that they functioned within,"[56] the new forms of reproducing social privilege began systematizing individuals according to the specialized needs of the new industries.

Although explicitly referred to as the basis for a meritocracy, the promise of mobility within these hierarchies depended on the performance of specialized tasks. Even as they affirmed the agency of the individual through competitive displays, individuals remained only components within the system. Even though kinship inheritance never disappeared, the reproduction of social privilege relied more and more heavily on the institutions that the Beauforts of the world built. This did not mean that society was homogenized to the degree that Wharton claimed—in fact, the new institutions often lent new authority to previous prejudices—but rather that bloodlines of the families needed to flow through the institutions to retain social privilege. Just as Beaufort produced Fanny outside of the family, corporate inheritance now reproduced social privilege outside of the family. From the perspective of the Larry Lefferts and Sillerton Jacksons, this form of indirect inheritance, with its disregard for family and social form, made each American just one more of Beaufort's bastards.

Notes

1. Edith Wharton, *The Age of Innocence* (New York: Collier Books, 1968), 19.
2. Melvin L. Adelman, "The First Modern Sport in America: Harness Racing in New York City, 1825–1870," *Journal of Sport History* 8:1 (1981): 5–32, see especially 5 and 8.
3. The term was first used by Josephson, but Licht has recently adopted it to distinguish those who integrated many different enterprises as opposed to those who were only involved in finance or promoted other enterprises. See Matthew Josephson, *The Robber Barons: The Great American Capitalists, 1861–1901* (New York: Harcourt Brace Jovanovich, 1934), Walter Licht, *Industrializing America: The Nineteenth Century* (Baltimore: The Johns Hopkins University Press, 1995), 138–45.
4. On the notion of middle management as an information processing system, see JoAnne Yates, "Evolving Information Use in Firms, 1850–1920: Ideology and Information Techniques and Technologies", in *Information Acumen: Understanding and Use of Knowledge in Modern Business*, ed. Lisa Bud-Freeman (London: Routledge, 1994), 26–50.
5. When the diagonal hooves of a horse (left front and right rear, for example) touch the ground simultaneously, the horse is trotting.
6. Peter C. Welsh, *Track and Road: The American Trotting Horse. A Visual Record 1820 to 1900 from the Harry T. Peters "America on Stone" Lithography Collection* (Washington, D. C.: Smithsonian Institute Press, 1967), 17.
7. Elliot J., Gorn and Warren Goldstein, *A Brief History of American Sports* (New York: Hill and Wang, 1993), 53.
8. Welsh, *Track and Road*, 49.
9. H. C. Merwin, *Road, Track, and Stable: Chapters About Horses and Their Treatment* (Boston: Little, Brown, and Company, 1892), 23.
10. Welsh, *Track and Road*, 51.
11. Oliver Wendell Holmes, *The Autocrat of the Breakfast Table* (New York: Popular Library, 1968), 25.
12. Edith Wharton, *A Backward Glance* (New York: Literary Classis of America, 1990), 783.
13. David C. Hammack, *Power and Society: Greater New York at the Turn of the Century* (New York: Columbia University Press, 1987), 36.
14. Wharton, *A Backward Glance*, 783.
15. Wharton, *The Age of Innocence*, 9–10.
16. Frederic Cople Jaher, *The Urban Establishment: Upper Strata in Boston, New York, Charleston, Chicago, and Los Angeles* (Urbana: University of Illinois Press, 1982), 249.
17. John Killick quoted in James R. Beniger, *The Control Revolution: Technological and Economic Origins of the Information Society* (Cambridge, MA: Harvard University Press, 1986), 127.

18. Alfred Chandler, *The Visible Hand: The Managerial Revolution in American Business* (Cambridge, Mass.: Belknap, 1977), 37.
19. Beniger, *The Control Revolution*, 129.
20. Wharton, *A Backward Glance*, 799.
21. Jaher, *The Urban Establishment*, 208, 279.
22. W. E. D. Stokes, *The Right to Be Well Born or Horse Breeding in its Relation to Eugenics* (New York: C. J. Obrien, 1917).
23. Although Beniger's scholarship is sound, I cannot agree with the basic premise of his argument, that information processing is a fundamental quality of life. A more-nuanced interpretation of his materials is that the rise of information technologies supplied a set of metaphors and values through which life was understood as an information processing system.
24. Hamilton Buseby, "The Evolution of the Trotting Horse," *Success* (1898), 278.
25. Dwight Akers, *Drivers Up: The Story of American Harness Racing*. New York: G. Putnam's Sons, 1938). 176.
26. Merwin, *Road, Track, and Stable*, 118.
27. Akers, *Drivers Up*, 176.
28. Stanford Papers, Special Collections, Leland Stanford Junior University, "Scrap Book #22", *Sc 33f*, 61.
29. Akers, *Drivers Up*, 168–9; Adelman, "The First Modern Sport," 21.
30. Norman E. Tutorow, *Leland Stanford: Man of Many Careers* (Menlo Park: Pacific Coast Publishers, 1971), 162.
31. John Bradburn, *Breeding and Developing the Trotter* (Boston: American Horse Breeder Publishing Company, 1906), 120–6.
32. Bradburn, *Breeding and Developing*, 120–6.
33. Bonner Papers, New York Public Library, "Scrap Book of Newspaper Clippings Vol. 1: The Trotting Horse."
34. Stanford Papers, "Scrapbook #22," *Sc 33f*, 33.
35. On Muybridge at Stanford, see Stanford University, Dept. of Art, *Edward Muybridge: The Stanford Years, 1872–1882* (Stanford, 1972).
36. Anson Rabinbach, *The Human Motor: Energy, Fatigue, and the Origins of Modernity* (New York: Basic Books, 1990), 102–3.
37. Rabinbach, *The Human Motor*, 104.
38. Busbey, "The Evolution."
39. Charles Marvin, *Training the Trotting Horse: A Natural and Improved Method of Educating Trotting Colts and Horses, Based on Twenty Years Experience* (New York: Avin Publishing Company, 1892), 200.
40. Stanford papers, "Scrap Book #22," *Sc 33f*, 95.
41. Marvin, *Training the Trotting Horse*, 210. Italics in the original.
42. David Starr Jordon,*The Days of a Man: Being Memories of a Naturalist, Teacher and Minor Prophet of Democracy* (Yonkers-on-Hudson: World Book Company, 1922), 370.
43. Stanford in conversation with friend Frank Pixley, quoted in George T. Clark, *Leland Stanford, War Governor of California, Railroad Builder and Founder of Stanford University* (Stanford: Stanford University Press; London: H. Milford, Oxford University Press, 1931), 417.
44. Stanford Papers, "Stanford Family Tree," box 3, folder 17.
45. Stokes, *The Right to Be Well Born*, 20–21.
46. See Lawrence Veysey, *The Emergence of the American University* (Chicago: The University of Chicago Press, 1965), 57–120.
47. Merle Eugene Curti and Roderick Nash, *Philanthropy in the Shaping of American Higher Education* (New Brunswick: Rutgers University Press, 1965), 114.
48. Veysey, *The Emergence of the American University*, 39.
49. Curti and Nash, *Philanthropy*, 112.
50. "Trotters of New York," *Harper's Weekly*, 15 May 1886, 3.
51. John Murray Forbes initially made his money in the East India trade as the Canton agent for his uncle, Thomas Handasyd Perkins. Wealthy Bostonians more frequently took up the new industrial businesses than their New York counterparts. See Jaher, *The Urban Establishment*, 23, 123–4.

52. Stanford Papers, "Scrap Book #22," *Sc 33f*, 65.
53. Hervey, *The American Trotter*, 314.
54. Livesay, *Andrew Carnegie*, 99–100.
55. Wharton, *The Age of Innocence*, 345.
56. JoAnne Yates, *Control through Communications: The Rise of System in American Management* (Baltimore: Johns Hopkins University Press, 1989), xvii.

PART II

Control and the New Bodies: Modes of Informational Experience

CHAPTER **4**

Desiring Information and Machines

MARK POSTER

The body is configured in practices.[1] It is constituted at the cultural level by ideas, attitudes, values; at the emotional level by inscriptions of desire in the unconscious; and at the physical level by movements, postures, spaces. The body is inserted into multiple spaces and times that are always already socially given (although changing and polyvalent as well), positioned in relation to material objects, machines, other humans, animals and plants, represented in cultural artifacts of many kinds, and territorialized by desire, awakened and repressed at different points and in different ways. Certainly the body has natural limits, capacities, deficiencies, being finite and subject to gravity, requiring oxygen and nutrients of various kinds. For millennia these limits defined the human experience in varying yet surprisingly stable ways.

Now things are changing. Life span increases. Fatalities increasingly occur from social causes like car accidents, wars, pollution, and environmental damage. Globalization invites marriages of previously distant individuals and groups, mixing bodies that in earlier epochs had little chance to commingle. Scientific advances in biochemistry open the body as a book to be rewritten, edited, and mixed with the "books" of parts of other bodies and even other species. Experiments in robotics and artificial intelligence redraw the line between the living and dead, combining carbon- and silicon-based materials in new mixtures of machines and flesh. Arguments from the 1990s that the body is shaped, in part, by culture appear quaint when confronted by somatic alterations on the horizon and already being put into practice. Many of these tendencies began centuries ago but seem to have reached, by the late twentieth century, a crisis point by which incipient changes in the body in the present and the future are likely far to surpass anything from the past, or indeed anything imagined in the past. Digital morphing programs, however trivial in themselves, represent this newfound plasticity to the human body.

In the context of such innovations, we consider in this chapter the relation of networked computing to the human body, placing the discussion in the context of the reconceptualization of the body initiated by poststructuralist feminist theory and continued by cultural studies, transgender studies, postcolonial studies, race studies and queer theory. Originated perhaps by Michel Foucault in *The Birth of the Clinic* (1963), Gilles Deleuze and Félix Guattari in *Anti-Oedipus*

(1972), Elaine Scarry in *The Body in Pain* (1985), Teresa de Lauretis in *Technologies of Gender* (1987), and Judith Butler in *Bodies that Matter* (1993), critical cultural studies have theorized the body both as resistant, material substrate and as culturally constituted. The body is here understood as represented in the discourse of medicine through the gaze instituted in the early clinic; as a surface of desire inscribed on, absorbing impressions, yet in the end undeterminable and capable of forming new patterns of energy and intensities; as gendered by binary heterosexual practices that bring to bear on it the vast resources of capital and the intimate privacy of the nuclear family; as racialized by hegemonic systems of prejudice; as brutalized, oppressed and hybridized by Western imperialism; as a locus of pleasure, multifarious without visible limit, a point of heterotopian transgression; as a site of self-fashioning and self-constitution. Even before the genome project and the Internet, intellectuals have explored the complex mutual determinations of nature and culture as they impinge on the frail but magnificent human body. My examination of the body's relation to new media departs from these stunning advances in the understanding of how the apparent naturalness of the body is also historically shaped.

I. Freud's Body

Before exploring the body in new media, I must return to what is still considered by many the major source comprehending the body, Freudian psychoanalysis. In Freud's writing and practice, we find all the conditions for thinking about the body: the overture to the cultural conception of the body and its denial and naturalization; the perception of the body as situated in determinate practices and the misrecognition of those practices; the subordination of the understanding of body in space to the need to grasp specific patterns of interior pain; the conflict of impulses between scientific critique and medical cure. One may assess the condition of the body immersed in new media only by first understanding how the body was understood without the least awareness of its imbrication with information machines. Let us not forget Freud's quip about the telephone and the progress of humanity it represented: that the telephone was like the pleasure received on a cold night by sticking one's leg out from under the covers in order to enjoy the warmth of returning it to where it was in the first place. Yes, one can with wired telephony enjoy conversations at a distance, he seemed to say, but without such devices one might not take such long trips in the first place. Conditions today for many no longer afford such humor.

Psychoanalysis theorized the body as a development, as a process of formation, through which zones of function and the pleasure inherent in the organs and nervous system were realized in particular patterns. The body for Freud is an input/output machine, taking in food, processing it for nutrients, and expelling waste. Associated with these functions are nervous excitation and release, pain and pleasure. At first, the body is dependent on others to supply the food. This is the natural condition of the newborn human. Later, the body would depend on the state of civilization for the same purpose. Freud, the scientist, understood three areas of the body as central to its early functioning: the mouth, the anus, and the genitalia. Highly intense nervous excitations beleaguer the three zones in course of their early functioning. Accompanying these excitations are energy forms he termed libido. Although undefined at first, libido undergoes a degree of organization in the course of the body's experience of its inputs and outputs. And although each individual develops somewhat unique patterns of the organization of libido, there are also general shapes of such organization. The question is how Freud understood the generalities at issue. I argue that in the interest of comprehending the specific organization of his patients and their pain associated with this organization, he played down the role of the situational practices of his day, his class, and his place in the world. To comprehend the contemporary situation of the body and its current libidinal organization, one must rediscover the parallel elements in the Freudian scene of Victorian Europe.

When Freud studied the sexual etiology of neurosis, the group that he encountered imposed quite specific practices upon the newborn child. The oral phase was controlled by the birth mother's breasts, and feeding was rigidly scheduled. The anal phase featured very early toilet training with severe threats of the withdrawal of love from the child for inadequate performance. The genital phase was carried out with threats of castration if the child practiced what was called masturbation. The pattern imposed on the child was maximal denial of the pleasure in its body, compensated for by a high degree of affection from adults. The affection was, however, limited to the smallest possible circle of others: the immediate parents. Thus the child's body matured in the following way: its own body was not a source of pleasure, but links were inscribed on the body with the proximate adults. These were highly intense and highly ambivalent, intense because of the small number of possible "objects," and ambivalent because these adults restrained severely the child's access to pleasure of the body, yet offered the child deep and sustained affection.

The social nexus of Freud's study then figured the child in relation to a small circle of adults, organizing the child's libido in relation to these adults. For Freud, the crucial aspect of the child's development was the emergence of the ego, which occurred after the body had been charged and marked. He theorized the child's options for ego development after the first three stages as consisting in a twofold path of libidinal discharge. Either the child could find objects toward which to orient the release of desire (object choice) or s/he could incorporate the object into its psyche (identification). Here at length is Freud's description of the situation of the child at the moment of the formation of the ego:

> The basis of the process is what is called "identification"—that is to say, the assimilation of one ego to another one, as a result of which the first ego behaves like the second in certain respects, imitates it and in a sense takes it up into itself. Identification has been not unsuitably compared with the oral, cannibalistic incorporation of the other person. It is a very important form of attachment to

> someone else, probably the very first, and not the same thing as the choice of an object. The difference between the two can be expressed in some such way as this. If a boy identifies himself with his father, he wants to *be like* his father; if he makes him the object of his choice, he wants to *have* him, to possess him. In the first case his ego is altered on the model of his father; in the second case that is not necessary. Identification and object-choice are to a large extent independent of each other; it is however possible to identify oneself with someone whom, for instance, one has taken as a sexual object, and to alter one's ego on this model.[2]

At stake in this process for Freud is the formation in the ego of a parental agency he calls the superego. Freud complicates the understanding of the process of superego formation by considering the place of earlier libidinal attachments (those of the first three stages discussed above) in relation to the superego. He writes, "the ego is formed to a great extent out of identifications which take the place of abandoned cathexes by the id; that the first of these identifications always behave as a special agency in the ego and stand apart from the ego in the form of a super-ego.... As the child was once under the compulsion to obey its parents, so the ego submits to the categorical imperative of its super-ego."[3] In this way, Freud relates the super-ego, the internalized parental agency, with the relatively systematized bits and pieces of body energy from early infancy. The ego forms and along with it the superego as a synthesis of the child's libidinal experience.

II. The Habitus of Psychoanalysis

There are many questions to ask about Freud's picture of the psyche and its relation to the body. First, and often overlooked, is the observation that "objects" for Freud are primary human beings (parents) or their body parts.[4] It is true that Freud discusses inert objects such as "the mystic writing pad" in relation to memory and the object thrown and retrieved in fascination by the child (the *fort-da* game) in connection with separation anxiety. Yet when writing about the basic formation of the personality, Freud gives overwhelming attention to living adults. This limitation, while perhaps sensible in the context of the people Freud encountered, will need to be revised when we discuss the child's development in the late twentieth century when information machines play a vastly more significant role in the life experiences of small children. We will then have to consider what revision to make in the description we inherit from Freud.

The second problem I raise in connection with Freud's understanding of the body is a more familiar one: the strong sense in his writing of heteronormativity. Identification occurs between the child and the parent of the same sex, leading to the passing on of heterosexual desire. Freud, as Teresa de Lauretis and others have noted, wrestles energetically with contradictory evidence. He recognizes the basic bisexuality of the human being but sees it as confusing the picture. He complains, "It is this complicating element introduced by bisexuality that makes it so difficult to obtain a clear view of the facts in connection with the earliest object-choices and identifications, and still more difficult to describe them intelligibly. It may even be that the ambivalence displayed in the relations to the parents should be attributed entirely to bisexuality and that it is not, as I have represented above, developed out of identification in consequence of rivalry."[5] For Freud one cannot "describe intelligibly" the basic parameters of the human psyche as long as one acknowledges, which one must, bisexuality. Here Freud negotiates with the heterosexual ideology of his day, recognizing its limits, yet conceding it pride of place. To render the bodily development of the child intelligible, the human, he thinks, must be heterosexual. Bisexuality gets lost in the shuffle of heterosexual science.

If heterosexuality is kept in place by Freud, so lamentably is patriarchy. The child's body, undergoing identification, now divides into two clear but unequal types: masculine and feminine. In Freud's words, "It is said that the influencing of the ego by the sexual object occurs particularly often with women and is characteristic of femininity . . ."[6] In short, for women, object choice often regresses to identification. They cannot keep separate their own ego from the objects they choose to love and so they reenact the process of identification from early childhood, something men, when making object choices, need not be bothered with. This of course is not the only way femininity and masculinity are distinguished by Freud, in every case to the advantage of the masculine, but it is enough to recognize that Freud generalizes what may have been the predominant (but certainly not universal) hierarchy of the sexes in his day into nothing less than the fate of all humankind.

The status of Freud's understanding of the child's body must then be understood as a description and analysis of the technology of power, the mechanisms through which a regime of the body is imposed. This regime—bourgeois, masculinist, heterosexist—must be taken as fragile and uncertain, historically contingent and ever at risk. Butler argues to this effect by pointing out that "the position of 'masculine' and 'feminine'. . . are established in part through prohibitions which *demand the loss* of certain sexual attachments, and demand as well that those losses *not* be avowed, and *not* be grieved."[7] As Freud argued, the crucial identification of the child with the parent of the same sex *substitutes for* the bodily attachments (cathexes) of the child's earlier emotional life. The father incorporated by the body of the child in a "successful" Oedipal resolution includes mourning for the loss of all the attachments to the mother as well. The body of the "normal" bourgeois boy is then an unstable amalgam of desires for both the mother and the father as they descend into unconsciousness, incorporated into the body from the desires for each of the parents. This incorporation, as Freud implies and as Butler insists (136), occurs through threats, prohibitions, and intimidations—all manner of force associated with the technology of Oedipal power.

Butler goes on to argue that the resulting masculine (and feminine) ego is so unstable that it exists only at the level of the imaginary. She writes,

> But where or how does identification occur? . . . Significantly, it never can be said to have taken place; identification does not belong to the world of events. Identification is constantly figured as a desired event or accomplishment, but one which finally is never achieved; identification is the phantasmatic staging of the event. In this sense, identifications belong to the imaginary; they are phantasmatic efforts of alignment, loyalty, ambiguous and cross-corporeal cohabitation; they unsettle the "I"; they are the sedimentation of the 'we' in the constitution of any "I," the structuring presence of alterity in the very formulation of the "I." Identifications are never fully and finally made; they are incessantly reconstituted and, as such, are subject to the volatile logic of iterability.[8]

Although I agree with Butler that the instability of the ego requires, for its analysis a "logic of iterability," I cannot agree that identification "has never taken place." Instead, I would argue that it takes place the way all historical events take place: as a mixture of the real and the imaginary. And that identification subsists in the way all cultural formations subsist: in their contingent iteration but one that for long periods, effectively cannot in practice be avoided. Victorian boys and girls became men and women in the regime of coerced heterosexuality, enduring it as they could, for better or for worse.

Yet another difficulty besets Freud's understanding of the child's body: the Eurocentrism of psychoanalysis is also recognized but disavowed by Freud. In the midst of a discussion of the formation of the child's personality in an obscure footnote in *An Outline of Psychoanalysis*, Freud acknowledges and dismisses the parochialism of his science: "No investigation has yet been made of the form taken by the events described above [childhood castration] among races and in civilizations which do not suppress masturbation among children."[9] The issue in this passage is the practice of Western parents to discourage with great vehemence, to say the least, the practice of children touching their genitals. The denial to the child of pleasure in the genital zone was a great battle of parents in their effort to impose proper behavior on their offspring. For Freud's theory of personality formation, this battle was crucial (even though he often intervened on the side of leniency). The castration threat was the emotional cauldron in which was brewed the fine drink of Oedipus, itself the discriminating taste of civilization, culture, and masculinity. Today, one avoids as an embarrassment talk of the threat, preferring to insist that the parents need only deny *something* for the operations of Oedipus to be effectuated. When I asked parents in Orange County in a questionnaire in the mid-1980s if they threatened their children with castration, one responded by warning me they would report me to the University administration for such a barbaric suggestion.[10] But Freud was not so timid. In the foregoing quote, he calls into question the scope of applicability of psychoanalysis "among races and in civilizations" where the threat is absent.

Imagine what might befall European culture if children freely touched their genitals, or even played with each others'.

Finally, Freud disavows forms of sexual preference that do not conform to the accepted views of his day. Homosexuality among women and men are simply not to be considered in understanding the process of identification and the consequent resolution of the Oedipus drama. He contends, in *An Outline of Psychoanalysis*, that these cases present confusions of the circumstances that can easily be dismissed as abnormal.[11] Ruling out sexual attraction among individuals with the same genital organs sets in place the rigidly defined personality types of masculinity and femininity as the only legitimate outcomes of bodily formation. Such models are not simply restrictive and hierarchical in their binary opposition: they also insist on one set of libidinal preferences for each individual, centering the body in a unity and excluding multiplicity both in bodily preferences and in relation to nonhuman objects.

The image of the body we inherit from Freud is therefore highly delimited. It is human centered, it is unitary, it is binary and it is hierarchical. Bodies of boys and girls are de-eroticized with certain zones designated for possible pleasures. Bodily desire is conscripted to the tasks of civilization in its bourgeois cultural form, with highly specialized, sex specific etiologies. The men sublimate desire into the economy and imperial adventures; the women into the home. Both hystericize flows of desire they cannot manage into anal and oral fixations. Compulsive heterosexuality develops into stylized masculinity and femininity. Non-European societies, as they are being devastated by imperialist politics and economics during the career of psychoanalysis under Freud's aegis, are exoticized as the lubricious other. We might recall that this Victorian body is constituted through relations of force primarily with other humans. As we investigate the fate of body in a context strewn with machines, it is well to note that a fully human organization of desire is not always and completely beneficent.

This image of the body rings true in relation to the circumstances of Freud's day but begins to lose its field of effectivity as we depart from those characteristic determinations. Although psychoanalysis remains one of the most complex and rich understandings of the human body available, it bears the traces of its historicity. As we confront the world of the contemporary technical conditions, we must rethink many of its pronouncements and tendencies. In this spirit, I raise the question of the body in relation to the Internet.

III. Media Bodies

Signs of difficulty with Freudian categories emerged clearly after 1945, a type of difficulty that also characterizes Marxist theory. Both theories position their truth value in relation to practices. In the case of Marx, the politics of the working class is a necessary component of the theory. If that politics differs from the expectations of the theory, at some point the theory must take this into account, either to explain away the anomaly or to alter itself to better connect with the circumstances. Neither Marxist nor Freudian theory is apodictic, claiming truth value transcendentally, in its own terms. With Freud, the problem emerges if patients express and display symptoms that have no relation to the theory. Freud acknowledged that his theory said little and could result in no effective therapy in the case of psychosis, for instance.

Increasingly since the mid-twentieth century, hysteria and neurasthenia have been displaced by a number of diagnoses and complaints ranging from identity confusions in Erikson's ego psychology, to schizophrenia in Laing's antipsychiatry, to narcissism in Kohut and Kernberg, to borderline disorders, multiple personality syndrome,[12] clinical depression, addiction and an array of other designations of trouble discovered/invented by other clinicians. In these circumstances, psychoanalysis must do more than insist on the truth of its theory and methods. Indeed, the rogue's gallery of mental disorders listed above is related to changing circumstances, changing patient groups, and changing cultural norms.

I argue that one way to understand these changes is to study the disciplinary practices of embodiment. In the past fifty years, the body has become subject to disciplinary practices that differ from those of the Victorian nuclear family. Patterns of feeding infants, toilet training them, and controlling their genital pleasures have drastically altered. Children of the middle class are no longer sequestered in apartments, away from the dangerous life in the streets, and confined to the tempestuous microsociety of parents and siblings. Instead, they are sent out to preschools within months of birth. Their bodies are tended by strange adults and engaged in play with their diminutive peers. Perhaps most significantly their surroundings in the home have been revolutionized. Parents are more attentive to the requirements of young children in the design and furnishing of their rooms and even the shared areas of the home. Color schemes, decorations and toys are selected to match the bodies of young people, even if the specific choices may be arguably no better suited to them than those of the past.

The child's space in the Victorian home lacked all the information technologies we assume today. It consisted of furniture, book cases perhaps, a bed, and few other articles. Decorating schemes concerned the education of the senses as well as reason. Parents were urged to fill the room (without overstuffing it) with prints from the great art of Europe and with books that reflected the highest values of that culture such as and especially Chaucer and *Morte d' Arthur.*[13] Children from families who might potentially find themselves on a psychoanalytic couch were being readied for the cultural work of European civilization. In reading descriptions of children's space from this era, one is struck by the degree to which bourgeois parents were able to design and control the world inhabited by their children. The meaning of privacy[14] here becomes clear. Although the parents were surely influenced by their social world in selecting the child's environment, the child could say little about it. No technology or youth culture available through its means interfered with the wish of the adults in these matters. No external forces penetrated the bourgeois interior to provide alternatives for the child to the dictates of adults. Each object, however trivial, that appeared before the child was there at the discretion of the child's parents. Perhaps only the window in the child's room afforded an unmonitored glimpse of the world beyond the privacy of the nuclear family.

The situation today is dramatically altered. The outer world—corporations, media providers, schools—designate children as a market. Beyond the family, institutions recognize children of various ages as part of society in a manner quite different from earlier times and with consequences of diverse sorts. At issue are not simply toy manufacturers, advertisers, television and film producers—who prey on children and young people with greedy expectations of monetary reward. There are also theme parks, large spaces devoted to attracting children, and restaurant chains like Chuck E. Cheese and other fast-food corporations; there are clothing manufacturers, from sneakers to jeans and T-shirts who have attended to and captured the fashion aspirations of increasingly younger persons. For better or worse, general social institutions cater to what they regard as the desires and bodily needs of young people from birth to majority. Whatever the motivations of the providers of these objects and amusements, one consequence is clear: youth cultures now exist that are beyond the control and in many cases the understanding of parents. One might hazard the speculation that our postmodern culture is characterized not simply by the multiculturalism of ethnicity, race, gender, and sexual preference but also and of equal importance by age. Society is splintered into microcultures of groups determined by age. But the important question for this discussion of the body concerns the specific mechanisms through which the child connects with and experiences cultural patterns that are unique to it. At the center of this question, I submit, are information machines.

Freud's clients took photographs, sent telegraphs, listened to phonographs and radios, answered their telephones, and, what is often overlooked in such lists, read books and deployed typewriters and other writing implements. From the mid-nineteenth century onward, information machines increasingly took their place in the lives and the homes of the Western bourgeoisie. These machines produce cultural objects (cameras, typewriters), store cultural objects (books, phonographs), transmit cultural objects (telegraph, radio, telephone), and disperse cultural objects throughout urban space (films, books). These objects of mechanical reproduction alter social relations by substituting machine mediation for face-to-face relations.[15] They reconfigure space by altering the relation of the public sphere and the private sphere,[16] altering the configuration of the subject. The solid walls and locked doors of apartments are invaded as if by an army of foreign cultures, upsetting the ability of the nuclear family to sustain its degree of separation from the wider world. The cultures made available through these information machines may be remote but also differentiated by age. In Freud's time, parents were still able to control the source and nature of these cultural objects. Books, phonograph records, prints and other media remained within their dominion. Children were able to obtain only those cultural objects purchased by their parents. In studying the development of the child's body, Freud was then still able to ignore the role of information machines and focus his attention on parents as the sole components of the family environment.

By the early twenty-first century, the number and variety of information machines installed in the home has increased remarkably. Televisions, answering machines, fax machines, computers, electronic games, network connections, and so many other devices have entered the residence. Each of these, along with the earlier technologies, has multiplied throughout the home and also become mobile with Walkmans, mobile phones, PDAs enabling the extension outward into the public sphere of the private world. Mechanical appliances, from refrigerators to microwave ovens, have added computer chips to their assemblage, rendering them "smart machines." The home has become infinitely permeable to the outside world with the result that the coherence of the culture of the nuclear family has been fragmented into what I call the segmented family.[17] Each member of the family now sustains their separate cultural world within the family. Media provide programming for each person with only a small proportion aimed at the family as a whole.

The Freudian body required as a necessary condition the privacy of the bourgeois home. The child needed to be nurtured by the parents in order for desire to implode and be sent reeling through its Oedipal course. The public world was a beyond, an other, to this child. As Thomas Keenan writes,

> The public is not the realm of the subject, but of others, of all that is other to—and in—the subject itself.... The public is not a collection of private individuals experiencing their commonality.... The public is the experience, if we can call it that, of the interruption or the intrusion of all that is radically irreducible to the order of the individual human subject, the unavoidable entrance of alterity into the everyday life of the "one" who would be human.[18]

When the windows onto the world opened by dint of information machines, the private space that cultivated the bourgeois body would be shattered forever.

If in the Victorian period, worker and bourgeois did not understand each, Catholic and Protestant felt distant, and regional cultures were barriers to communication, today men and women have separate subcultures and children, if they do not understand adults, are equally opaque to their elders. Media serve to heighten and to intensify what Durkheimian sociologists term the structural and functional differentiation of modern society.

IV. Media Discourse for Children

A good example of the ability of children to find alternatives to the culture of their parents, to look outside the window of the family, is the story of a boy who developed his gay sexual orientation through the Internet. Growing up in a Southern Baptist town and finding only condemnation for homosexuality, Jeffrey, at age 15, wrestled with his sense of liking boys. His only outlet beyond his home environment was the Internet. With a simple search on the Web for "gay" and "teen," he was able to contact a community of like-minded youths. In his words, "The Internet is the thing that has kept me sane.... I live constantly in fear. I can't be my true self. My mom complains: 'I can see you becoming more detached from us. You're always spending time on the computer.' But the Internet is my refuge."[19] Jeffrey's life online was in sharp opposition to his face-to-face world. Yet he was able to sustain his emotional life for some time through the relations he was able to have on the Internet. Jennifer Egan draws the implications from the case of Jeffrey: "For homosexual teenagers with computer access, the Internet has, quite simply, revolutionized the experience of growing up gay. Isolation and shame persist among gay teenagers, of course, but now, along with the inhospitable families and towns in which many find themselves marooned, there exists a parallel online community—real people like them in cyberspace with whom they can chat, exchange messages and even engage in (online) sex" (113). Here information technology provides young people with connections to subcultures that are absent from their environment and scorned by their parents. One might compare this situation with the efforts, dating back to the 1960s, of parents in such communities to censor books in school libraries like J. D. Salinger's, *Catcher in the Rye*, for representing youth rebellion with only the suggestion of homosexual content. Today, the Internet offers considerably more than Salinger: communicating with communities of youth rebels, gays, and other cultures shunned by many parents. Children are now able to participate in cultures of their own choosing, regardless of parental values. And this is so for young people at a surprisingly early age.

Cultural contacts available to children through information technologies are not always in direct opposition to parental views. My argument is not so much about resistance to parental authority in these situations. Instead, I maintain that the child's world is now composed of media as well as adults. The child learns about the world through machines as well as people. If the content of the child's experience is broader than the objects the parents might select, the more important issue is that media play a central role in the constitution of child as a cultural self. The child now is formed through a media technology of power, a set of discourses, images, and sounds that arrive at the child's awareness through stereos, televisions, telephones, game consoles, and networked computers. The child's body is libidinized not exclusively through object-choices and identifications made with parents. In fact, object-choice and identification, ego functions that shape desire, do not grasp well the relation of the child to the information machine.

Young children watching television are not in a mirror stage. For the mirror simply reflects. It produces a logic of self and other. As Lacan claims, the small child enters an imaginary misrecognition with others (especially the mother) in which they reflect back to the child an image the child takes as itself.[20] Screens on televisions and computers are technologies more complex than mirrors, however metaphorical. The logic of the screen is one of incorporation: of the child by the screen and the screen by the child. This is its fusion effect which is captured in films like Gary Ross' *Pleasantville* (1998), where characters jump into the screen and partake of the diegesis of television from the 1960s, and David Cronenberg's *Videodrome* (1983) where Max Renn, the James Woods character, incorporates a television into his stomach. In both cases, the relation of viewer to screen is one of fusion. The screen is thus a liminal object, an interface between the human and the machine that invites penetration of each by the other.

The machinic mediation sets up desire in immediate and indirect paths that, I suggest, follow a logic of fusion and distraction specific to information machines. First, an oneiric consciousness relaxes the ego, blurring its intentionality, while focusing the eyes on the screen and the ears to the speakers. The child's televisual gaze changes the body and prepares for a form of desiring unmatched in relations with parents and other children. This state of consciousness is different from the child's response to adults in the room. Next the simulated reality of the small screen's diegesis draws the child into the object, forming a bond of desire in an imaginary register. On the television there may be humans, animations or advertisements. In each case, the character solicits attention in a mediascape of desire that inscribes the child differently from the domain of face-to-face relations. Television discourse is a one-way flow of images, sounds, and symbols to which the child responds in an imaginary without feedback. The world on the screen is constructed always differently from the world outside the screen, with cultural forms combined in ways impossible in real life. The child comes to desire mediated events and, in the case of television, is on its way to become what is called in the vernacular "a couch potato." Desire has been diffracted through the screen and an unconscious constructed in an imaginary register of the body that is outside the oral, anal, and genital phases of libidinal organization.

Take the example of *Teletubbies*, a show broadcast for young children daily on public television beginning in April 1998 in the United States. The show, originally created by the BBC in 1997, is highly successful with an estimated 500,000 viewers in more than fifty countries. It is considered the first program televised in the United States specifically produced for one-year-olds. Teletubbies are four friendly figures with the names Tinky Winky, Dipsy, Laa-Laa, and Po. They are furry, colorful, round in the midriff, and speak with voices of young children. Some might consider them aliens with antennae. The mood of the show is thoroughly warm and cuddly. Teletubbieland, where they reside, is a pastoral of rolling hills, flowers, and bunnies. The children play with balls and the like, laughing frequently in a continuous state of mild enjoyment. The show begins with a sunrise. The sun has a face of a baby in its middle, one who is smiling and laughing, inviting the child-viewer into the show in a pre-ego fantasy of identification. *Teletubbies* offers a nonthreatening, gentle visual environment that is pleasant in every way.

This apparently benign program has nonetheless been the source of much controversy. Reverend Jerry Falwell, for instance, accused *Teletubbies* of celebrating homosexual life styles to infants.[21] Other critics argue that the program seduces infants to watch television and initiates them into the world of commodities through spinoff products. Susan E. Linn and Alvin F. Poussaint, in "The Trouble With Teletubbies," write in outrage:

> What's worrisome about *Teletubbies* is that, to date, there is no evidence to support its producers' claims that the program is educational for one-year-olds. There is no research showing that the program helps babies learn to talk. There's none to suggest that it facilitates motor development in 12-month-olds. There is no data to substantiate the claim that young children need to learn to become comfortable with technology. In fact, there is no documented evidence that *Teletubbies* has any educational value at all. When asked about research, people associated with *Teletubbies* respond that studies show how much children and parents like the program. That may be so. The fact that children like something, or parents think they do, does not mean that it is educational, or even good for them. Children like candy, too. Given the lack of research, why would PBS import a television program for one-year-olds that has no proven educational value?[22]

Linn and Poussaint suspect that something is amiss in this harmless-looking television series. Just because children like something, they intone, "does not mean that it is . . . good for them." Parents, they advise, ought to beware or their children will "become comfortable with technology." These are sage warnings of a new discipline of the body.

One thing that is amiss is the complete absence of adults in Teletubbieland,[23] a most unusual characteristic of children's television programs or books. Children's culture almost always mirrors the social world where adults are ubiquitous and prevail. Along with the absence of adults go the absence of parental demands, rules, and punishments. *Teletubbies* does not warn children to abstain from touching genitals. In fact, it rarely forbids them anything. The childlike figures play and romp without restraint, except for an adult voiceover that announces when it is time for the teletubbies to return home. Their departure is slow and proceeds with interruptions and returns. An indication that the creators of the program are aware of separation anxiety, teletubbies play a *fort-da* game with the audience, disappearing down a hole only to return once or twice before finally bidding the audience adieu. Objects on the show also behave in a hyperreal manner: appearing and disappearing all of a sudden, in magical performances that defy the rules of the material world. The show obeys no obvious narrative logic but that of a one-year-old's fantasy. When the screen emerges in a teletubbie's stomach and runs a skit, this may be repeated. Here, the child's emotional demand for repetition, familiar and perplexing enough to parents, guides the flow of images. *Teletubbies,* as one critic writes, is devoted to the pleasure of infants:

> It is this structure of play and the validation of pleasure that has caused many American critics to label this show "pointless."... Parents want more from their children's entertainment than mere pleasure. Gratification that does not advance scholarship or skill can strike parents as hedonistic and escapist, especially when involving the consumption of television—already a low status activity.[24]

Teletubbies presents to infants the structure of their own desire—through the mediation of the machine. A fused machinic desire solicits the child and prepares a body outside the Oedipal paradigm.

In the midriff of each teletubbie is a television screen which the teletubbies view, usually once during each broadcast. On the screen are children who perform various acts like fingerpainting, and greeting and saying goodbye to the teletubbies. This doubling of the screen erects a fantasy world of television within television, information machine within information machine. It suggests that beings and televisions are one assemblage, that the viewing child might be a television. The child's mildly attentive consciousness focuses first on the image of the baby in the sun, then on the teletubbies, then on the screen in the teletubbies' stomachs. The imaginary merger between the characters on the screen and the child could not be more complete. The media technology of power structures a scene of fusion between the child and the machine. Within the fused position, the child's body becomes one with mediatized culture. Child and other are constituted in a cyborg assemblage. The family has been complicated by cultural objects from outside and the self of the child has been multiplied and dispersed, cathecting not only to the Oedipal triangle but to the mediascape.

Conclusion

Parents, of course, remain central figures in the child's emotional life. The Oedipus complex is not eradicated just complicated by information machines, as well as by other practices of contemporary families. Sending children to preschool at very early ages diffuses their desire to a wide variety of people. Transformations in family patterns alter the libidinal inscription of the body: mixed or blended families, single-parent families, and same-sex families all have increased dramatically since the 1970s, rendering the classic nuclear family a minority. If we add the entry of mothers into the workforce to this list of changes, the picture that emerges is very different from that of the Victorian middle class. What I call the segmented family of the late twentieth- and early twenty-first centuries multiplies the cultural forms in the home. Above all, the formation of the body through identifications and object choices has altered by dint of information machines. Oneiric assemblages of child and machine inscribe the body with a structure of desire whose shape may be difficult to characterize but whose difference from the Oedipal child is certain.

Notes

1. See Judith Butler, *Bodies That Matter: On the Discursive Limits of "Sex"* (New York, Routledge, 1993).
2. Sigmund Freud, *New Introductory Lectures* (New York: Norton, 1965), 62–3.
3. Sigmund Freud, *The Ego and the Id* (New York: Norton, 1960), 38.
4. Melanie Klein, *Contributions to Psycho-Analysis: 1921–1945* (New York: McGraw-Hill, 1964).
5. Freud, *The Ego and the Id*, 22–3.
6. Freud, *New Introductory Lectures*, p. 63.
7. Judith Butler, *The Psychic Life of Power: Theories of Subjection* (Stanford: Stanford University Press, 1997) 135.
8. Butler, *Bodies that Matter*, 105.
9. Sigmund Freud, *An Outline of Psychoanalysis* (New York: Norton, 1949), 93 note.
10. Mark Poster, "Narcissism or Liberation?: The Affluent Middle-Class Family," in *Postsuburban California: The Transformation of Orange County since World War II*, eds. R. Kling, S. Olin and M. Poster (Berkeley: University of California Press, 1991) 190–222.
11. Freud, *An Outline*, 121.
12. On the vexed question of the relation of psychological diagnosis of multiple personality to the postmodern concept of multiple selves, see L. Layton, "Trauma, Gender Identity and Sexuality: Discourses of Fragmentation" *American Imago* 52:1 (1995): 107–25.
13. E. Wharton, *The Decoration of Houses* (New York: Arno Press, 1975), 177ff.
14. On the profound importance of privacy to the bourgeois nuclear family, see J. Lukacs, "The Bourgeois Interior" *American Scholar* 29:4 (1970): 616–630.
15. Walter Benjamin, "The Work of Art in the Age of Mechanical Reproduction" in *Illuminations* (New York: Schocken, 1969), 217–51.
16. Joshua Meyrowitz, *No Sense of Place: The Impact of Electronic Media on Social Behavior* (Oxford: Oxford University Press, 1986).
17. Poster, "Narcissism or Liberation?"
18. Thomas Keenan, "Windows: of Vulnerability" in *The Phantom Public Sphere*, ed. B. Robbins (Minneapolis: University of Minnesota Press 1993), 133.
19. Jennifer Egan, "Lonely Gay Teen Seeking Same," *New York Times Magazine*, 113.
20. Jacques Lacan, "The Mirror-phase as Formative of the Function of the I," *New Left Review* 51 (1968): 71–7.
21. S. Woodson, "Exploring the Cultural Topography of Childhood: Television Performing the 'Child' to Children" *Bad Subjects* 47 (2000).
22. S. E. Linn and A. F. Poussaint, "The Trouble with Teletubbies," *The American Prospect* 10: 44 (1999): 18.
23. The one exception is the adult voiceover at the beginning and ending of each segment, perhaps reflecting a transition from the "normal," adult world to the exclusively children's world of *Teletubbies*, or covering over the anxiety of this "separation" for the child.
24. Woodson, "Exploring the Cultural Topography."

CHAPTER **5**

LSDNA: Consciousness Expansion and the Emergence of Biotechnology

RICHARD DOYLE

> *I had to struggle to speak intelligibly.*
>
> —Albert Hofmann on his self-experiment with LSD-25

> Finding a place to start is of utmost importance. Natural DNA is a tractless coil, like an unwound and tangled audio tape on the floor of the car in the dark.
>
> —Kary Mullis on the invention of polymerase chain reaction

I. Undoing Life

It often came on paper, where, a certain novelist wrote, all the real fucking happens these days. What happened is this: Once upon a time there was a narrative of vitality. OK, not a narrative, but an apprehension, a fear, a vision. *It* did not have a beginning, middle, or end. *It* was. Its contours were no more determinable than those of consciousness. Indeed, our consciousness, as humans, was our sole alleged difference from *it*. From something called "life." We were *more than* life, but were also, tragically, confined to it. You know the story all too well. Some of its major authors were Cuvier, Lamarck, Darwin, Bichat. Even Shelley. But not, surprisingly enough, Hegel.[1]

Just because this sudden transformation happened on paper does not mean it was easy to endure. Enormous numbers of humans went on tranquilizers just to deal with the effects. Others, elsewhere, knew nothing but rumors. Nonetheless, they became increasingly implicated in its effects, the effects of understanding and experiencing both "life" and "consciousness" as *informatic* events. Suddenly—and it is sudden, a real surprise—both our vitality and our thought were distributed, scattered across a network and nowhere in particular. Timothy Leary and Francis Crick were speaking the same language, the language of information where the organic and the machinic enfold each other helically and, sometimes, the capacity for replication goes through the ceiling. This new language of information would introduce a novel response and abyss: the pleasures and hells of eternal replication. The distribution of both life and consciousness enabled a dream of immortality from which we have not yet awakened, as the spectre of infinite clonal replicants provokes entropies of identity even as it preserves the self from onslaughts of difference.

In at least one of its forms, the effect of this shift toward an informatic understanding of life is perhaps not really so novel. The seductions of control are hardly, historians of science tell us, foreign to the practices and effects of technoscience. Evelyn Fox Keller makes this point concisely and precisely in her not quite eponymous *From Secrets of Life, Secrets of Death,* itself a self-professed sequel to her 1986 "Making Gender Visible in the Pursuit of Nature's Secrets." Here the scientific "impulse" is underwritten by a

> perennial motif that underlies much of scientific creativity—namely, the urge to fathom the secrets of nature and the collateral hope that, in fathoming the secrets of nature, we will fathom (and hence gain control) of our own mortality.[2]

Surely Keller is correct that the nascent practices of molecular biology were cryptograhic in character, an espionage project unknowingly launched by Erwin Schrödinger during World War II. Surely it is crucial to this moment that while Schrödinger's 1943 Dublin lectures were introducing audiences to the concept of the code-script, Alan Turing again brilliantly decrypted a modified U-boat code, giving the Allies code superiority for the remainder of the war.[3] But in this segment on that strange growth called "biotechnology," I want to focus on what Keller has called the "collateral" of hope in this cryptographic impulse, namely the collateral equation between *inquiry* and control, specifically the control of mortality.

For is "fathoming" a secret identical to the exercise of control? The very cadence and rhetorical encryption of Keller's formulation here—a general parenthetical syllogistic conclusion within and alongside a hyphentated invocation of particularity—suggests that concepts and practices of secrecy are more unstable than Keller implies with her invocation of physicist Richard Feynman's "disease" of lock picking: "One of my diseases, one of my things in life, is that anything that is secret, I try to undo" (40). For Keller, Feynman's confession is emblematic of a desire to control life and death.

Yet the impulses to "undo" secrecy and the exercise of control can operate as extraordinarily divergent vectors. Yes, the appropriately named Feynman carries out the oh-so-human desire to reveal and expose, the garish and obscene revelation through which control sometimes emerges. Nietzsche associated this nihilism and its will to truth—undoing falsehood, exposing it in a negative revelation—with the unhappy conscience of a science that killed a god who had never *lived*. Hence the question of science's proximity to life:

> Even if *language*, here as elsewhere, will not get over its awkwardness, and will continue to talk of opposites where there are only degrees and many subtleties of gradation; even if the inveterate Tartuffery of morals, which now belongs to

> our unconquerable "flesh and blood," infects the words even those of us who know better—here and there we understand it and laugh at the way in which precisely science at its best seeks most to keep us in this simplified, thoroughly artificial, suitably constructed and suitably falsified world—at the way in which, willy-nilly, it loves error, because, being alive, it loves life.[4]

Nihilism was in part, for Nietzsche, a product or symptom of a will to reveal, the undoing of the secret that Keller writes of. On the other tentacle, Nietzsche cultivates and laughs with a *fröhliche Wissenschaft* that lives not through the bottomless plumbing of interiority—the revelatory practice of the secret—but in *life*. Here, laughter sprouts out of an "understanding" that is not necessarily correlative to knowledge: those of us who should know better, here and there, here as elsewhere, the *fröhlicher Wissenshaftler, understands*. This "understanding" results less from a will than from a hosting, a sudden encounter or recognition of a differential. Less a capacity to steadfastly endure the respective differences of entropy and time than a flexible hospitality to a chaotic difference, this understanding finds itself in love with life, overtaken by it. Here, in catastrophic laughter, we notice that life and death are less opposites than differences, non sequiturs whose only image is the outbreak of laughter. These changes in kind challenge usual identity complexes, hailing vertiginous capacities for differentiation as well as the disciplined stabilities of knowledge.

Willy-nilly, here and there, science is at its best when as inquiry, its direction cannot be predicted. In its very effort to embed humans in a thoroughly falsified but unfortunately simplified vision, science encrusts or "infects" the living with a mechanical revelation. What Bergson characterized as "the encrustation of the mechanical upon the living"—the sudden folly of the machinic/biological difference—is less an analysis of the uncanny than a reminder that any specifically *biological* difference can never be known in advance. Such sudden revelations of artifice are not a matter of epistemology but of self-experiment: we do not know what an organism can do! In the context of biotechnology, such open-ended life forms augur, perhaps, changes in kind not just of species but of *life itself*.

This unpredictable character of science and life implies not relativism but an extraordinary capacity for surprise, a sudden difference between knowledge and life which is periodically associated with laughter. Laughter often responds to the specificity of *life* as an evolutionary force, an unfolding or becoming whose logic is not identical to knowledge. Here, the undoing of one knot is the creation of another, not a bringing of secrets into the outside of knowledge, but endless and essentially unpredictable wander from knot to knot, problem to problem. Laughter, then, becomes for Nietzsche an ecology of human "agency" through which this understanding can emerge. As such, this "agency" is a paradoxical one, as it is fundamentally implicated in an activity which, for the most part, humans can neither instigate nor control.[5]

The random walk that instigates laughter is at play with more than asteroids or other dice. Nietzsche links this "willy-nilly" itinerary to life itself, in some sense putting him in accord with molecular biologist Jacques Monod's absurdist notion of life as "frozen accident." Here error becomes once again contextualized as an "absurd" loss rather than the ecstatic differential of surprise that accompanies much laughter.

In short, such laughter sprouts from a sudden difference. Bergson, whom most contemporary biologists regard as representative of a telelogical vitalism, also notices that it is through laughter that the difference between the mechanical and the living often is expressed, a practice grounded not in telos but in imageless errancy. According to late millenium writers such as Stuart Kauffman, we humans find this erratic understanding of life too desolate and inconsolable to embrace:

> we find ourselves on a tiny planet, on the edge of a humdrum galaxy among billions like it scattered across vast megaparsecs, around the curvature of space-time back to the Big Bang. We are but accidents, we are told.[6]

And yet, for Nietzsche, gay scientists love error precisely to the extent that they love life. And the love of life *appears to be fundamentally a symptom of being alive.* Both love and laughter, as activities, are not fundamentally human agencies that emerge from consciousness. For both, the involvement of a self is necessary but impossible: for both the self is at play and, indeed, *at stake,* in love and laughter. The love of life—a symptom of being alive—is therefore a symptom not of knowledge or consciousness but of the force of life, a force exterior to, beyond or quite simply other than consciousness and the self.

Thus there is at least one other valence of this "undoing" that was Feynman's disease—not revelation and its alleged collateral, control, but a pulling apart or deterritorialization, an undoing of the self that is usually the creation of another. Sudden laughter puts us in stitches. It is side splitting, undoing our usual sense of autonomy and will. With the logic of a science fiction monster confronted by a death ray, the arrival of laughter only intensifies while we attempt to obliterate it.

In the schematic argument that follows, I want to map the rather willy-nilly itinerary of molecular biology's rhetorical and conceptual evolution and its debts to those forms of agency best exemplified by laughter but available to many extraordinary affects "proper" to even the scientific will: laughter, terror, ecstasy. Such modes of response, I will argue, were crucial to a conceptual evolution whose feedback loop arrives at cloning and tends toward a nanotechnological impasse: DNA information, at first understood by molecular biology as a fundamentally stable semantic phenomenon or "secret," becomes a spectacularly mutable technology of replication and differentiation by the early 1980s. This undoing of life, and the concomitant "loss" of integrity in the organism—wrought first by recombinant DNA and then polymerase chain reaction—seems to occur in response to an undoing and doing of identity sculpted by that most tabooed and double entendred scientific enterprise: the self-experiment. In particular, the necessary role of the self-experiment in the scientific study of hallucinogens—an inquiry not into life but into consciousness—provides the ecology for the emergence of these innovative and even ecstatic modes of interaction.

My discussion will respond to the common self-experiments at play in the seemingly diverse ecologies of the hallucinogenic "expansion" of "consciousness" and the engineering of evolution, biotechnology. These ecologies will be treated as twin or replicated domains where an informatic desire distributes and disperses both consciousness and life into inhuman, inorganic, and extraterrestrial realms.

II. Taking DNA, or Tripping Over the Organism

Thus far, I have claimed a shift from an emphasis on the interiority of organisms and the associated revelation of their essence to a harnessing of DNA's capacity to replicate and its subsequent "distribution" of life. No longer simply the attribute of a sovereign organism, life now emerges out of the connections of a network, involving an essential impropriety—it is life's habit of refusing containment that becomes interesting for biotechnology and capital.

These replicants are extraordinarily different models of living systems, both of which take place under the sign of "DNA" and molecular biology. One, the cracking of the code, looked to expose vitality as an attribute of a "periodic crystal," an orderly rather than mysterious enterprise best apprehended by physics. Another, a model of living systems we associate with biotechnology and its collateral capital and publicity markets, is less interested in what DNA might "mean" than with what it can *do*, and the relation between what it can do—replicate—and its production of value. Indeed, an emphasis on the primordial importance of *copying* reminds us of the impossibility of keeping secrets, as replication allow alleles and ideas to travel to multiple contexts, some intended and some not. Contemporary life science, despite all the chatter of God associated with the recent rough draft of the human genome, is interested less in predictability than experimentation and mutability, capacity for deterritorialization that generates value in the economy.[7]

While molecular biology was busy on its eighth day of creation, discovering, decoding and analyzing the "secret" of life, LSD-25 was also proffered in the laboratories and then the communes and other crowds of the world as the secret of "consciousness." Contrary to its usual representation as a seamless technology of unveiling associated with the instant gratification allegedly sought by an entire decade—the 1960s—LSD was continually treated as an enormously powerful but equally unreliable tool for the probing, revelation and "expansion" of consciousness. Writers, researchers, and experimentalists such as Timothy Leary, Humphrey Osmond, Richard Alpert, and Stanislov Grof all sought to study the function of "set and setting" in the instantiation of hallucinogenic practices and their capacity to transform human consciousness. These writers *took DNA* and used it to frame and articulate hallucinogenic sessions as programmable but not controllable events, experiments with and on the self. As such, theirs was a fundamentally *pragmatic* rather than *semantic* relation to the "information" of DNA, more recipe than message.

How did these writers *take* DNA? In fact, there were many experiments among researchers attempting to ingest DNA and RNA itself as a hallucinogen, sometimes in the hope of developing a "learning lozenge" that would inscribe the experience of LSD onto the brain. But nucleic acids were also crucial *rhetorical* vectors composing hallucinogenic discourse of the 1950s and 1960s, the talk, thought experiments, manuals, and technical papers that resulted from variously intentional and unintentional ingestions.

Hallucinogenic discourse, both of scientific and "recreational" nature, faced a similar rhetorical dilemma as the rest of the ecstatic traditions it responds to: it must report on an event that is *in principle* impossible to communicate. Writers of mystic experience from St. Theresa to William James have treated the unrepresentable character of mystic events to be the very hallmark of ecstasis. Hallucinogenic discourse faced a similar struggle in the effort to report on the knowledge beyond what Aldus Huxley (and Jim Morrison . . .) described as the "doors of perception." To deal with this struggle, many researchers had recourse to the rhetoric of nucleic acids—DNA and RNA became privileged characters in the stories and practices of hallucinogenic science. Nucleic acids were more, though, than "content providers" for the channeling of hallucinogenic knowledges into quasi-scientific protocols. As carriers of the news of molecular biology's informatic vision, rhetorics of nucleic acids were also set and setting for hallucinogenic sessions themselves: more than reporting devices, these rhetorics of nucleic acids helped to suggest that these sessions were themselves, like DNA, programmable.

III. Double Take

Problems of reportage troubled the discourse of LSD almost from its very inception, and certainly from its very first ingestion. In his fundamental ergot studies in 1938, Albert Hofmann first synthesized lysergic acid diethylamide, abbreviated LSD-25 (lyserg-saure-diathylamid) for laboratory usage. This novel molecule was primarily noted for its strong effects on the uterus, but as Hofmann retroactively reports in his autobiography named for a molecule, *LSD: MY Problem Child,*

> The research report also noted, in passing, that the experimental animals became restless during the narcosis. The new substance, however, aroused no special interest in our pharmacologists and physicians; testing was therefore discontinued.[8]

Hofmann continued his work in the ergot field, but LSD-25 was thought to have little pharmacological value, so between 1938 and 1943 "nothing more was heard of the substance LSD-25."

And yet Hofmann still had ears for the crying of LSD-25. According to his own account, LSD would not leave him alone:

> And yet I could not forget the relatively uninteresting LSD-25. A peculiar presentiment—the feeling that this substance could possess properties other than those established in the first investigation—induced me, five years after the first synthesis, to produce LSD-25 once again so that a sample could be given to the pharmacological department for further tests. This was quite unusual; experimental substances, as a rule, were definitely stricken from the research program if once found to be lacking in pharmacological interest.... Nevertheless, in the spring of 1943, I repeated the synthesis of LSD-25. As in the first synthesis, this involved the production of only a few centigrams of the compound (14).

Although Hofmann's attribution of presentiment must be placed within its context as an autobiographical confession, it nonetheless well names a peculiar agency that often adheres to those self experiments that are survived: the inability to forget. The memory of LSD and, as we shall see, LSD-25 itself, seems to have little truck with the usual operations of will. Indeed, according to Hofmann, his response to the crying of LSD-25—synthesis—resulted paradoxically in an interruption or a dissolution:

> In the final step of the synthesis, during the purification and crystallization of lysergic acid diethylamide in the form of a tartrate (tartaric acid salt), I was interrupted in my work by unusual sensations (15).

This interruption of the I, rather than ending an experiment, instigates one. In a report to a superior, Hofmann did his best to offer a description of the phenomenon that ensued after the interruption, but he only "surmised" a connection with the LSD-25.

Hofmann's analysis of the cause of the interruption itself was interrupted by the question of ingestion. Given his method of synthesis, Hofmann reasoned that his accidental passage must have been through the skin, and that the substance must therefore be of extraordinary—indeed, unprecedented—potency. Thus Hofmann's causal analysis of the unusual sensations seemed to offer two extraordinarily unlikely—that is, unprecedented—alternatives. On the one hand, the cause could remain unknown, and the tasteless and odorless LSD-25 had merely been associated with the experience rather than causing it. In this instance the strange interruption retained its enigmatic status, a nonsequitur of Hofmann's experience. On the other hand, if the minute quantity of LSD-25 was the causal agent of the interruption, then Hofmann was faced with the equally unlikely scenario that he had discovered the most potent compound known to history. To resolve the situation, Hofmann had recourse to an extraordinary nonsequitur, itself seemingly emerging without cause: "There seemed to be only one way of getting to the bottom of this. I decided on a self-experiment" (16).

In what sense can one "decide on a self-experiment? What warrants this decision? If it seems obvious that indeed, Hofmann did decide on such a course of action, it must also be noted that such a decision is of necessity itself an experiment, one that emerges not from any deliberative logic but from an incalculable action, a breakage in a chain of reasoning: just do it. Hofmann's deliberation on his possible responses to the interruption was itself not subject to anything like a procedure, an algorithm shorter than repetition by which he could arrive at a resolution of the two equally enigmatic if thoroughly differentiated outcomes.[9] Thus this decision to self-experiment is itself a testing of a hypothesis: that repetition is the most hi fidelity and compressed procedure for resolving the matter. The outcome of this experiment—Hofmann's implication of himself into the research—cannot itself be meaningfully differentiated from the experimental dosing of LSD-25. Only an additional experiment—the synthesis and ingestion of LSD-25—will retroactively provide this experiment in decision making with anything like a result. As if to mark the extreme danger that the implication of a self and body into the experiment entails, Hofmann writes oxymoronically of a self-experiment embarked upon with "caution":

> Exercising extreme caution, I began the planned series of experiments with the smallest quantity that could be expected to produce some effect, considering the activity of the ergot alkaloids known at the time: namely, 0.25 mg (mg = milligram = one thousandth of a gram) of lysergic acid diethylamide tartrate (16).

The danger, here, however, is not only the risk of an unknown compound of apparently extraordinary psychic potency. Instead, the instance or event of danger emerges coincident to the interruption of work itself: how to proceed? For as Hofmann notes after the now deliberate ingestion, LSD-25 is nothing if not the incessant and yet irregular arrival of the question: *How to go on?* This is a question of endurance for the experimental self: usual experimental protocol demands that everything is involved in an experiment *except* the self, and yet here it is precisely only the self and its responses that are the very assay of LSD-25. Only the variable examples of history provided anything like a protocol for self-experiment, calibration for the assay.[10]

This assay had great difficulty generating any readout:

> 4/19/43 16:20: 0.5 cc of 1/2 promil aqueous solution of diethylamide tartrate orally = 0.25 mg tartrate. Taken diluted with about 10 cc water. Tasteless.
>
> 17:00: Beginning dizziness, feeling of anxiety, visual distortions, symptoms of paralysis, desire to laugh (16).

LSD-25 did little to present itself for inscription. Without flavor, within forty minutes it produces predominantly anticipatory symptoms, events which were about to make themselves more fully known. The "desire to laugh," as such an experimental anticipatory symptom, was particularly difficult to assay. For by what means would any desire to laugh be registered by an observing self, except by laughter itself and its subsequent attempted blockage? What agency would interrupt said desire and, interrupted, in what sense did one desire to laugh?

Obviously, anyone can wish for laughter: this is the unlikely hope marked by the sitcom laugh track. But what is named by Hofmann is perhaps less a wish than a tantalizing inclination, a becoming laughter that is neither a cackling or its absence, a meanwhile in which the proximity of the future—I am about to laugh—is unbearable to the self of the present. For Hofmann, this desire indeed becomes unbearable, the doing of the experiment veritably undone:

> Supplement of 4/21: Home by bicycle. From 18:00–ca.20:00 most severe crisis. (*See special report.*)
>
> Here the notes in my laboratory journal cease. I was able to write the last words only with great effort. By now it was already clear to me that LSD had been the cause of the remarkable experience of the previous Friday, for the altered perceptions were of the same type as before, only much more intense.
>
> I had to struggle to speak intelligibly (16).

Hofmann's certainty regarding the causal role of LSD-25 in the visions and disturbances he experienced was equaled by his inability to communicate the character and nature of the struggle. While under the variable influence of LSD-25, Hofmann periodically ceases to be capable of even an *attempt* at communication and thus, an attempt at experimentation. In place of the usual and invisible expectation of communicability that is the rhetorical arena of scientific observation, the struggle suggests Hofmann's reliance on another mode of knowing altogether: the struggle of the ordeal, an event in which knowing is not separable from an irreducible participation. "*I had to struggle to speak intelligibly.*"

This inability to communicate is not a deficit in the hallucinogenic experience, but a symptom of it. In this sense, the self-experiment is both failure and success: as an experiment *with* the self, the outcome is close to null. No meaningful report can be generated, and therefore the knowledge of the hallucinogenic experience can in no way be gathered or repeated. As an assay, the self is found wanting. If the experiment is an occasion at which, strangely, the self is to be present as the very apparatus through which the inquiry is to made visible and replicable, then the apparatus has faltered and the experiment is nothing but artifact.

And yet later, Hofmann would nonetheless prepare a special report to his supervisor, Professor Stoll. Here Hofmann was struck by the *capacity to remember* the experience with great, even machinic precision:

> What seemed even more significant was that I could remember the experience of LSD inebriation in every detail. This could only mean that the conscious recording function was not interrupted, even in the climax of the LSD experience,

> despite the profound breakdown of the normal world view. For the entire duration of the experiment, I had even been aware of participating in an experiment, but despite this recognition of my condition, I could not, with every exertion of my will, shake off the LSD world. Everything was experienced as completely real, as alarming reality; alarming, because the picture of the other, familiar everyday reality was still fully preserved in the memory for comparison (20).

As an experiment *on* the self, the ingestion of LSD-25 was indeed a resounding success—the experimental object was unmistakably transformed, alteration extending even to the agency of Hofmann himself. As a recording, Hofmann could not avoid remembering as, for the purposes of LSD-25, Hofmann became what Merlin Donald describes as "external symbolic storage" device. Combined with Hofmann's own memory of LSD's agency—"And yet I could not forget the relatively uninteresting LSD-25"—Hofmann's transformation by the experiment is subtly but nonetheless inescapably inscribed in the confession: LSD-25 is not easily erased from the experience and memory of the experimental subject, a subject who, like it or not, is recording. Indeed, in some sense, Hofmann is both recorder and recording here, as he must respond exegetically to the demands of memory.

Hence, although the conclusions to Hofmann's exercise of extreme caution remained to be determined, there could be no argument but that the risk had yielded interesting data, namely, the transformation of Hofmann himself into a being seemingly incapable of forgetting. On this factor alone, Hofmann could determine that he had happened on a substance of unprecedented potency. Its usefulness and character would, of course, call for further research, but Hofmann's understanding of the causal nature of LSD-25 in his experience was certainly an important result capable of representation to his colleagues.

But if Hofmann had great faith in the splitting capacities of LSD as a molecule that allows for an almost unique position as a retroactive observer and real time participant, his supervisors, at least at first, did not:

> As expected, the first reaction was incredulous astonishment. Instantly a telephone call came from the management; Professor Stoll asked: "Are you certain you made no mistake in the weighing? Is the stated dose really correct?" Professor Rothlin also called, asking the same question. I was certain of this point, for I had executed the weighing and dosage with my own hands. Yet their doubts were justified to some extent, for until then no known substance had displayed even the slightest psychic effect in fraction-of-a-milligram doses. An active compound of such potency seemed almost unbelievable (21).

Although the incredulous questions focused repeatedly on issues of quantitative importance, clearly it was also the qualitative transformations wrought by LSD-25 that inspired disbelief. The missing special report, whatever it says, was evidently unable to carry out its task of translating the delirium of the self-experiment into the allegedly intersubjective space of scientific communication.

So little convinced were Hofmann's colleagues of his report that they took the almost unbelievable step of repeating the self-experiment:

> Professor Rothlin himself and two of his colleagues were the first to repeat my experiment, with only one third of the dose I had utilized. But even at that level, the effects were still extremely impressive, and quite fantastic. All doubts about the statements in my report were eliminated (21).

That is, LSD-25 was in this instance both the object of scientific inquiry and the medium for the communication of the results of that inquiry, the translation of a solo experiment into the general equivalent of truth. While Hofmann's wife returned from Lausanne on hearing reports that Albert had had some sort of "breakdown," his fellow scientists willingly and immediately ingested LSD in order to eliminate any doubts fostered by Hofmann's report, a psychedelic republic of letters.

It is as an experiment *on* the self that Hofmann's discoveries are replicated by the community. Only by encountering a veritable undoing of self—a submission to the possible transformation one is in fact testing for—can interesting data from this novel pharmacological agent be gathered, evaluated and transmitted.

IV. Informatic Prayer

The combination of ineffability common to many mystic traditions and the necessity of communication proper to scientific practice continued to pose problems for the study of hallucinogens as they migrated from Sandoz Pharmaceuticals, where Hofmann worked, to places like the pre-1962 psychology department of Harvard University, the working home of Timothy Leary, Ralph Metzner, Richard Alpert, and Michael Horowitz. Among a parade of intellectuals and artists that moved through Leary's burgeoning circle was writer William S. Burroughs, whose cut-up techniques had, a few years earlier, been used to strip a written text of its meaning—what Burroughs called the virus of the Word—and to allow such texts to interrupt normal consciousness. After a visit with Burroughs, Leary hit upon the idea of using Burroughs's cut-up technique as a framework for reporting the hallucinogenic experience. In this framework, both Burroughs's texts and LSD were experiences one less understood than underwent and, in undergoing them, recorded them. These recordings—as with Hofmann's precise recall of his LSD trip—were not, however, communicative in the usual sense: they could be understood only retroactively, after the subject had his or herself encountered LSD-25. With the cut-up method, Leary hoped to interrupt the grip of the authorial ego that might interfere with the more direct recording of the LSD experience. In other words, the LSD experience was treated less as an event to be reported on than an experience to be assayed by writing—the cut-up was a symptom and not a description of the encounter of LSD. In this sense, the information gathered about LSD was understood less as a semantic production than a pragmatic one: all data concerned not an *understanding* of LSD but instead consisted of testing LSD for what it could *do.* Writing, in this context, becomes less a struggle for intelligibility than a prolix signature of the LSD experience.

Indeed, the apparent need to write in response to LSD provoked a graphomania of sorts among Leary's crowd. By 1962, they had been removed from their positions at Harvard University and were now leading itinerant seminars from Zihuatanejo, Mexico, several islands in the Caribbean, and then finally the Hitchcock brothers mansion in Millbrook, New York. The *Psychedelic Review* regularly published the group's prolific work and in 1963 Leary, Metzner, and Albert prepared a psychedelic "manual" for use in association with LSD entitled *The Psychedelic Experience.* Published early in 1964, *The Psychedelic Experience* was presented as a source of protocols for the management and study of psychedelic sessions, protocols "based on the Tibetan Book of the Dead." As such, much of the writing, including *The Psychedelic Experience,* was oriented less to "understanding" LSD experience than with dealing with it, enduring it. These texts were intended to be part of the very interface of psychedelia, algorithms for attaining and prolonging particular states:

> One may want to pre-record selected passages and simply flick on the recorder when desired. The aim of these instruction texts is always to lead the voyager back to the original First Bardo transcendence and to help maintain that as long as possible.[11]

These recipes and techniques for ecstasy were repeatedly and explicitly linked by Leary to the writing and execution of a sequence of steps in a computer environment, programming:

> A third use would be to construct a "program" for a session using passages from the text. The aim would be to lead the voyager to one of the visions deliberately, or through a sequence of visions. . . . One can envision a high art of programming psychedelic sessions, in which symbolic manipulations and presentations would lead the voyager through ecstatic visionary Bead Games (98).

In this framework, hallucinogenic subjects become both authors of and platforms for "symbolic manipulations and presentations," interactive wetware of infinite experiment and transformation. In the preface to a later work, Leary would write that "*The Psychedelic Experience* was our first attempt at session programming."[12]

Crucial to this vision was that the function of information here was to "program." Less an activity of understanding or even communication than of repetition and transformation, these programs had to be endured. Indeed, the writings "based" on the Book of the Dead were often not understandable or recognizable until, like Hofmann's colleagues, one had oneself undergone the encounter with LSD. "The most important use of this manual is for preparatory reading. Having read the Tibetan Manual, one can immediately recognize symptoms and experiences which might otherwise be terrifying, only because of lack of understanding as to what was happening. Recognition is the key word" (97). Thus, the Tibetan Book of the Dead was treated as a source book not to be decoded as much as deployed, deterritorialized and deployed in divergent contexts, used more than understood.

Leary made this pragmatic understanding of language (which, for Leary, was indistinguishable from information) even clearer in *Psychedelic Prayers,* his 1966 "translation" of the Tao Te Ching, what Leary called a "time tested psychedelic manual":

> like all great biblical texts (sic), the Tao has been rewritten and re-interpreted in every century and this is how it should be. The terms for Tao change in each century. In our times, Einstein rephrases it, quantum theory revises it, the geneticists *translate* it in terms of DNA and RNA, but the message is the same.[13]

Translation, for Leary, is both universally available—"the message is the same"—and utterly variable and reliant upon context, what Leary will refer to as "set and setting" for the LSD experience but which can be extended to his own work of translation as well: "these translations from English to psychedelese were made while sitting under a bamboo tree on a grassy slope of the Kumoan Hills overlooking the snow peaks of the Himalayas" (38).

Like the drugs and plants with which they are to be used, psychedelic manuals are catalysts for transcendent experiences—"or can be, given the appropriate preparation, attitude, and context [the "set and setting" in Leary's felicitous phrase]" (10). Crucial to this facility for novel set and settings is not, strictly speaking, its universal message, but rather its capacity to be translated and travel into novel contexts: "The advice given by the smiling philosophers of china to their emperor can be applied to how to run your home, your office, and how to conduct a psychedelic session" (38).

Note here that the psychedelic manual is not concerned with the communication or an elucidation of a meaning, although it also does so. Instead, the text is seen as how-to tool for managing and transforming diverse contexts with the help of exotic and yet thoroughly debugged techniques. Its translation is less the production of an equivalent meaning than the "porting" of code to a different platform: "It became apparent that, in order to run exploratory sessions, manuals and programs were necessary to guide subjects through transcendental experiences with a minimum of fear and confusion" (36). This lack of interest in the semantic operation of such manuals is itself pragmatic: reportage continually fails, even while context is itself a powerful constituent of the psychedelic experience. Hence, repetitive failure would be tend to be avoided as a negative feedback loop, and other models of abstraction would be experimented with. The incommunicability of the psychedelic experience was taken to be a measure of the complexity of human consciousness as well as the insufficiency of most concepts to it: "No current philosophic or scientific theory was broad enough to handle the potential of a 13-billion-cell computer" (36).

As "prayers," Leary's translations also remind the reader of the active rhetorical register involved in the LSD context: prayers, above all, demand prayer. For Leary, their effects emerge out of their very utterance, a whispered utterance that needs to be said more than heard: "they should be read *very* slowly and in a serene voice. They should be considered prayers to be whispered" (33). Ideally, then, these rhetorics should approach pure action. Only a trace of the utterance will persist, a whispered scar of language's ecstatic embodiment.

The prayer program was divided by Leary into six parts, and the very center of the sequence is occupied by a twelve-part sutra entitled "Homage to DNA." Here Leary, "translating" and transforming the Book of the Dead into epideictic praise of the book of life, instructs the reader/tripper to "contact cellular consciousness" via the utterance of these meditations. The first, "The Serpent Coil of DNA," invokes the old and now familiar image of the ouroborus, a figure also used by Waddington in his discussion of feedback:

> we meet it everywhere , but we do not see its front.... When we embrace this ancient serpent coil, we are masters of the moment, and feel no break in the curling back to primeval beginnings. This may be called unraveling the clue of the life process. (67)

Rather than containing a "message," DNA is hailed as a molecule of ceaseless activity whose very embrace leads to an unraveling or, to translate, undoing. But even as a master of the moment, this "undoing" reveals no central wisdom other than this: our implication in an ancient coil of repetition without beginning or end. Thus rather than triumph, Leary's doggerel suggests an affirmation of primeval complicity, a complicity owed to the very self replicating ouroborus of DNA, and perhaps, a complicity with LSD and the prayer itself. The universal message alluded to by Leary is none other than repetition itself.

Contrasted with the molecular biology of the same period—the transmitter of the rhetorics of nucleic acids that were the set and setting of hallucinogenic knowledge—research into psychedelics, both in the lab and the commune, were strikingly pragmatic in their understanding of information. Both Hofmann and Leary wrote of the need for endurance, a practice that would enable repetition within and beyond the hallucinogenic experience even while the very assay of the experience—the self—was becoming variable and even breaking down. Indeed, it was the practice of repetition itself that seem to emerge as crucial to the acquisition and transmission of hallucinogenic knowledges. For Leary and the programmers of psychedelic practice, sessions were sequences of information to be actualized differentially—no two sessions could be enacted in precisely the same fashion, precisely because of the transformative effects of LSD and its teachings, repetitive teachings of undoing—"unraveling the clue of the life process." While molecular biology continued to write and talk of decoding the book of life and practicing a hermeneutics of DNA, hallucinogenic discourse suggested that information was less a phenomenon to be understood than it was a potent mutagen of human experience, mutagens that could only be understood retroactively.

V. Highway 128, Revisited

> If I had not taken LSD ever would I have still been in PCR? I don't know, I doubt it, I seriously doubt it. (Kary Mullis)

In his 1998 autobiography *Dancing Naked in the Mind Field,* Nobel Prize winner Kary Mullis details his invention of polymerase chain reaction, or PCR. While both the topic—the invention of a veritable Xerox machine for nucleic acids—and the genre—scientific autobiography with a hint of scandal, the promise of a forbidden, "naked" truth—encourage the telling of a heroic tale of innovation, Mullis's narrative continually highlights the thoroughly contingent and ungovernable arrival of a concept. Far from simply inflating the role of a lone scientist struggling to know life and the cosmos, *Dancing Naked in the Mind Field* offers a testimony to the thoroughly other, even alien, character of a scientific vision.

By every account, Mullis was himself an unpredictable creature. A biochemist by training, Mullis's most important publication in graduate school treated time reversal and its cosmological implications. In love with the craft of producing new compounds in a more and more efficient fashion, Mullis was every bit as much tinkerer as theorist.

This tinkering extended to Mullis's own consciousness. Like many of his UC-Berkeley chemistry colleagues, Mullis considered the potent new hallucinogens to be experimental objects as real and interesting as any other in their practice. Indeed, after one particularly difficult episode of incorrect dosage, Mullis underwent an entire personality change: overtaken by amnesia, Mullis had the sense that indeed he had become someone else as a result of the encounter.

Nor was Berkeley Mullis's sole connection to hallucinogens. During the period of PCR's emergence as a concept, Mullis visited with Albert Hofmann at the house of his friend, Ron Cook. Indeed, Mullis later compares Hofmann's three-lettered discovery with his own:

> The famous chemist Albert Hofmann was at Ron's that night. He had invented LSD in 1943. At the time he didn't realize what he had done. It only slowly dawned on him. And then things worked their way over the years as no one would have predicted, or could have controlled by forethought and reason. Kind of like PCR.[14]

This conjunction of LSD and DNA is of course strikingly resonant with Leary's treatment of both above: for Mullis, the main connection between the two was contingency itself, a practice "no one would of predicted, or could have controlled by forethought and reason." Only the unfolding of events allows for a retroactive ascription of narrative and reason—no description of the events could have been compressed into anything shorter than the events themselves. In short the practice of translating LSD and PCR into respectively technoscientific objects involved nothing less than the transformation of knowledge without precedent, a non sequitur not available to prophecy. There appeared to be no shortcut to either, with only the narrative of accident and coincidence making retroactive sense of the event. Unforeseeable, the effect of both could only be endured, responded to like the surfer Mullis is.

But Mullis himself goes further in his account of the connection between PCR and LSD. Retaining the contingency of the concept, Mullis nonetheless credits his experience with LSD with a certain capacity for the sheer strangeness of the idea of PCR. In a BBC interview Mullis hesitantly ascribed a paradoxical agency to LSD in the invention of PCR:

> PCR's another place where I was down there with the molecules when I discovered it and I wasn't stoned on LSD, but my mind by then had learned how to get down there. I could sit on a DNA molecule and watch it go by and I didn't feel dumb about that, I felt I could, I mean that's just the way I think is I put myself in all different kind of spots and I've learned that partially I would think, and this is again my opinion, through psychedelic drugs. If you have to think of bizarre things PCR was a bizarre thing. It changed an entire generation of molecular biologists in terms of how they thought about DNA.

Mullis's idea was a remarkably simple one. Indeed, in his cabin off of Highway 128 in Mendocino County, California, the idea struck Mullis as far too simple to be workable:

> If the cyclic reactions that by now were symbolized in various ways all over the cabin really worked, why had I never heard of them being used? . . . Why wouldn't these reactions work?[15]

Key to Mullis's understanding was the cyclic or iterative character of DNA itself. Unlike earlier molecular biologists who had sought the "meaning" of the genetic code, Mullis sought only to replicate enormous quantities of any given sequence. Mullis writes that because of his knowledge of computer programming, "I understood the power of a reiterative mathematical procedure" (5).

It is precisely as an iteration machine that Mullis treats DNA. Although H. G. Khorana and his team had provided the foundation for PCR as early as 1965, they did so within a cryptographic paradigm of discovering the nature and meaning of the genetic code. Francis Crick, writing the lead article for a volume devoted to the triumph of the newly decrypted code, declared that the "historic occasion" was the understanding of the four-letter alphabet of ATCG and its triplet combinations that produce the twenty amino acids, which in turn fold up into proteins: "From all of this we were able to work out the *meaning* of several of the remaining doubtful triplets."[16] Khorana's article appears thirty-six pages later in the volume, and it is clear from its frame and its content that above all Khorana's group sought understanding and even proof: "It was therefore clearly desirable to try to *prove* the total structure of the genetic code by this method."[17] As Paul Rabinow points out in *Making PCR,* the problem of generating DNA sequences was solved by cloning methods, so Khorana's group left the techniques they had discovered for another decade and perspective.

Although Rabinow writes that Mullis's approach was the "opposite" of Khorana's, it is perhaps more precise to notice that each had a fundamentally different understanding of information. For Khorana, the genetic code was a space of revelation—the production of a gene was in the service of an understanding of the entire genetic code, an understanding attentive to the rhetorical and epistemological genres of a proof. By contrast, Mullis sought less to understand DNA—for that, after all, was accomplished—than to transform it. It is this pragmatic understanding of genetic information through which PCR emerged.

Consider, for example, Mullis's account of the PCR process:

> If I could arrange for a short synthetic piece of DNA to find a particular sequence and then start a process whereby that sequence would reproduce itself over and over, then I would be close to solving the problem . . . The concept was not out of the question because in fact one of the natural functions of DNA molecules is to reproduce themselves.[18]

This treatment of DNA as an ourborous—not unlike that of Leary's—looked to the pragmatic capacities of nucleic acids: the natural capacity to replicate. And although Rabinow, in his anxiety to avoid a heroic vision of scientific innovation, seeks to distance his own account of PCR from Mullis's, he perhaps overlooks the fact that Mullis was not hero but host, as his instantiation of an iterative, pragmatic understanding of information replicated not just DNA but psychedelic culture itself, as the notion of "expansion" and decontextualization that dislocated consciousness in the 1960s now deterritorialized life, allowing it to become just another sample. Where to begin on the effects of such a deterritorialization or "expansion" of life? Perhaps one must begin with replication. And if any molecule was replicated, perhaps that molecule was LSD.

Notes

1. In the *Philosophy of Nature,* Hegel often treats vitality in terms easily articulable by a logical system of "division" and "unification." But his parsing of the ecological becoming of organisms in terms of the cut or its repair is hardly itself one based "negation" in any logical—rather than ecological—sense. In proximity to a plant, for example, Hegel writes a terrifying description of exteriority as constituting knots and folds torn out of themselves in a strange topology of retroactivity: "Since the plant does not hold itself back in inner, subjective generality against outer individuality, it is equally torn out of itself by light, from which it takes the specific confirmation and individualisation of itself knotted and multiplied into a multiplicity of individuals" (§ 269). This exfoliation motif of life—a growth torn out of itself not unlike that of the Alien—proliferates most

readily in low budget horror and science fiction films, an ecology from which we could learn much concerning "life."

2. Evelyn Fox Keller, *Secrets of Life, Secrets of Death: Essays on Science and Culture* (New York: Routledge, 1992), 39. For her earlier essay, see Keller, "Making Gender Visible in the Pursuit of Nature's Secrets" in *Feminist Studies/Critical Studies,* edited by Teresea de Lauretis (Bloomington: University of Indiana Press, 1986), 66–77.
3. See http://www.dcs.shef.ac.uk/~u7ss/enigma1.html.
4. Friedrich Nietzsche, *Beyond Good and Evil: Prelude to a Philosophy of the Future,* trans. Walter Kaufmann (New York: Vintage Books, 1966), 35.
5. This contingency associated with laughter can perhaps be best exemplified by the following scenario of Armageddon: Asteroid hits earth, producing enormous quantities of nitrous oxide, or laughing gas: death by laughter. The willy-nilly itinerary of an asteroid is a potent producer of laughter indeed, although it is perhaps not as comic as current schemes for using missiles to defend the Earth from such alien projectiles, what will be henceforth known as "gasteroids."
6. Stuart Koffmann, *At Home in the Universe: The Search for the Laws of Self-Organization and Complexity* (Oxford, UK: Oxford University Press, 1995), 4.
7. Of course, the "code script"—and its subsequent rhetorical transformations of code, program, and network—were all elements of a system to DO something: uncover the mysteries of life or, as Francois Jacob put it, "to triumph over death." In this sense, the emphasis on the replication activity of DNA rather than its communicative capacities is the return of the repressed of Schrödinger's machinic and ultimately cryptographic, rendering of life. I have discussed this topic in greater length in *On Beyond Living: Rhetorical Transformations of the Life Sciences* (Stanford: Stanford University Press, 1997).
8. Albert Hofmann, *LSD: My Problem Child,* trans. Jonathan Ott (New York: McGraw-Hill, 1980), 12.
9. It is not the case, for example, that Hofmann had no options: as his own account details, many more ways of fathoming LSD-25's effects were available, including the animal testing alluded to earlier.
10. The question of caution is an interesting one here, seemingly oxymoronic to any notion of self-experiment, as the very status of the self is at stake. Hofmann writes that during the terrifying peaks of his hallucinogenic panic, he worried over the possibility of ever communicating the fact of LSD-25's unforeseeable effects to his family. This retroactive recognition of Hofmann's highlights the true experimental conclusion of his work—that there is no preparation, decided or otherwise, for the encounter with the qualitative difference of LSD-25: "Would they ever understand that I had not experimented thoughtlessly, irresponsibly, but rather with the utmost caution, and that such a result was in no way foreseeable? My fear and despair intensified, not only because a young family should lose its father, but also because I dreaded leaving my chemical research work, which meant so much to me, unfinished in the midst of fruitful, promising development. Another reflection took shape, an idea full of bitter irony: if I was now forced to leave this world prematurely, it was because of this lysergic acid diethylamide that I myself had brought forth into the world" (18).
11. Timothy Leary, Ralph Metzner, and Richard Alpert, *The Psychedelic Experience: A Manual Based on the Tibetan Book of the Dead* (New Hyde Park, NY: University Books, 1964), 97.
12. Timothy Leary, *Psychedelic Prayers and Other Meditations* (Berkeley: Ronin, 1997), 37.
13. Leary, *Psychedelic Prayers,* 37.
14. Kary Mullis, *Dancing Naked in the Mind Field* (New York: Pantheon Books, 1998), 11.
15. Mullis, *Dancing in the Mind Field,* 8–9.
16. F. H. C. Crick, "The Genetic Code: Yesterday, Today, and Tomorrow," *Cold Spring Harbor Symposium on Quantitative Biology* 31 (1966): 3.
17. H. G. Khorana, in *Cold Spring Harbor,* 39.
18. Mullis, *Dancing in the Mind Field,* 6.

CHAPTER 6

$ell: Body Wastes, Information, and Commodification

ROBERT MITCHELL

In one of the more striking scenes in David Fincher's film *Fight Club* (1999), the narrator and his double steal plastic bags of discarded human body wastes from a dumpster behind a liposuction clinic. The two then convert the fats from this corporeal garbage into a commodity, soap, that they sell to fashion boutiques for $20 a bar. The narrator positions this commodification of body waste as a subversive maneuver, claiming that "[i]t was beautiful. We were selling rich women their own fat acids [or "asses"; the pun is deliberate] back to them." The viewer is no doubt supposed to find this scenario hyperbolic, the satire depending on the notion that we, the viewers, have not yet gone so far that we would pay good money to buy back our own bodily waste after it had been processed through the infrastructure of capitalism and commodity culture.

Yet this process is no wild imagining of the Hollywood entertainment industry, for the commodification of body wastes is big business. From the sale by hospitals of infant foreskins (used to manufacture artificial skin) and aborted embryo gonads (a source of stem cells), to the patenting of infected cell lines as research materials, to the sale of celebrity DNA-rich saliva, body wastes are fully integrated into the structures of late capitalism, and many of us "buy back" these wastes in the form of placenta-enhanced shampoos and drugs developed from patented cell lines.[1] We might be inclined to bemoan this, and see the quip of *Fight Club*'s narrator as a joke on all of us who live in an era in which the human body is decomposed within the field of capital, only to be reconstituted, and sold, under the aegis of money. Within this reading, the integrity of the human body is under attack, and the commodification of body wastes is seen as the first sign of the complete erasure of dignity itself. Dignity, the argument goes, demands that the human body stand distant from both economic value and our own wastes.

However, the status of body wastes as commodities is uncertain and puzzling terrain, and in this chapter I contend that the discourse of "human dignity" is a problematic strategy with which to engage productively the contemporary logic of body reterritorialization and commerce. Although the discourse of dignity could probably never "protect" the human body in the ways that its adherents supposed, the increasing dominance of information as a mediating term between bodies and money has altered the terrain fundamentally. More specifically, the introduction of notions of information to discussions of body commerce has created a strange merger between the discourses of dignity and garbage, with both being used as wedges to divide individuals from the gains of commerce, while at the same time allowing ownership by corporate bodies. In the first part of this chpater, I outline the key assumptions of this discourse of dignity that for most of the latter half of the twentieth century has framed discussions of the relationship between commodity exchange and the human body. I then investigate the role of dignity in the California Supreme Court case of *Moore v. Regents of the University of California*, a case in which the California judicial system attempt to protect "dignity" by denying an individual (Moore) property rights in human cells while at the same time allowing corporate entities to reap enormous profits from the cultivation of this corporeal waste. Like a number of previous commentators, I find this case particularly revelatory of contemporary relationships between human bodies, information, and commodities. However, I focus on a relatively underdiscussed aspect of the case: the importance of notions of bodily waste in the court's attempt to link arguments about dignity and information. I conclude by arguing that the court's reasoning highlights the problematic hygienic logic of the discourse on dignity, which positions the legal authorities as parental arbiters of the proper uses of body parts. Moreover, it points to the need to reexamine the critique of commodification, and especially the commodification of body wastes. Instead of understanding commodification as an intrinsically evil process, we ought to turn to the work of commodity theorists such as Appadurai, Bourdieu, and Douglas and Isherwood to construct a more productive understanding of the relationship between body parts and commodities.

I. The Discourse of Dignity

The twentieth century has witnessed a massive reterritorialization of individual bodies, as physicians, hospitals, and corporate entities have developed and enabled practices such as tissue banking (of blood, eyes, and the like) and organ transplantation, and attending these processes has been a persistent concern about the emergence of "markets" for these body parts. Numerous medical practitioners, governmental bodies, and individual citizens have expressed discomfort with the idea that body parts such as blood or organs could be bought and sold.[2] Some critics assert that existing "gift" and "donation" systems of body part transfer enable modes of communal cohesion that will be lost in the movement to a commodity system of exchange. Others point to the exacerbation of individual, class, and global economic inequity that might result from the establishment of organ markets. Opponents of commercialization suggest that the supply of organs might be diminished following the establishment of body part markets, as people formerly willing to donate organs find themselves loath to sell them (or to give them when others are allowed to sell them). Critics also point to the legal problems that emerge when body parts are treated as commodities (most notably, questions concerning "product liability").[3] The "marketing" of body parts is almost invariably presented as "black marketing," that is, a form of commerce outside the bounds of human values and decency.[4]

Although critics thus have employed a number of different rationales to oppose the transformation of body part transfer into commodity exchange, many, if not most, of these objections depend upon, or at least make recourse to, the discourse of "dignity." Within this logic, the human body and human life are represented as possessing an innate dignity that is threatened as soon as a part of the body is assigned an economic value. The value of the human body is understood as absolute, incomparable to any other value. As a result, it cannot be "owned," for ownership would displace the absolute value of dignity into the realm of commercial values that can be exchanged for one another. Parts of the human body, so the argument goes, should have only use values, and never exchange values. Thus, the authors of a recent report to the French government on bioethics conclude that "[h]uman dignity forbids that man be given a right to own his own body."[5]

As Paul Rabinow has noted in *French DNA: Trouble in Purgatory*, the basic structure of this logic of dignity was outlined by Kant, although it seems to have been fully integrated into biomedical discourses following World War II. In his *Grounding for a Metaphysics of Morals*, Kant argued that rational agents were characterized by "dignity," which could not be related to the realm of values: "In the kingdom of ends everything has either a price or a dignity. Whatever has a price can be replaced by something else as its equivalent; on the other hand, whatever is above all price, and therefore admits of no equivalent, has a dignity."[6] Following the disclosure of the atrocities committed during World War II, a version of this logic was adopted and incorporated into the United Nations charter as well as the French "Déclaration Universelle des Croits de l'Homme (Universal Declaration of Human Rights)," both of which employ the term "dignity" as definitional of humans.[7] More recently, Kant has been explicitly invoked in discussions of body part transfers. Stephen R. Munzer, for example, uses Kant to argue that since "[e]ntities with dignity differ sharply from entities that have a price on the market," if individuals were granted "property rights in bodies and exercised those rights, they would treat parts of their bodies in ways that conflict with their dignity. They would move from the level of entities with dignity to the level of things with a price."[8] In all these examples, dignity is positioned as bulwark against the threat of the absolute annihilation of communal values and human life. Protecting dignity by opposing body part sales is presented as equivalent to protecting humanity itself against the maelstrom of exchangeable values and the nihilism that attends the revaluation of all non-commercial values in commercial terms.

Within this logic, the only way to transfer body parts (for example, blood or organs) from one person to another without loss of dignity is through "gift" or "donation."[9] Gifts and donations, in this view, are understood as nonutilitarian (one does not donate for gain) and divorced from exchange (one does not give item *x* for item *y*). Understood in this way, body part gifts and donations are able to skirt the realm of exchangeable values while still traversing the distance from one body to another. Authors committed to this view contend that these gifts allow for direct human contact that is unsullied by any passage through the dehumanizing field of commodities and commerce. I will question below whether this is a viable understanding of either body part transfer or commodities, but in any case, this understanding of the gift has underwritten most attempts to protect the dignity of humans by, on the one hand, preventing body parts from being commercially alienated while, on the other hand, still allowing for the movement of body parts from one individual to another.

At the same time, however, this logic of dignity coexists uneasily with the fact that in most Western countries, several types of body "parts" are, in fact, alienable. The alienability of labor, for example, is especially problematic to the extent that philosophical justifications for this right are often traced back to Locke, who justified the exchange of labor for goods through an appeal to an individual's relation of proprietorship to their own body. Within this understanding, an individual may sell her labor *because* she owns herself.[10] More concretely, most Western countries allow for the individual sale of either nonliving or nonorgan body parts such as hair, fingernails, and milk (and an increasing number of states in the United States allow the sale of sperm and eggs).[11] Although adherents of the logic of dignity have more or less ignored the problem of the alienability of labor, the sale of body parts such as hair and fingernails often have been explained by recourse to a logic of waste. These body parts, it is argued, are "abandoned" by the living, functioning body, and may thus be sold.[12] These abjected parts thus can be easily and usefully distinguished from organs (which are much more difficult and dangerous to separate from the body). The sale of waste tissue is acceptable; the sale of integral organs is not. Yet, however much sense that distinction between "integral" and "abandoned" tissue might have made in the past, the situation has become much more complicated lately, as "information" has come to confuse the distinction between dignity and waste.

II. Sacrality, Profanity, and Moore's Unclean Cells

In order to understand the increasing importance of information to the discourse of dignity, it is useful to recognize that the distinction between the realms of dignity and relative value is our modern version of the dichotomy between the sacred and the profane. This is quite clear in the language of the California Supreme Court decision of *John Moore v. The Regents of the University of California* (1990).[13] This case went through three levels of the California juridical system, but in the final California Supreme Court decision, the regents of the UC system were allowed to maintain exclusive patent rights to a cell line generated from infected hairy-cell leukemia tissue taken from UCLA patient John Moore. The court acknowledged that Moore's doctors had not informed him that they were performing extra tests in order to develop, and patent, this cell line, but it nevertheless denied Moore any proprietary rights to his diseased cells, the patent on his cell line, or its proceeds (properties worth $3 billion, according to Moore's lawyers). Moore was not entitled to ownership in these properties, argued one judge, in part because to do so would be to mingle two spheres that must be kept separate. Moore, wrote the justice,

> has asked us to recognize and enforce a right to sell one's own body tissue for profit. He entreats us to regard the human vessel—the single most venerated and protected subject in any civilized society—as equal to the basest commercial commodity. He urges us to commingle the sacred with the profane.[14]

To allow Moore to sell a part of oneself, the court suggested, would initiate a process of contagious mingling that would threaten the integrity of all sacred human vessels.

Excluding Moore from property rights to his diseased cells required considerable legal dexterity on the part of the court. In the (divided) majority opinion, the court attempted to preserve the distinction between the sacred and the profane by distinguishing between the right to convert a body part into property and other rights that it did want to protect, such as privacy rights. The court observed that Moore's lawyers had attempted to join privacy and property: "[l]acking direct authority for importing the law of conversion [i.e., property] into this context, Moore relies, as did the Court of Appeal, primarily on decisions addressing privacy rights." But the court concluded that "[n]o party has cited a decision supporting Moore's argument that excised cells are 'a species of tangible personal property capable of being converted.'" The right to privacy and the right to property, the court argued, were distinct and should not be confused.

More specifically, the court suggested that the link between privacy and property that both Moore and the California Court of Appeal (which had agreed with Moore) had attempted to establish was, quite literally, shit. Moore's lawyers had cited in support of their claim *Venner v. State,* a case originating in a police seizure of a criminal defendant's feces for evidence. In that case, the Maryland justices had argued that the defendant had the right to control the search of his "abandoned" feces. The Maryland court had reasoned that just as people often assert "ownership, dominion, or control" over body parts such as "hair, fingernails, toenails, blood and organs" so too ought one be able to exert these powers over "excrement, fluid waste, [and] secretions."[15] The California Supreme Court justices noted that from this case, the California Court of Appeal had concluded in their earlier decision "that '[a] patient must have the ultimate power to control what becomes of his or her tissues. To hold otherwise would open the door to a massive invasion of human privacy and dignity in the name of medical progress.'" "Yet," continued the justices of the California Supreme Court,

> one may earnestly wish to protect privacy and dignity without accepting the extremely problematic conclusion that interference with those interests amounts to a conversion of personal property. Nor is it necessary to force the round pegs of "privacy" and "dignity" into the square hole of "property" in order to protect the patient, since the fiduciary-duty and informed-consent theories protect these interests directly by requiring full disclosure.[16]

In other words: not only had Moore and his lawyers attempted to mingle the sacred and the profane, but they had misunderstood the true nature of shit (it is something private, but not property), and had thus failed to make obvious qualitative distinctions (they had mistaken round pegs for square holes).

In order to clarify for Moore how he should have dealt with body wastes such as feces and his own diseased cells, the court emphasized its point by supplementing its logic of dignity with another line of reasoning, focused on garbage. The majority justices were no doubt aware that it was a bit peculiar to base their decision on the sacred nature of the human body, since what was at issue was the status of *diseased* tissue that had to be removed if Moore wished to remain an integral "human vessel."[17] Thus, within the court's supplementary logic, Moore was forbidden from receiving money from his body parts not just because they were sacred, but also because they were profane. The court argued that Moore had effectively "abandoned" his diseased cells when he consented to their removal, and thus agreed to their disposal in the biological garbage system of the UCLA hospital facility. This garbage system is regulated by federal and local codes designed to safeguard public health, and "virus-infected cells, such as Moore's T-lymphocytes," noted the court, "fit reasonably within the statute's definition of 'infectious waste.'" The court thus positioned itself as a concerned parent, hygienically safeguarding Moore (and all of us) from two types of unholy messiness: on the one hand, the assault on dignity that occurs when we try to value *any* body part; on the other, the assault on dignity and public health that occurs when individuals try to hold and possess their own wastes.

Yet even as the court ruled that *individual* bodies ought to keep their hands off their offal, they did not require the same behavior of corporate bodies, and "information" served as the magic term that provided the distinction between the two sorts of behavior. Luckily for the UC regents—and, the court argued, the collective good—the UCLA body waste disposal system allowed for scientific research designed to extract information from biological garbage (and, if warranted, patenting of material derived from that information). In this particular case, scientists had created a cell line, rather than information per se, but the court suggested that this act of creation could only occur within an ecology of scientific information flows. The court contended that this ecology of information flows, and the acts of biomedical creation within this ecology, required strenuous work on the part of scientific researchers in general, and the doctors who had labored to turn Moore's infected cells into a research cell line, in particular. Moore, by contrast, had simply dumped his wastes into this sewage system, and any proceeds from his leavings rightly accrued to those who took on the unpleasant and difficult task of locating and extracting nuggets of data in the streams of waste. Moreover, not only did Moore fail to labor in support of information and biomedical creation, but his property claims posed a threat to very existence of this ecology. The court argued that allowing Moore property rights in his cells would impede two sorts of flows: the free flow of information between researchers and the free flow of capital from investors to biomedical corporate entities. "At present," the court contended,

> human cell lines are routinely copied and distributed to other researchers for experimental, usually free of charge. This exchange of scientific materials, which still is relatively free and efficient, will surely be compromised if each cell sample becomes the potential subject matter of a lawsuit.[18]

The progress of science, in other words, requires an efficient and unrestricted flow of information and cell lines from institution to institution. At the same time, however, the production of information requires a constant flow of capital from the outside to the inside of the research institution, and "the theory of liability that Moore urges us to endorse threatens to destroy the economic incentive to conduct important medical research." Why, the court asked, would researchers bother to produce any useful research and information if they weren't able to derive financial profit from that work?[19]

In *Shamans, Software, and Spleens,* James Boyle argues that these two rationales seem to work at cross purposes to one another, for the court finds itself claiming, on the one hand, that property rights in cells will impede the flow of information, but on the other hand, that the production of information requires that someone be granted property right in cells.[20] However, I am not so sure that this is a contradiction, for the court seems to be implying a complicated ecology between three different types of bodies. First, there is the individual "dignified" human body, and, second, a dignified "body of information" that is produced by research institutions, both of which must be kept pure. But they can maintain their dignity only if *corporate* bodies (who have already submitted to the profane economic world) labor in the sewers of commerce. The body of information mediates between these other two bodies (individual; corporate), with "waste" serving as the rhetorical technology that keeps individual bodies dignified.

The biomedical dream that motivates this understanding of the complicated series of dispersals, transformations, and multiplications outlined above is that waste, entering the field of economics, can be magically transformed into valuable information and products; diseased tissue leaves one body as waste so that healthy tissue can be produced in another body somewhere down the line. But where exactly does the magical economic field begin and end? The lawyers for the UC Regents argued—and the California Supreme Court agreed—that individuals such as Moore stood outside of this field, for in the case of the patented cell line, his original tissue had become so fragmented and abstracted that the end product bore no real economic relationship to the original "donation."

As this line of reasoning indicates, the economics of body waste are far more difficult to track and think about than those involving healthy tissues. Body part transfers such as organ transplants involve a relatively direct relationship between two parties (the donor and the recipient), though it is true that this relationship is mediated by a complicated array of other players, such as families, doctors, hospitals, and insurance companies. Yet these complications are negligible in comparison with the economics of profitable body waste, for in a case such as Moore's, infected tissues leave one body and are dispersed, altered, and multiplied within a research institution, before then being translated into an economic field of patents, copyrights, and trademarks.

There is no doubt some element of hypocrisy in the court's decision, but much more importantly, this case illustrates that the discourse of dignity is probably not capable of serving as a useful strategy of social justice in our contemporary situation. The discourse of dignity always has had to ignore the very complicated economic relationships that have enabled blood, egg, and semen banking, as well as organ donation, but recent changes in the legal status of body parts have outstripped this discourse entirely. The U. S. Supreme Court made clear in *Diamond v. Chakrabarty* (1980) that commercial agents may patent genetically engineered organisms, and recent decisions have supported the rights of economic entities to patent even "naturally occurring" sequences of DNA. We may want to oppose elements of this process, but, as the Moore case highlights, "dignity" is probably not viable ground from which to mount this resistance. The discourse of dignity is, at bottom, an hygienic logic, and it thus implicitly positions "the authorities" (e.g., the courts) as arbiter of disputes concerning how messes such as Moore's infected cells shall be cleaned up. Perhaps even more importantly, the discourse of dignity simply obscures the fact that at stake is not the question of body part ownership per se so much as the question of *who* shall own these parts, and—even more importantly—the question of what "ownership" and "commodification" should mean.

III. Commodities and Exchange

To return briefly to *Fight Club*, we can note that the film registers this same commercial–hygienic logic of dignity through the image of the transformation of body wastes into expensive soap. Yet although *Fight Club* exposes this logic, it does not finally break outside of it, for the film positions the transformation of body wastes into consumer products as an index of the intolerable degradation that characterizes contemporary society. The film's protagonists reestablish personal (and, presumably, collective) dignity only by deriving glycerine from human fats to make bombs, with which they then destroy the world's financial institutions. Yet I suggest that instead of following this approach—for example, using cases such as Moore's as an opportunity to retreat even deeper into the hygienic logic of dignity—we ought to rethink the processes and meanings of commodification itself. As Paul Rabinow has noted in a discussion of blood and organ transfers, the prevailing belief that body parts should be "donated" (rather than sold) is not in fact a way of protecting these transfers from the realm of commerce; instead, it is simply one strategy for structuring "how body materials, social solidarity, public health, and money should function together."[21] The organ donation system does not prevent organs from entering the realm of commercial value; it simply dictates a specific form of entry and emergence. The research hospital waste system, through which Moore's cells traveled, organizes the intersection of money and body parts in another manner.

Several important efforts to rethink the connections between bodies, information, and commerce have focused on redefining the term "property." Commentators in this camp note that acknowledging property rights in bodies or body parts is not equivalent to saying "anything goes," nor will it necessarily transform bodies into commodities. Although, in common parlance, "property" denotes a *thing*, in legal discourse it refers to a "bundle of rights" (some collective, some individual), and for every kind of property, rights are subject to regulations.[22] Property rights are thus understood as ways of establishing sites of control and responsibilities over things. Andrews and Nelkin, for example, suggest that recognizing personal property interests in human tissue might lead to greater individual control over uses of the flesh.[23] Feminist theorists such as Rosalind Pollack Petschesky have argued that property rights do not have to be understood through a Lockean paradigm (in which a property right is equivalent to a right to alienate) but rather as a right to control access to bodies.[24] If understood this way, she argues, individuals *should* be granted the right to own their bodies.

Another approach, closely aligned with this understanding of property as a question of access, has centered on the idea of establishing a "commons" of body parts and information, within which individual ownership claims would be tightly regulated (if not excluded completely). This approach would agree with the California Supreme Court decision that individuals should not own body parts or information derived from bodies, but it would also forbid commercial entities from exercising ownership claims. Fox and Swazey trace the emergence of the notion of the biomedical commons back to the late 1960s, and the rhetoric of the commons has become increasingly dominant in the last two decades.[25] So, for example, in *The Human Body Shop: The Cloning, Engineering, and Marketing of Life*, attorney Andrew Kimbrell bemoans the fact that "[w]e have begun the process of commercially exploiting our common genetic heritage."[26] Vandana Shiva has argued that patenting and commodification of elements of the bioworld are akin (and can be traced back) to colonization of non-Western cultures. She notes that recent international intellectual property agreements exclude the "kinds of knowledge, ideas, and innovations that take place in the 'intellectual commons'—in villages among farmers, in forests of tribespeople and even in universities among scientists."[27] In *Shamans, Software, and Spleens: Law and the Construction of the Information Society,* James Boyle advances the notion of an "information commons," to which individuals would contribute as research subjects and from which they would receive back relatively cheap drugs and therapies. The notable failure of eighteenth- and early nineteenth-century critics of enclosure to protect the British commons might not seem to bode well for contemporary analysts committed to resuscitating this terminology and strategy, but supporters point to the success of the commons approach in the 1960s and 1970s in helping protect outer space and the deep-sea bed from exclusive property claims.[28] Yet, one wonders, along with Manuel Castells, whether language that dates back to the transition from an agricultural to an industrial age will prove to be fully adequate to help solve the problems of body part commerce in the information age.

Thus, in addition to developing strategies that seek to redefine property to distinguish it from commodities, we also ought to revalue our notion of the commodity itself. The recent emergence of commodity and consumption theory offers a number of important considerations that can help us to reformulate the connections between body materials, social solidarity, public health, and money. The approaches to commodification and consumption exemplified by texts such as Mary Douglas and Baron Isherwood's *The World of Goods: Towards and Anthropology of Consumption,* John Brewer and Roy Porter's *Consumption and the World of Goods,* and Daniel Miller's *Acknowledging Consumption: A Review of New Studies* offer numerous possibilities for more fully understanding (and controlling) the processes through which bodily wastes are "magically" transformed into values.[29] The discussion and definition of "the commodity" in Arjun Appardurai's introduction to *The Social Life of Things: Commodities in Cultural Perspective* is particularly useful. He notes that "commodities" are generally identified with mass production, marketing, and advanced monetary systems, but suggests that this is a particularly limited understanding of the term. He proposes instead that we focus our understanding of the commodity around the fact of exchange, and advances the notion of the "commodity situation." "*The commodity situation in the social life of any 'thing',*" Appadurai writes, is "*the situation in which its exchangeability (past, present or future) for some other thing is its socially relevant feature.*"[30] Commodities, in Appadurai's view, thus encompass the phenomena of barter and gifts, as well as monetary transactions, but at stake in his version of commodity analysis are the conditions and elements of exchange. Appadurai also reminds us of Marx's distinction between capitalist commodity flows, which employ commodities as a middle term to generate money (Money-Commodities-Money) and noncapitalist flows, which use money as a mediating term between commodity exchange (Commodities-Money-Commodities).

This understanding of commodities and commodification allows one to avoid seeing the "black" market and money as the only possible linkage between exchange and the deterritorealization of body parts. It allows us to recognize, for example, that James Boyle's notion of an "information commons" (described above) really has very little to do with traditional notions of "the commons" (which generally focus on questions of access rather than exchange), but is rather more productively understood as an attempt to redefine the commodity flows between patients and therapeutic products.[31] The question is not: does everyone have *access* to body parts and information, but rather, in what ways will individuals or groups *receive something in exchange* for their contributions of body parts or information? This reformulation of the notion of the commons can be theorized through the theoretical framework outlined by Douglas and Isherwood in *The World of Goods.* Although Douglas and Isherwood focus on consumption, rather than commodities per se, they articulate a vision of commodity flows as a "live information system," in which commodities are "both the hardware and the software, so to speak, of an information system whose principal concern is to monitor its own performance."[32] They suggest that different information systems of commodity exchange can be assessed morally by their inclusiveness: that is, do systems of commodity exchange hinder or facilitate ever more inclusive interpersonal interactions?

Reimagining the commodity in this way also allows one to avoid a problem that has plauged so many recent discussions of biomedical "gifts" and "donations"; that is, the tendency of adherents of this latter discourse to ignore questions of exchange. Within the discourse of dignity, the gift is supposed to represent a way of moving human tissues from one body to another while at the same time avoiding the realm of commutable values. In other words, body part donations are not commodities because they are not exchanged for anything. Yet this is an extremely restrictive understanding of the gift, and one that is at odds with sociological and anthropological investigations of the gift phenomena, most of which deal with gift *exchanges.* In their discussion of organ donation, for example, Fox and Sweeny perform the near-miraculous feat of attempting to support their argument with Marcel Mauss's inaugural text *The Gift: The Form and Reason for Exchange in Archaic Societies* while at the same time voiding all elements of exchange from their discussion.[33] Thus, the dynamic nature of the "obligations" that Mauss found so characteristic of "archaic" gift exchange (one is obligated to *reciprocate* gifts) is stalled in Fox and Swazey, as they parcel out the various obligations to the different people involved in organ donation. The "obligation to give," for example, devolves exclusively on the body part donor, whereas the "obligation to receive" is focused solely on the organ recipient.[34] Yet while it is true that organ recipients are unlikely to donate organs themselves, this does not mean that exchange is absent from this process.[35] One might apply the framework developed by Pierre Bourdieu to Fox and Swazey's analysis to uncover the various forms of cultural and symbolic, as well as biological, capital that are exchanged in organ donation.[36]

These theoretical failures of gift models suggest that perhaps one ought to attempt to subvert the language of the market from within, rather than simply trying to oppose it head on. This is in part a practical necessity, for the discourse of property tends to co-opt its rhetorical opposition. E. Richard Gold, for example, has astutely noted that to argue about a good (for example, Moore's cells) in terms of "property" is automatically "to subject [this good] to property discourse," where that latter term is understood as the "combination of conceptions, assumptions, and language used by legal practitioners—judges, attorneys, and legislators—to decide to whom and in what circumstances we ought to grant rights of control over a good."[37] Gold argues that in our current situation, "property discourse" is so firmly wedded to notions of "the market" that non-market values such as "dignity, charity, sharing, life, and community . . . do not find voice within property discourse" (156–7). Although he acknowledges that property discourse in principle could be expanded to include consideration of nonmarket values, he concludes that this is unlikely to happen swiftly enough to make this a useful strategy. Thus, property discourse threatens to be the sole arbiter of who shall have "the right to determine whether materials derived from the body will be treated as commodities (rather than treated with respect) or consumed (rather than preserved)" (x). Yet Douglas and Isherwood's approach suggests that our choices are not commodities *or* respect, consumption *or* preservation. To shift the discussion from "property" to "commodity exchange" may be one of the few ways to alter property discourse from within, employing some elements of the language of the market while at the same time highlighting for courts and legislators the continuity of that system with other modes of exchange.

As an example of what an "internal" subversion of market and property discourse might look like, I conclude with the example of PXE, International. This nonprofit corporation was established by Patrick and Sharon Terry, whose children suffer from pseudoxanthoma elasticum, a rare progressive disease that causes premature aging of cells (with symptoms such as vision impairment, skin lesions, and an extremely abridged life span).[38] Frustrated by the lack of interest, on the part of the pharmaceutical industry, in developing therapies or cures for this disease, the Terrys took two radical measures. First, in cooperation with researchers at the University of Hawaii, they were able to isolate, and then patent, the gene that causes pseudoxanthoma elasticum. Second, they established a nonprofit corporation responsible for licensing use of this gene to interested researchers, and created institutions and protocols that allow PXE, International to direct its own international flows of tissues and information related to this disease (so, for example, they have created their own blood and tissue banks, epidemiological studies, and MRI databanks). Rather than relying on the largesse of the families of afflicted individuals (as is often the case for fund-raising activities by patient groups afflicted by, for example, cystic fibrosis), the Terrys have refashioned the commodity form (in this case, a patented gene) to create new flows of body tissues and information. The Terrys stress that this refashioning was impossible without possession of a patent. This is true in part because patent ownership has allowed them access to venture capital in ways that were not possible before, for although pharmaceutical companies are still uninterested in a disease that afflicts a relatively small number of people, they are well aware that the pseudoxanthoma elasticum gene may play a partial role in other, more lucrative, diseases, and they thus will need to negotiate with PXE, International in order to access this gene legally. PXE, International is willing to license use of the gene under a number of different conditions: for example, free use for researchers working on therapies for pseudoxanthoma elasticum, and profit-sharing agreements for corporations not interested in those sorts of therapies. In this case, then, PXE, International has pushed the commodity form from within, rather than attempting to locate "alternatives" in the form of gifts or donations, and they have accomplished this by focusing on the question of exchange.

Moore v. the Regents of the University of California highlights the ways in which the rhetoric of dignity tends to slip into an hygienic and authoritarian discourse that corporate interests and the courts have found so useful in recent years, whereas the example of the PXE, International gene patent highlights a possible solution for understanding the dynamics and possibilities for body part exchanges in the information age. Both examples help us to see that exchange is not necessarily or intrinsically problematic (and is inevitable, in any case). Moreover, they might allow us to develop strategies that turn intellectual property rights to more expansive ends. Douglas and Isherwood's suggestion that we understand consumerism as an information system, in which goods are used as ways to include and exclude people from certain sorts of social participation and solidarity, is particularly helpful, and the example of PXE, International suggests how one which create more inclusive realms of exchange. Rather than naively opposing the integration of body parts and wastes into systems of value, we ought instead to investigate how the logic of value can be used to encompass a greater number of people in the goods that arise from the exchange of body parts and wastes.

Notes

1. For on overview of different forms of commerce in bodily wastes, see Lori Andrews and Dorothy Nelkin, *Body Bazaar: The Market for Human Tissue in the Biotechnology Age* (New York: Crown Publishers, 2001), 2–3.
2. For a discussion of arguments against the commercialization of blood, see Douglass A. Starr, *Blood: An Epic History of Medicine and Commerce* (New York: Quill, 2000), 186–206; for an overview of the objections to organ commercialization, see Pranlal Manga, "A Commercial Market for Organs? Why not" *Biothetics* 1:4 (1987), 321–38. For a discussion of arguments against commodification of body parts in general, see

Paul Rabinow, *French DNA: Trouble in Purgatory* (Chicago: University of Chicago Press, 1999). My analysis in this paper is indebted to Rabinow's book, as well as to his shorter and earlier article, "Severing the Ties: Fragmentation and Dignity in Late Modernity," in *Essays on the Anthropology of Reason* (Princeton: Princeton University Press, 1996), 129–52.

3. Starr outlines these concerns with respect to attempts to classify blood as a commercial product in the United States in the 1950s; see Starr, *Blood*, 193–206.
4. Mangal also notes that "[i]mplicit in much of the literature [on the sale of organs] is the idea of a thorough-going private market as the only form of commercialization, in which the donor 'sells' an organ for a price to a recipient or an intermediary and the organ recipient (patient) 'pays' a direct price to the donor or intermediary," in Mangal, "A Commercial Market," 323–4.
5. *Sciences de la Vie: De l'ethique au droit* (1988), cited in Rabinow, *French DNA*, 98.
6. Immanuel Kant, *Grounding for a Metaphysics of Morals*, trans. James W. Ellington (Indianapolis: Hackett Publishing Company, 1981), 40.
7. See Rabinow, *French DNA*, 103.
8. Stephen R. Munzer, "An Uneasy Case against Property Rights in Body Parts," *Social Philosophy and Policy* 11:2 (1994): 266. For a useful discussion and contextualization of Munzer's argument, see "The Ethics of the Organ Market: Llyod R. Cohen and the Free Marketeers," in *Biotechnology and Culture: Bodies, Anxieties, Ethics*, ed. Paul E. Brodwin (Indianapolis: Indiana University Press, 2000), 224–37.
9. The biomedical literature on gift exchange is vast and continuing to grow. For the classic account of blood donation as a mode of gift giving, see Richard Titmuss's famous comparison between French and Anglo/American blood banking practices, *The Gift Relationship: From Human Blood to Social Policy* (New York: Pantheon Books, 1971). In *Spare Parts: Organ Replacement in American Society* (New York: Oxford University Press, 1992), Renée C. Fox and Judith P. Swazey describe organ transplantation as a form of gift giving, while at the same time criticizing the expansion of organ donation practices. In *The Human Body Shop: The Cloning, Engineering, and Marketing of Life, Second Edition* (Washington, D.C.: Gateway, 1997), Andrew Kimbrell ends with a plea to give up all forms of commercial body part exchange and "return" to a gift system of exchange. For a more general discussion that situates organ donation against a larger field of community-bonding gift practices, see Lewis Hyde, *The Gift: Imagination and the Erotic Life of Property* (New York: Vintage Books, 1983).
10. For a useful discussion of Locke's notion of property, see C. P. MacPherson, *The Political Theory of Possessive Individualism* (Oxford, UK: Clarendon Press, 1962). For a contextualization of Locke's understanding against subsequent theories of property, see *Property: Mainstream and Critical Positions*, ed. C. P. MacPherson (Toronto: University of Toronto Press, 1978). It is important to note, however, that *legal* definitions of property, at least in the United States, generally do not make explicit appeals to this Lockean doctrine. Moreover, in 1914, in order to prevent the possibility that labor unions would be brought to court under antitrust laws for conspiring to control access to a product (i.e., labor), the U.S. Congress declared that "the labor of a human being is not a commodity or article of commerce" (U.S. Code October 15, 1914, ch. 223, sec. 6, 38 Stat. 731).
11. In the United States, state policies concerning organ and nonorgan body part exchange are an absolutely bewildering hodgepodge of state-directed regulations. Many states distinguish between organs (which may not be sold) and nonorgans (which may be sold). For a useful state-by-state overview of regulations of human egg sales, see Kenneth Baum, "Golden Eggs: Towards the Rational Regulation of Oocyte Donation," *Brigham Young University Law Review* 107 (2001): 127, note 54. Rabinow notes that in France, the *source* of body parts is vital in determining its movement into, or exclusion from, the realm of commodity exchange. So, for example, "[a]lbumin changes legal status in France depending on whether it is taken from blood or from the placenta. In French law, the placenta is *un déchet* [a waste product]. Consequently, it can enter into the commercial law, and it does," in Rabinow, *French DNA*, 86. Albumin from blood, however, cannot be sold in France.
12. Munzer provides a slightly more complicated version of this argument, suggesting that the acceptability of body part sales is a function of risk: the greater the risk to the health of an individual, the more the sale of that part threatens the dignity of the individual.

Removing and selling body parts such as hair and nails pose little if any health risk, and, therefore, their sale has little impact on human dignity. See Munzer, "An Uneasy Case."

13. My discussion of Moore's case is heavily indebted to James Boyle's extremely useful analysis in *Shamans, Software, and Spleens: Law and the Construction of the Information Society* (Cambridge, MA: Harvard University Press, 1996), 21–4, 97–107 and E. Richard Gold's discussion in *Body Parts: Property Rights and the Ownership of Human Biological Materials* (Washington, D.C.: Georgetown University Press, 1996), 23–40. Also very useful are Rabinow, "Severing the Ties" and John Frow, *Time and Commodity Culture: Essays in Cultural Theory and Postmodernity* (Oxford: Clarendon Press, 1997), 152–79. For a more comprehensive list of the extensive secondary material on this case, see the sources cited in Gold, *Body Parts*, 181–2 note 15, as well as the more updated list in the notes to Keith Sealing, "Great Property Cases: Teaching Fundamental Learning Techniques with Moore v. Regents of the University of California," *Saint Louis University Law Journal* 46 (Summer 2002): 755–74.
14. *Moore v. Regents of the University of California*, 51 Cal. 3d at 143. This statement was made by Justice Arabian, who concurred with the majority, but wrote his own opinion. For a nuanced discussion of the four different opinions in this case (the majority opinion, one concurring, and two dissenting), see Gold, *Body Parts*, 26–38.
15. *Venner v. State*, 354 A.2d at 483, 498 (Md. App. 1976).
16. *Moore v. Regents of the University of California*, 51 Cal. 3d at 140.
17. It would be interesting to see how a critic of body part commerce such as Munzer (who positions the acceptability of commerce as a function of the risk necessary to remove the part) would deal with the status of Moore's tissue. In his case, *not* to remove the diseased tissue would have resulted in a higher risk than to keep it in his body.
18. *Moore v. Regents of the University of California*, 51 Cal. 3d at 144–5.
19. Whether or not this is true is still an open question. In *Body Bazaar*, Andrews and Nelkin point to several studies that suggest that proprietary claims in tissue types, research procedures, and genetic information in fact have cut down considerably on the flow of information between institutions and laboratories. See Andrews and Nelkin, *Body Bazaar*, 5, 55.
20. Boyle, *Shamans, Software, and Spleens*, 24.
21. Rabinow, *French DNA*, 71. Mangal also makes this point, noting that "[v]irtually every proponent of the market approach [to organ transfers] points out that medicine is so suffused with commercialism that it is contradictory or inconsistent not to permit at least the selling if not the buying of organs" (Mangal, "A Commercial Market," 330). He also notes the substantial international trade that exists in body parts destined for academic purposes.
22. See MacPherson, *Property*, 2. Boyle notes that the court in Moore's case seems to have forgotten this fact insofar as it suggested that extensive state or federal regulation of a substance precluded property claims. See Moore, *Shamans, Software, and Spleens*, 23.
23. See Andrews and Nelkin, *Body Bazaar*, 179, who note that "[t]he property approach was successful in *York v. Jones*, 717 F. Supp. 421 (E. D. VA. 1989), as a way of giving a couple control over decisions about their in vitro embryo" (237 note 28). See also Lori B. Andrews, "My Body, My Property," *The Hastings Center Report* 16:5 (October 1986). At the same time, however, Nelkins and Andrews note that commodification is not necessarily an unqualified good for individuals, for if "donors" are transformed into "sellers," they might be subjected to prosecution under liability laws, or body parts might be considered taxable parts of one's estate (*Body Bazaar*, 165–8).
24. Rosalind Pollack Petchesky, "The Body as Property: A Feminist Revision," in *Conceiving the New World Order*, eds. F. D. Ginsburg and R. Rapp (Berkeley: University of California Press, 1995), 387–406.
25. See Fox and Swazey, *Spare Parts*, 73.
26. Kimbrell, *The Human Body Shop*, viii.
27. Vandana Shiva, *Biopiracy: The Plunder of Nature and Knowledge* (Boston, Massachusetts: South End Press, 1997), 10.
28. See, for example, "The Bellagio Declaration," a document advocating the use of the rhetoric of the commons in resisting expansion of intellectual property claims; included as Appendix B in Boyle, *Shamans, Software, and Spleens*, 192–200.

29. See Mary Douglas and Baron Isherwood, *The World of Goods: Towards an Anthropology of Consumption* (New York: Norton, 1979); *Consumption and the World of Goods*, eds. John Brewer and Roy Porter (New York: Routledge, 1993); Daniel Miller, *Acknowledging Consumption: A Review of New Studies* (New York: Routledge, 1995).
30. Arjun Appadurai, "Introduction: Commodities and the politics of value," in *The Social Life of Things: Commodities in Cultural Perspective*, ed. Arjun Appadurai (Cambridge, UK: Cambridge University Press, 1986), 13.
31. In "An All-Consuming Experience: Obstetrical Ultrasound and the Commodification of Pregnancy," (in Brodwin, *Biotechnology and Culture*, 147–70), Janelle S. Taylor develops a fascinating discussion of the relationship between commodification and the construction of fetuses as "persons" in the experience of pregnancy. Although her approach differs slightly from mine (for Taylor, commodification still seems to be positioned as intrinsically problematic), the essay nevertheless provides very useful material for a reconsideration of body parts and commodification.
32. Douglas and Isherwood, *The World of Goods*, xiv, 49.
33. See Marcel Mauss, *The Gift: The Form and Reason for Exchange in Archaic Societies*, edited by W. D. Halls (New York: W. W. Norton, 1990), 1–18.
34. Fox and Swazey, *Spare Parts*, 38–42. In the section entitled "Obligations to Repay the 'Gift of Life' and the 'Tyranny of the Gift'," Fox and Swazey attempt to deal with the recipients felt need to reciprocate the gift, but seem to conclude that recepients find themselves incapable of doing so.
35. The nonnegligible likelihood of reciprocal blood donation (a recipient becomes a donor) allows Titmuss, in his discussion of blood gift giving, to see the presence of exchange more fully than Fox and Swazey. See Titmuss, *The Gift Relationship*.
36. Throughout his vast *oeuvre*, Bourdieu's fundamental principle has been that "practice never ceases to conform to economic calculation even it gives every appearance of disinterestedness by departing from the logic of interested calculation (in the narrow sense) and playing for stakes that are non-material and not easily quantified"; cited in Pierre Bourdieu, *Outline of a Theory of Practice*, trans. Richard Nice (Cambridge, UK: Cambridge University Press, 1977), 171. For Bourdieu, all forms of capital (economic, symbolic, cultural, and the like) potentially can be converted into one another, depending on the rules of a particular social formation.
37. Gold, *Body Parts*, 43.
38. For more information about this corporation and its activities, see their website at http://www.pxe.org/. My claims are based primarily on personal communications with the Terrys, but all of these points are substantiated in the Terrys' presentations about PXE, International archived at the Duke University Center for Genome, Ethics, Law and Policy Fall 2002 Symposium, "Commercialization of Human Genomics: Consequences fro Science and Humanity," at http://www.law.duke.edu/conference/gelp/program.html.

CHAPTER 7

The Virtual Surgeon: New Practices for an Age of Medialization

TIMOTHY LENOIR

Surgery is experiencing rapid change in its culture, institutions, and material practices due to what we might best characterize as an accelerating technological revolution in medialization. Media inscribe our situation. We are becoming immersed in a growing repertoire of computer-based media for creating, distributing, and interacting with digitized versions of the world, media that constitute the instrumentarium of a new epistemic regime. In numerous areas of our daily activities, we are witnessing a drive toward fusion of digital and physical reality; not the replacement of the real by a hyperreal, the obliteration of a referent and its replacement by a model without origin or reality as Baudillard predicted, but a new playing field of ubiquitous computing in which wearable computers, independent computational agent-artifacts, and material objects are all part of the landscape. To paraphrase William Gibson's character Case in *Neuromancer*, "data is being made flesh."

Surgery provides a dramatic example of a field newly saturated with information technologies. In the past decade computers have entered the operating room to assist physicians in realizing a dream they have pursued ever since Claude Bernard: to make medicine both experimental and predictive. The emerging field of computer-assisted surgery offers a dramatic change from the days of individual heroic surgeons. Soon surgeons will no longer boldly improvise modestly preplanned scripts, adjusting them in the operating room to fit the peculiar case at hand. Increasingly, to perform surgery surgeons must use extensive 3D modeling tools to generate a predictive model, the basis for a simulation that will become a software surgical interface. This interface will guide the surgeon in performing the procedure.

I. The Minimally Invasive Surgery Revolution

These developments in surgery date back to the 1970s when the first widely successful endoscopic devices appeared. First among these were arthroscopes for orthopedic surgery, available in most large hospitals by 1975, but at that point more a gimmick than a mainstream procedure. Safe surgical procedures with such scopes were limited because the surgeon had to operate while holding the scope in one hand and a single instrument in the other.

What changed the image of endoscopy in the mind of the surgical community and turned arthroscopy, cholecystectomy—removal of the gallbladder with instruments inserted through the abdominal wall—and numerous other endoscopic surgical techniques into common operative procedures? The introduction of the small medical video camera attachable to the eyepiece of the arthroscope or laparascope was a first major step. French surgeons were the first to develop small, sterilizable high-resolution video cameras that could be attached to a laparoscopic device. With the further addition of halogen high-intensity light sources with fiberoptic connections, surgeons were able to obtain bright, magnified images that could be viewed by all members of the surgical team on a video monitor rather than just by the surgeon alone. This technical development had consequences for the culture of surgery; it contributed to greater cooperative teamwork and opened the possibility for surgical procedures of increasing complexity, including suturing and surgical reconstruction done only with videoendoscopic vision.[1] French surgeons performed the first laparoscopic cholecystectomy in 1989. A burgeoning biomedical devices industry sprang up almost immediately to begin providing the necessary ancillary technology to make laparoscopic procedures practical in your local hospital, such as new, specialized instruments for tissue handling, cutting, hemostasis, and many more.

Due to their benefits of small scars, less pain, and a more rapid recovery, endoscopic procedures were rapidly adopted after the late 1980s and became a standard method for nearly every area of surgery in the 1990s. Demand from patients has had much to do with the rapid evolution of the technology. Equally important have been the efforts of health care organizations to control costs. In a period of deep concern about skyrocketing healthcare costs, any procedure that improved surgical outcomes and reduced hospital stays interested medical instrument makers. Encouraged by the success of the new videoendoscopic devices, medical instrument companies in the early 1990s foresaw a new field of minimally invasive diagnostic and surgical tools. Surgery was about to enter a technology-intense era that offered immense opportunities to companies teaming surgeons and engineers to apply the latest developments in robotics, imaging, and sensing to the field of minimally invasive surgery. Although path-breaking developments had occurred, the instruments available for such surgeries allowed only a limited number of the complex functions demanded by the surgeon. Surgeons needed better visualization, finer manipulators, and new types of remote sensors, and they needed these tools integrated into a complete system.

II. Telepresence Surgery

A new vision emerged, heavily nurtured by funds from the Advanced Research Projects Agency (ARPA), the NIH, and NASA, and developed through contracts made by these agencies to laboratories such as Stanford Research Institute (SRI), Johns Hopkins Institute for Information Enhanced Medicine, University of North Carolina Computer Science Department, the University of Washington Human Interface Technology Laboratory, the Mayo Clinic, and the MIT Artificial Intelligence Laboratory. The vision promoted by Dr. Richard Satava, who spearheaded the ARPA program, was to develop "telepresence" workstations that would allow surgeons to perform telerobotically complex surgical procedures that demand great dexterity. These workstations would recreate and magnify all of the motor, visual, and sensory sensations of the surgeon as if he were actually inside the patient. The aim of the programs sponsored by these agencies was eventually to enable surgeons to perform surgeries, such as certain complex brain surgeries or heart operations not even possible in the early 1990s, improve the speed and surety of existing procedures, and reduce the number of people in the surgical team. Central to this program was telepresence-telerobotics, allowing an operator the complex sensory feedback and motor control he would have if he were actually at the work site, carrying out the operation with his own hands. The goal of telepresence was to project full motor and sensory capabilities—visual, tactile, force, auditory—into even microscopic environments to perform operations that demand fine dexterity and hand–eye coordination.

Philip Green led a team at SRI that assembled the first working model of a telepresence surgery system in 1991, and with funding from the NIH Green went on to design and build a demonstration system. The proposal contained a diagram showing the concept of workstation, viewing arrangement, and manipulation configuration used in the surgical telepresence systems today. In 1992, SRI obtained funding for a second-generation telepresence system for emergency surgeries in battlefield situations. For this second-generation system, the SRI team developed the precise servomechanics, force-feedback, 3D visualization, and surgical instruments needed to build a computer-driven system that could accurately reproduce a surgeon's hand motions with remote surgical instruments having five degrees of freedom and extremely sensitive tactile response (see Figure 7.1).

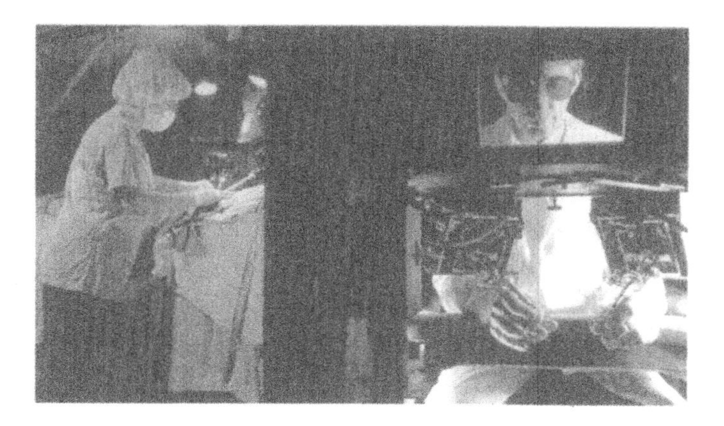

Fig. 7.1 Philip Green, force-reflecting surgical manipulator, *Time* Magazine, Special Issue, Fall 1996.

Fig. 7.2 Intuitive Surgical DaVinci Computer Assisted Robotic Unit, from Intuitive Surgical promotional material, Intuitive Surgical, Palo Alto, CA, 1999.

In late 1995, SRI licensed this technology to Intuitive Surgical, Inc. of Mountain View, CA. Intuitive Surgical furthered the work begun at SRI by improving on the precise control of the surgical instruments, adding a new invention, EndoWrist^{TM}, patented by company cofounder Frederic Moll, which added two degrees of freedom to the SRI device—inner pitch and inner yaw (inner pitch is the motion a wrist performs to knock on a door; inner yaw is the side-to-side movement used in wiping a table)—allowing the system to better mimic a surgeon's actions; it gives the robot ability to reach around, beyond and behind delicate body structures, delivering these angles right at the surgical site. Through licenses of IBM patents, Intuitive also improved the 3D video imaging, navigation, and registration of the video image to the spatial frame in which the robot operates. The system employs 250 megaflops of parallel processing power (see Figure 7.2).

A further crucial improvement to the system was brought by Kenneth Salisbury from the MIT Artificial Intelligence Laboratory who imported ideas from the force-reflecting haptic feedback system he and Thomas Massie invented as the basis of their PHANTOM™ system, a device invented in 1993 permitting touch interactions between human users and remote virtual and physical environments. The PHANTOM™ is a desktop device which provides a force-reflecting interface between a human user and a computer. Users connect to the mechanism by simply inserting their index finger into a thimble. The PHANTOM™ tracks the motion of the user's finger tip and can actively exert an external force on the finger, creating compelling illusions of interaction with solid physical objects. A stylus can be substituted for the thimble and users can feel the tip of the stylus touch virtual surfaces. The haptic interface allows the system to go beyond previous instruments for minimally invasive surgery (MIS). These earlier instruments precluded a sense of touch or feeling for the surgeon; the PHANTOM™ haptic interface, by contrast, gives an additional element of immersion. When the arm encounters resistance inside the patient, that resistance is transmitted back to the console, where the surgeon can feel it. When the thimble hits a position corresponding to the surface of a virtual object in the computer, three motors generate forces on the thimble that imitate the feel of the object. The PHANTOM™ can duplicate all sorts of textures, including coarse, slippery, spongy, or even sticky surfaces. It also reproduces friction. And if two PHANTOM™ s are put together a user can "grab" a virtual object with thumb and forefinger. Given advanced haptic and visual feedback, the system greatly facilitates dissecting, cutting, suturing and other surgical procedures, even those on very small structures, by giving the doctor inches to move in order to cut millimeters. Furthermore, it can be programmed to compensate for error and natural hand tremors that would otherwise negatively affect MIS technique.

The surgical manipulator made its first public debut in actual surgery in May of 1998. From May through December 1998 Professor Alain Carpentier and Dr. Didier Loulmet of the Broussais Hospital in Paris performed six open-heart surgeries using the Intuitive™ system. In June of 1998, the same team performed the world's first closed-chest video-endoscopic coronary bypass surgery completely through small (1-cm) ports in the chest wall. Since that time, more than 250 heart surgeries and 150 completely video-endoscopic surgeries have been performed with the system. The system was given approval to be sold throughout the European Community in January 1999.

III. Computer Modeling and Predictive Medicine

A development of equal importance to the contribution of computers in the MIS revolution has been the application of computer modeling, simulation, and virtual reality to surgery. The development of various modes of digital imaging in the 1970s, such as CT (which was especially useful for bone), MRI (useful for soft tissue), ultrasound, and later PET scanning, have made it possible to do precise quantitative modeling and preoperative planning of many types of surgery. Because these modalities, particularly CT and MRI, produce 2D "slices" through the patient, the natural next step (taken by Gabor Herman and his associates in 1977) was to stack these slices in a computer program to produce a 3D visualization. Three-dimensional modeling first developed in craniofacial surgery because it focused on bone, and CT scanning was more highly evolved. Another reason was that in contrast to many areas of surgery where a series of 2D slices—the outline of a tumor for example—give the surgeon all the information he needs, in craniofacial surgery the surgeon must focus on the skull in its entirety rather than on small sections at a time.

Jeffrey March and Michael Vannier at Washington University in St. Louis, Missouri, pioneered in the application of 3D computer imaging to craniofacial surgery in 1983. Prior to their work, surgical procedures were planned with tracings made on paper from 2D radiographs. Frontal and lateral radiographs were taken and the silhouette lines of bony skull edges were traced onto paper. Cutouts were then made of the desired bone fragments and manipulated. The clinician would move the bone fragment cutout in the paper simulation until the overall structure approximated normal. Measurements would be taken and compared to an ideal, and another cycle of cut-and-try would be carried out. These hand-done optimization procedures would be repeated until a surgical plan was derived that promised to yield the most normal-looking face for the patient.

Between 1983 and 1986 March, Vannier and their colleagues computerized each step of this 2D optimization cycle. The 3D visualizations overcame some of the deficiencies in the older 2D process. Two-dimensional planning, for instance, is of little use in attempting to consider the result of rotations. Cutouts planned in one view are no longer correct when rotated to another view. Volume rendering of 2D slices in the computer overcame this problem. Moreover, comparison of the 3D preoperative and postoperative visualization often suggested an improved surgical design in retrospect. A frequent problem in craniofacial surgery is the necessity of having to perform further surgeries to get the final optimal result. For instance, placement of bone grafts in gaps leads to varying degrees of resorption. Similarly a section of the patient's facial bones may not grow after the operation, or attachment of soft tissues to bone fragments may constrain the fragment's movement. These and other problems suggested the value of a surgical simulator that would assemble a 3D interactive model of the patient from imaging data, provide the surgeon with tools similar to engineering computer-aided design tools for manipulating objects, and allow him to compare "before" and "after" views to generate an optimal surgical plan. In 1986 March and Vannier developed the first simulator by applying commercial CAD software to provide an automated optimization of bone fragment position to "best fit" normal form. Since then, customized programs designed specifically for craniofacial surgery have made it possible to construct multiple preoperative surgical plans for correcting a particular problem, allowing the surgeon to make the optimal choice.

These early models were further extended in an attempt to make them reflect not only the geometry but also the physical properties of bone and tissues, thus rendering them truly quantitative and predictive. R. M. Koch, M. H. Gross, and colleagues from the ETH Zurich, for example, applied physics-based finite element modeling to facial reconstructive surgery. Going beyond a "best fit" geometrical modeling among facial bones, their approach is to construct triangular prism elements consisting of a facial layer and five layers of epidermis, dermis, subcutaneous connective tissue, fascia, and muscles, each connected to one another by springs of various stiffness. The stiffness parameters for the soft tissues are assigned on the basis of segmentation of CT scan data. In this model, each prism-shaped volume element has its own physics. All interactive procedures such as bone and soft tissue repositioning are performed under the guidance of the modeling system, which feeds the processed geometry into the finite element model program. The resulting shape is generated from minimizing the global energy of the surface under the presence of external forces. The result is the ability to generate highly realistic 3D images of the postsurgical shape. Computationally based surgery analogous to the craniofacial surgery described above has been introduced in eye surgeries, in prostate, orthopedic, lung, and liver surgeries; and in repair of cerebral aneurysms.

Equally impressive applications of computational modeling have been introduced into cardiovascular surgery. In this field, simulation techniques have gone beyond modeling structure to simulating function, such as blood flow in the individual patient who needs, for example, a coronary bypass surgery. Charles A. Taylor and colleagues at the Stanford Medical Center have demonstrated a system that creates a patient-specific, three-dimensional finite element model of the patient's vasculature and blood flow under a variety of conditions. A software simulation system using equations governing blood flow in arteries then provides a set of tools that allows the physician to predict the outcome of alternate treatment plans on vascular hemodynamics. With such systems predictive medicine has arrived.

IV. Medical Avatars: Surgery as Interface Problem

Such examples demonstrate that computational modeling has added an entirely new dimension to surgery. For the first time the surgeon is able to plan and simulate a surgery based on a mathematical model that reflects the actual anatomy and physiology of the individual patient. Moreover, the model need not stay outside the operating room. Several groups of researchers have used these models to develop "augmented reality" systems that produce a precise, scaleable registration of the model on the patient so that a fusion of the model and the 3D stereo camera images is made. This procedure has been carried out successfully in removing brain tumors and in a number of prostatectomies in the Mayo Clinic's Virtual Reality Assisted Surgery Program (VRASP), headed by Richard Robb.

In addition to improving the performance of surgeons by putting predictive modeling and mathematically precise planning at their disposal, computers are playing a major role in improving surgical outcomes by providing surgeons opportunities to train and rehearse important procedures before they go into the operating theater. By 1995, modeling and planning systems began to be implemented in both surgical training simulators and in real-time surgeries. One of the first systems to incorporate all these features in a surgical simulator was developed for eye surgery by MIT robotics scientist Ian Hunter. Hunter's microsurgical robot (MSR) system incorporated features described above such as data acquisition by CT and MRI scanning, use of finite element modeling of the planned surgical procedure, a force-reflecting haptic feedback system that enables the perception of tissue cutting forces, including those that would normally be imperceptible to the surgeon if they were transmitted directly to his hands.

A distinctive feature of Hunter's MSR was its immersive virtual environment that fused video, touch, and even sound into a virtual reality experience (see Figure 7.3). The haptic environment in Hunter's system was fused with 3D stereo camera images fed to a head-mounted display. As if in a flight simulator, the surgeon could rehearse his procedure on the model of the individual patient he had constructed. In addition, the model could be used as a training site for student surgeons, copresent during a practice surgery, sharing the same video screen

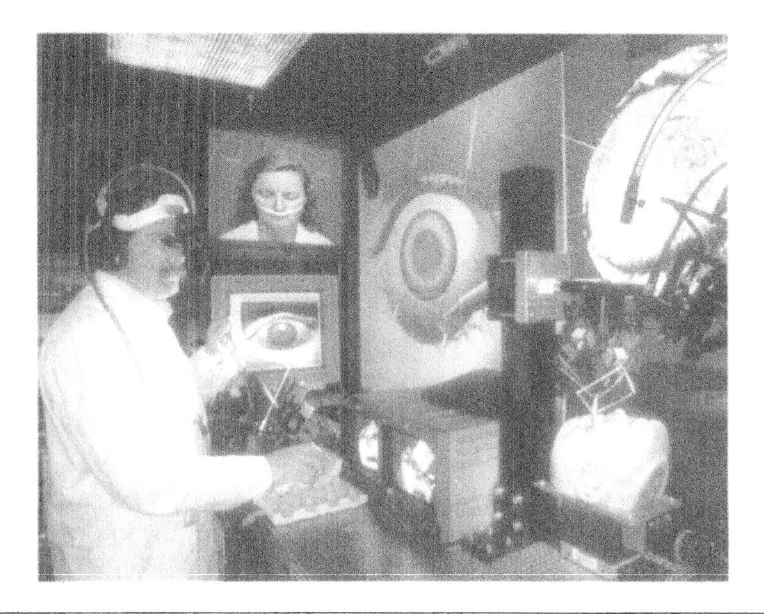

Fig. 7.3 Ian Hunter's microsurgical robot, *Presence: Teleoperators and Virtual Environments*, vol. 2, 1993.

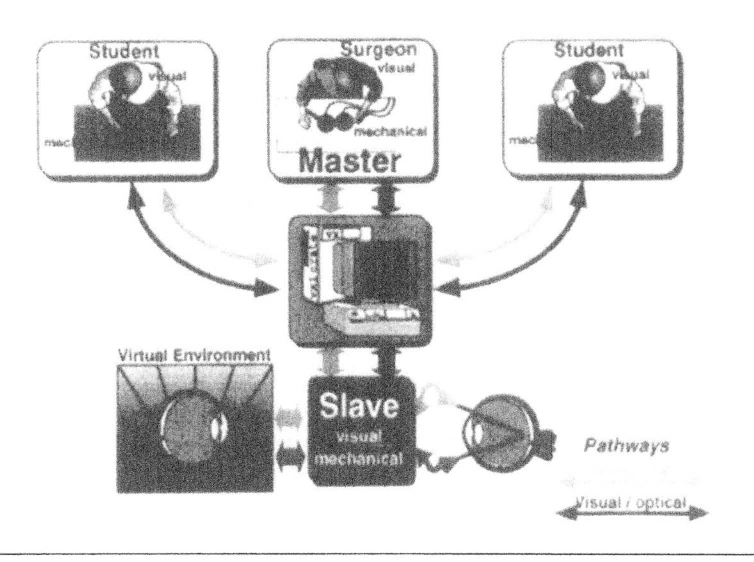

Fig. 7.4 Ian Hunter et. al, *Presence*, vol. 2, 1993, showing "fade-in" of student surgeons.

and feeling the same surgical moves as the master surgeon (see Figure 7.4). But such systems can also be deployed in a collaborative telesurgery system, allowing different specialists to be faded in to "take the controls" during different parts of the procedure. Indeed, a "collaborative clinic" incorporating these features was demonstrated at NASA-Ames on May 5, 1999 with participants at five different sites around the US.

V. Discourse Networks 2000: Regimes of the Robot

The microsurgical systems I have sketched above are by no means wild fantasies of technoenthusiast surgeons. After little more than a decade of serious development, many of these systems are already in use in select areas in Europe and several have been approved for clinical trials in the United States. To be sure, these developments are by no means a large movement in contemporary medicine; they constitute a fraction of funds spent on medical development. Nevertheless, it is intriguing to ponder the conditions that would lead them to be implemented more widely and the consequences entailed for both patients and surgeons were these technologies to become widely adopted. Let us begin by considering the arguments of proponents of the systems and the economic and political pressures that support their efforts.

Proponents of these new systems advance arguments based on claims for cost-saving measures the new technologies permit as a result of less invasive procedures and improved recovery chances of patients due to limitation of blood loss during surgeries that are more accurately planned and more precisely executed. Proponents also point to more efficient use of costly facilities through telepresence and the improvement of training regimes for surgeons. Such arguments question our tolerance for high error rates in surgeries (greater than 10 percent in some areas) whereas in other areas of risk, such as pilot training for commercial airlines, we would find even a 2 percent error rate intolerable. In the case of pilot error, one reason for the low incidence of error is arguably the availability of high-quality simulation technology for training.

A salient feature of contemporary health care is its attention to designing health care plans, diagnoses, and therapies targeted for the individual patient. This coincides with the demand for greater involvement by individuals in decisions related to their own health. The new surgical techniques map onto these concerns for individually tailored therapies. As I have suggested, the new modeling and simulation tools enable the design of procedures based on actual patient data rather than on generic experience with a condition—procedure "x" is what you do in situation "y." Dynamic simulation and modeling tools enable surgeons to construct alternative surgical plans based on actual anatomic and physiological data projected to specific outcomes in terms of lifestyle and patient expectations. Proponents argue that the new surgical tools take the guesswork out of choosing a procedure specific to the case at hand. Such outcomes not only increase patient satisfaction but also reduce costly repetition of procedures that were not optimized on the first pass.

The downside of this greater precision for the patient, of course, is increased surveillance. It is strangely ironic that while the new technology brings the capability to design therapies—including drugs—specifically targeted for the individual, and hence freeing the individual from infirmity and disease in a way never before imagined, it does so most efficiently and cost effectively by instituting a massive system of preventive health care from genome to lifestyle. In the age of medialization, your lifestyle is medicalized.

It is not difficult to see how the surgical systems explored here would mesh with such a system. The systems I have discussed deploy anatomical overlays and patient-related data as aids to the surgical procedure, but other layers of augmentation can be foreseen. Analogous to the insertion of material constraints, cost factors, and building code regulations in current CAD–CAM design tools, surgical simulators could be augmented with the list of allowable procedures the patient's HMO authorizes, and within this list various treatment packages could be prescribed according to benefit plan. Currently in a number of states, hospitals and managed care facilities that receive reimbursement from Medicaid dollars are required to treat patients with a prioritized list of diagnoses and procedures, ranked according to criteria such as life expectancy, quality of life, cost effectiveness of a treatment, and the scope of its benefits. The Oregon Health Plan, which first implemented this system, ranked 700 diagnoses and treatments in order of importance. Items below line 587 are disallowed.[2] Currently in facilities such as emergency rooms a staff supervisor examines the treatment prescribed by staff physicians, and physician-decisions to ignore the guidelines require the prescribing physician to produce a formal written justification. Physicians are reluctant to confront this additional layer of bureaucracy, particularly since the financial risks incurred by denial of Medicaid funding can be a potential source of friction with the management of the HMO employing them. In the future, the appropriate constraints and efficiency measures could be preprogrammed into the surgical treatment-planning simulator.

The new computer-intensive, highly networked surgical systems I have explored also carry consequences for the discipline of surgery and for the agent we call "surgeon." In the age of heroic medicine, the days before the advent of the corporate health care system, surgeons were celebrated as among the most autonomous of professional agents. Society granted these demigods of the surgical wards great status and autonomy in exchange for their ability to bring massive amounts of scientific and medical knowledge to bear in a heartbeat of surgical practice.[3] These guys had the proverbial "right stuff," agency par excellence. But in the telerobotics systems examined here, the surgeon function dissolves into the ever more computationally mediated technologies of apperception, diagnosis, decision, gesture, and speech. The once autonomous surgeon-agent is being displaced by a collection of software agents embedded in megabits of computer code. How is this possible?

Consider the surgeon planning an arterial stent graft before the advent of real-time volume rendering. He used a medical atlas—or perhaps more recently a 3D medical viewer—in combination with echocardiograms, CAT scans, and MRI images of his patient. At best, the surgeon dealt with a stack of 2D representations, slices separated by several millimeters. These were mentally integrated in the surgeon's imagination and compared with the anatomy of the standard human. Through this complex process of internalization, reasoning and imagining, the surgeon "saw" structures he would expect to be seeing as he performed the actual surgery, a quasi-virtual surgical template in his imagination. The surgeon worked as the head of a team in the operating room with anesthesiologists and several surgical assistants, but the surgeon mentally planned and executed the surgery alone. The position of the surgeon as an autonomous center of agency and responsibility was crucial to this system.

In the new surgical paradigm, the surgeon first begins with the patient dataset of MRIs, CAT scans, and other physiological data. The surgeon enters that data into a surgical model using a variety of software and data management tools to construct a simulation of the surgery to be performed. An entire suite of software tools enables the construction of such a simulation depending on the type of procedure to be performed. The Virtual Workbench, Cyberscalpel, and various systems for interfacing anatomical and physiological data with finite element modeling tools are all elements of this new repertoire of tools for preparing a surgery. A surgical plan is constructed listing the navigational coordinates, step-by-step procedures, and specific patient data important to keep in mind at critical points. The simulation is, in fact, an interactive hypermedia document.

Voxel-Man provides is a particularly clear illustration of this hypertextualization of the surgical body. The key idea underlying the approach is to combine in one single framework a computer-generated spatial model to which is attached a complete atlas with textual description of every detail necessary for every volume element in all the anatomical structures along the path of a surgery. These constituents differ for the different domains of knowledge such as structural and functional anatomy. The same voxel (volume pixel element) may belong to different voxel sets with respect to the particular domain. The membership is characterized by object labels that are stored in "attribute volumes" congruent to the image volume, including features like vulnerability or mechanical properties, which might be important for the surgical simulation. Patient-specific data for that particular region, such as the specific frames of MRI or CAT data used to construct the simulation, can also be included.

Such intelligent volumes are not only for preparing the surgery, or later for teaching and review. Built into the patient-specific surgical plan, the hypertext atlas assumes the role of surgical companion in an "augmented reality" system. In Hunter's surgical manipulator, for example, various pieces of information—patient-specific data, such as MRI records, or particular annotations the surgical team had made in preparing the plan—appear in the margins of the visual simulation indicating particular aspects of the procedure to be performed at the given stage of the surgery. The surgical and the procedures they design are thus inscribed in a vast hypertext narrative of spatialized scripts to be activated as the procedure unfolds.

Well before we enter the operating room of the future, it is clear that the surgeons are going to be significantly reconfigured in terms of skills and background. Two processes are driving that reconfiguraton: medialization and postmodern distributed production. Key to medialization is the externalization of formerly internal mental processes, the literalization of skill in an inscription device. Consider the virtual surgeon. Will that new techno supersurgeon be an upgrade on last generation's heroic surgeon? Such a surgeon would undoubtedly have background knowledge in the texts and practices of anatomy, biochemistry, physiology, and pathology, including some traditional practices from earlier generations. But they will require familiarity, if not hands-on experience in new fields such as biophysics, computer graphics and animation, biorobotics, mechanical, and biomedical engineering. They will also need to be aware of the importance of network services and bandwidth issues as enabling components of their practice. Obviously, it is unrealistic to assume that last generation's heroic surgeon is going to come packaged with all these features, anymore than next year's undergraduates are going to show up to math class with slide rulers. If we have learned anything about postmodern distributed production, it is to expect flat organizational structures, distributed teamwork, and modularization. Thus, given the complexity of all these fields, surgical systems will likely come packaged as turnkey systems. Many surgeons will be operators of these systems, performing "routine" cardiac bypass surgeries implementing predesigned surgical plans from a library of stored simulations owned by the company employing them. I am not saying that surgeons will simply become technicians or that surgery will cease to be a highly creative field. What I am saying, however, is that creativity will be of a different sort as many of the functions now internalized by surgeons get externalized into packaged surgical design tools, computer aided design packages such as Autocad, 3D Studio Max, or Maya have reconfigured the training, design practices, and the creativity of architects. Some surgeons with access to resources will undoubtedly engage in high-level surgical design work, but that process will be highly mediated in teamwork involving software engineers, robotics experts, and a host of others.

The radiologist has been crucial to the surgeon's ability to carry off such a complex surgery prior to the age of medialization. Like the surgeon, the radiologist has been a highly valued and relatively autonomous agent. As a key professional in the surgical design process, the radiologist would make X-rays and more recently CAT, MRI, or various other types of scanning modalities appropriate to the diagnosis of a suspected disease. Examining a dozen or so images, or more recently, a hundred or so slices of a CAT or MRI scan, the radiologist would prepare a diagnostic report for the surgeon. Like the similar skill of the physician, the radiologist's diagnosis was heavily dependent on acquiring keen mental skills of observation for detecting artifacts and spotting lesions or other abnormalities that would be the subject of the report. But the relative autonomy of the radiologist and his or her relationship to the diagnostic and surgical design process will certainly change in the near future. As real-time computer generated imaging becomes the norm, software tools for visualization and automated segmentation of tissues will displace the radiologist as interpreter of the data. Indeed, pressures are already mounting in this direction as the manufacturers of imaging systems such as GE, Siemens, and Brücke install systems that rapidly generate over 1,000 images rather than a few dozen slices. Radiologists are currently under siege by an explosion of new data. Given the cardinal rule of data processing, that valuable data should not go unused, the segmentation of this data into tissues, organs, and other anatomical structures, together with the detection of abnormalities is becoming a problem for software automation. As automated tools for handling the explosion of imaging data arrive, the radiologist will undoubtedly reorient his or her professional activity and training to focus on new problems, such as construction of surgical simulations. To do so, radiologists will work closely with computer programmers and software engineers. Needless to say, if radiology as a medical specialty survives, the background, types of knowledge, and training of its practitioners will be radically different.

In the virtual reality (VR) surgical theater we are describing, a much more concrete discursive setting appears, one of stainless steel clamps, CPUs, and video monitors. The imagining surgeon and imagining colleagues brought together by the technologies of telepresence share this networked theater. But the theater is filled also with hyperreal colleagues, constructed by programmers from a constellation of software companies. Conditions of hyperreality are created by the projection of models and data visualizations back onto the patient's body and into the surgeon's computer-mediated vision. But, in our case, the "imaginary" functions are all literal. They are completely externalized in computational algorithms for data segmentation, volume rendering, and graphical presentation.

The products of these imaging modalities differ from their predecessors in important ways. Much less expensive predecessor systems for representing 3D anatomy, such as stereoscopic viewers and wax models dating back to the eighteenth century, also allowed the surgeon to rotate the structure and examine it from many angles. However, these previous systems were all generalized anatomies rather than patient-specific, living anatomies, generated in real time. Moreover, ray-tracing volumetric rendering methods produce physically accurate representations of the internal volumes and not just the surfaces as with meshed surface triangle methods like marching cubes.

VI. The Surgical Body as Hypertext Narrative

The scenario envisioned five to ten years in the future by the National Research Council's Committee on Virtual Reality Research powerfully illustrates the shifting ontology of the virtual surgeon. In a discussion of use of VR in training heart surgeons, VR researchers describe how haptic augmentation can correct the tremors of the hand as it guides a scalpel over a beating heart:

> Jennifer Roberts... is training to become a surgeon and is at her SE (surgical environment) station studying past heart operations. She previously spent many hours familiarizing herself with the structure and function of the heart by working with the virtual heart system she acquired after deciding to return to medical school and to specialize in heart surgery. This system includes a special virtual-heart computer program obtained from the National Medical Library of Physical/Computational Models of Human Body Systems and a special haptic interface that enables her to interact manually with the virtual heart. Special scientific visualization subroutines enable her to see, hear, and feel the heart (and its various component subsystems) from various vantage points and at various scales. Also, the haptic interface, which includes a special suite of surgical tool handles for use in surgical simulation (analogous to the force-feedback controls used in advanced simulations of flying or driving), enables her to practice various types of surgical operations on the heart. As part of this practice, she sometimes deliberately deviates from the recommended surgical procedures in order to observe the effects of such deviations. However, in order to prevent her

medical school tutor (who has access to stored versions of these practice runs on his own SE station) from thinking that these deviations are unintentional (and therefore that she is poor material for surgical training), she always indicates her intention to deviate at the beginning of the surgical run.

Her training also includes studying heart action in real humans by using see-through displays (augmented reality) that enable the viewer to combine normal visual images of the subject with images of the beating heart derived (in real time) from ultrasound scans. Although there are still some minor imperfections in the performance of the subsystem used to align the two types of visual images, the overall system provides the user with what many years ago (in Superman comics) was called X-ray vision. In this portion of her training, Jennifer examines the effect of position, respiration, exercise, and medication on heart action using both the see-through display and the traditional auditory display of heart sounds.

Today, Jennifer is studying recordings of a number of past real heart operations that had been recorded at the Master Surgical Center in Baltimore. In all of these operations, the surgery was performed by means of a surgical teleoperator system. Such systems not only enable remote surgery to be performed, but also increase surgical precision (e.g., elimination of hand tremor) and decrease need for immobilization of the heart during surgery (the surgical telerobot is designed to track the motion of the heart and to move the scalpel along with the heart in such a way that the relative position of the scalpel and the target can be precisely controlled even when the heart is beating).

The human operator of these surgical teleoperator systems generally has access not only to real-time visual images of the heart via the telerobotic cameras employed in the system, but also to augmented-reality information derived from other forms of sensing and overlaid on the real images. Some of these other images, like the ultrasound image mentioned above are derived in real time; others summarize information obtained at previous times and contribute to the surgeon's awareness of the patient's heart history.

All the operations performed with such telerobotic surgery systems are recorded and stored using visual, auditory, and mechanical recording and storage systems. These operations can then be replayed at any time (and the operation felt as well as seen and heard) by any individual such as Jennifer, who has the appropriate replay equipment available. Recordings are generally labeled "master," "ordinary," and "botched," according to the quality of the operation performed. As one might expect, the American Medical Association initially objected to the recording of operations; however, they agreed to it when a system was developed that guaranteed anonymity of the surgeon and the Supreme Court ruled that patients and insurance companies would not have access to the information. This particular evening, Jennifer is examining two master double-bypass operations and one botched triple-bypass operation.

During her training time on the following day, she is going to monitor a heart operation in real time being performed by a surgeon at the Master Surgical Center in Baltimore on a patient in a rural area of Maryland roughly 200 miles away. Although substantial advances have been made in combating problems of transport delay in remote surgery (by means of new supervisory control techniques), very few heart operations are being conducted remotely at ranges over 500 miles.[4]

We find several remarkable features in this scenario, which builds its vision of the future from Hunter's microsurgical robot. The Committee focuses on the utility of the system for teaching purposes. In Hunter's system, multiple participants can be "faded in" and "faded out" so that they actually feel what the surgeon directing the robot feels. But here, a reverse video effect seems to set in: it is difficult to determine who is in control, robot system or human. A human team clearly programs the robot, but the robot enhances perception and actually guides the hand of the surgeon, correcting for errors due to (human-generated) hand tremor. The guiding hand of the microsurgical system "trains" Jennifer's erratic movements.

More than vision is at stake in computer-mediated surgery. Consider a recent development in the field: gesture macros and speech macros. Under the pressures of documenting medical procedures for reimbursement by HMOs, some physicians have sought ways to say more with less. Using speech-recognition systems, the physician needs to utter only certain phrases into a recognizer, from which the system constructs many pages of "written" documentation of the medical performance according to the templates of legal language.[5] These speech macros mediate, via computational logic, the surgeon's self-presentation, and they inscribe the surgeon into the legal and authorial structures of the new collective medical agent.

Notes

1. See J. Perissat, D. Collet, and R. Belliard, "Gallstones: laparoscopic treatment—cholecystectomy, sholecystostomy, and lithotripsy: our own technique, *Surgery Edoscopy*, 4 (1990): 1–5; F. Dubois, Icard, G. Berthelot, and H. Levard, "Coelioscopic cholecsystectomy: preliminary report of 6 cases," *Annals of Surgery*, 211 (1990): 60–2.
2. See Jerome Kassirer, "Managed Care and the Morality of the Marketplace," *The New England Journal of Medicine*, 333:1 (July 6, 1995): 50–2; Thomas Bodenheimer, "The Oregon Health Plan—Lessons for the Nation, Part One," *The New England Journal of Medicine*, 337:9, (August 28, 1997): 651–5; Idem., "The Oregon Health Plan—Lessons for the Nation, Part Two," *The New England Journal of Medicine*, 337:10, (September 4, 1997): 720–3.
3. The classic sources on this point are Eliot Freidson, *The Profession of Medicine* (New York: Dodd, Mead, 1970); Magali Sarfatti Larson, *The Rise of Professionalism* (Berkeley; University of California Press, 1977); Charles Rosenberg, *The Care of Strangers* (New York: Basic Books, 1987); Paul Starr, *The Social Transformation of American Medicine: The Rise of a Sovereign Profession and the Making of a Vast Industry* (New York; Basic Books, 1982).
4. Nathaniel I. Durlach and Anne S. Mavor, eds., *Virtual Reality: Scientific and Technological Challenges* (Washington, DC: National Academy Press, 1995), 25–6.
5. See T. W. Dillon and A. F. Norcio, "User performance and acceptance of a speech input interface in a health assessment task," *International Journal of Human-Computer Studies* (UK) 47: 4 (1997): 591–602; G. Kahle, et al., "Evaluation of digital speech processing system in clinical practice," *Current Perspectives in Healthcare Computing Conference* 1996: 79–84; L. M. DeBrujin et al., "Speech interfacing for diagnosis reporting systems: an overview," *Computer Methods and Progress in Biomedecine*, 48: 1–2 (1995): 151–6; John Gosbee and Michael Clay, "Human factors problem analysis of a voice-recognition computer-based medical record," *Proceedings of the 6th Annual IEEE Symposium on Computer-Based Medical Systems*, 1993, 235–40.

CHAPTER 8

The Bride Stripped Bare to Her Data: Information Flow + Digibodies

MARY FLANAGAN[1]

In a class I taught in 2001 entitled "New Technologies and Communications Media," I presented excerpts from a twentieth-century film in order to foster discussion of the cultural position of information historically within the Western imaginary. The romantic comedy, *Desk Set* (1957), depicts a research library in the center of the large "Federal" corporation, and is perhaps the first film depicting an IBM-like "EMERAC" machine as it spews out the credits.[2] Katherine Hepburn plays Bunny Watson, the head of the all-female reference division of the company, a woman who can spontaneously answer any question asked of her or find the answer almost as quickly. From reciting "By the shores of Gitche Gumee" by Longfellow to answering inquisitive calls about which issue of the *New York Times* contained which report, she is an information maven. However, Spencer Tracy arrives as the character Richard Sumner, an "efficiency engineer," seemingly determined to replace the human knowledge of the research division with the overstuffed, ballroom-sized IBM computer.

Bunny's troubled romantic relationship with her boss, Mike Cutler, puts gender politics of the workplace at the forefront of the romantic comedy, and the film is rife with troubling sexist remarks and behavior. Every time the women know the answer to a difficult query, the narrative counters this "untraditional" knowledge by reinscribing the feminine: the female characters seem to be always applying lipstick, ordering dresses, watering plants, or showing each other their new clothes for the upcoming dance. In this way, although there is an attempt to depict a "single-woman work culture" in the 1950s, this culture is focused around their interactions with men.[3]

Unexpectedly, it is Bunny's methodological and meticulous, almost machine-like command of knowledge that allows her to beat the very machines sent to replace her. Her knowledge is vast, crisp, and helpful at all times. At the same time the EMERAC is installed in the reference area (with a new female operator), one is installed in Payroll and promptly generates pink slips for everyone in the company. Still, Bunny beats the machine in an uncanny way, saving the day with her genuine human knowledge, her way of connecting events and facts in a sensible order; she becomes a metaphoric "bride" defeating the efficient machines of her bachelor suitor. The machine spins out of control, while Bunny remains cool and knowledgeable. In the end, Bunny shares her space with the machine yet controls it; she is then enfolded into a new and seemingly more equitable heterosexual relationship.

Desk Set is one of many media examples that depicts data, technological change, and information in ways which tie them directly to women's bodies. Although Bunny is allowed to possess knowledge and information, her dangerous knowledges are tamed by bouts of overemphasized femininity. Bunny repeatedly costumes herself in the film; she holds up glamorous gowns, hoping to transform herself literally at "the ball"—while underneath, keeping her "suits" of knowledge for the domain of the anonymous reference call. Although *Desk Set*'s Bunny ultimately triumphs over technology, images of women today produced by our technologically driven media forms are not necessarily invoking such images of equity and mastery, and are still conspicuously bound up in a heterosexual system of representation.

Representing data is not only a big business and the focus of countless classic professions; as we see from this example of filmic representation of the role of data, the practice has political and cultural implications and has historically been tied to the representation of women's bodies. With the emergence of contemporary ways of gathering, storing, and defining information, the complex relationship between information and the human body has evolved in interesting ways. Virtual bodies tend toward the hyperreal—characters such as (my favorite), Lara Croft from the *Tomb Raider* series, Kasumi in *Dead or Alive 2*, or Sarah Kerrigan in the *Starcraft* games are all, in one way or another, hyperreal, exaggerated "hyperbodies."[4] These hyperbodies seem to be slowly giving way to even more "realistic" representations as technology permits and audiences consume. Digital media are now working toward building a unique aesthetic for perfection—note the most suprahuman model yet created, the "down-to-the eyelash-follicle-perfect" rendering of the brilliant and beautiful Dr. Aki Ross in the 2001 film *Final Fantasy: The Spirits Within*. From virtual sets for television which create 3D backgrounds in real time, to virtual newscasters and characters themselves, the field of information design as an area of graphic design has quickly evolved into an area of biological design. Information today is enacting its own etymology: information, from the Greek word *morph;* in Latin "morph" became "form," to the Latin *informare:* to bring something into form.

As mentioned in previous essays by myself and others, computationally rendered graphics and the systems, machines, and traditions that produce them are powerful yet problematic for a number of reasons.[5] Virtual spaces are conscious creations produced by a numeric process, a process produced primarily through programming code. N. Katherine Hayles has noted that "even though information provides the basis for much of contemporary society, it is never present in itself."[6] But a symbolic system that has allowed computer code to come into an existence of its own is indeed taking shape. The merging of data production with digital media has much significance and immediately conjures up questions from the domain of semiotics. In fact, the dictionary definition of "virtual" was possibly written by the founder of the American branch of modern semiotics, Charles Sanders Peirce.[7] Bodies are acting as conduits for information; with the general move to make bodies perfect yet believably "realistic," strange eruptions occur, signaling gaps within our current system of signs.

This chapter explores the evolution of electronic media representation of information, from charts and graphs which represent data, to information's new form: virtual and organic flesh. Our connection with and interfaces to the computer change daily. Among human–computer interface (HCI) researchers, computer-generated "personalities" are seen as the evolutionary form of the icons used in next-generation graphical user interfaces.[8] In this essay, I want to explore the nodal points around which information bodies can be organized, and their system of signs. However, the essay focuses not on language but on the embodied code of virtual characters. I map this embodied data through phenomena—the world's first virtual newscaster, Ananova, a character-based live information interface, Motorola's Mya data service, and Syndi, the "celebrity portal" search engine that purports a subjective search experience through a character—to explore the ramifications of data embodiment. Specifically, in this chapter, I look to virtual characters developed to represent data and news in online environments as the site of a monumental shift in digital cultural consciousness. Where older models such as those represented in *Desk Set* contain data in the female form, present embodiments have left the real female body behind in a significant way. Embodied data through phenomenon such as virtual newscasters and other data agents are becoming commonplace, but what are the implications of making information biological? If our data must now be embodied, what is the impact on how we understand data in the form of human bodies, especially female-shaped bodies that act as conduits for information flow? What is the role of the female in this representation, and what are the social implications of embodied data?

I. Duchamp, Technology, and Gender

A representational review is in order to reflect on the visual coupling of woman and technology, code and bodies, for it is at the female body that the formation and contestation of digibodies is occurring. Notions of the machinistic successes of the nineteenth-century industrial revolution resulted in shifts in art, such as that produced by cubists and futurists, to the industrio-sexo-mechanical metaphors churned out by Marcel Duchamp. Duchamp, one of the influential thinkers of the time, represented the shift from the industrial to postindustrial Western cultures repeatedly in his work through the figure of woman. After his groundbreaking "Nude Descending a Staircase #2" (1912), Duchamp expressed ideas about time, work, and technological innovation in his artwork, and throughout his career veered away from representational forms (in this, he furthered Delaroche's proposal, for Duchamp believed that representational forms themselves were dead).[9] Problematic as Dadaist and Surrealist representations of women were, there are a number of similarities between contemporary cultural change and the shift that occurred in the earlier part of the twentieth century. In trying to read our current semiotic shift, I look to Duchamp because he also briefly broke the system of representation between gender, the body, and technology. If de Saussure established semiotics as a system of signs within society and focused meaning as defined by the relationships of one sign to another, and if Peirce described the science of signification and the sign as an "object," Duchamp was the radical who turned popular semiotic conventions upside down, moving away from both representation and cultural references, breaking the link between his works and the "objects" they supposedly represented.

Scholars have debated the nature of historical and contemporary linkages between the figure of woman and technology. For example, as Andreas Huyssen notes in his book *After the Great Divide: Modernism, Mass Culture, Postmodernism*, technology and woman have historically been linked: "As soon as the machine came to be perceived as a demonic, inexplicable threat and as the harbinger of chaos and destruction . . . Woman, nature, machine had become a mesh of signification which all had one thing in common: otherness."[10] We can also look ahead and see how negative images of women played out in Weimar Germany's technological and industrial shift.[11] In twentieth-century culture struggling to find a place for photography and film within the traditional arts, it was in part through Duchamp's challenges to artistic definitions and institutions that it became possible to contextualize the creative act using mechanical reproduction paradigms—especially in the 1920s, when there was a major shift in both technology and in visible/virtual culture.

At around nine feet high and over five feet wide, Duchamp's famous 1923 painting *The Bride Stripped Bare by Her Bachelors, Even* depicted "the bride" as an engine to symbolize twentieth-century progress (see Figure 8.1). "The Large Glass," as it is also called, is considered to be his most major work and took eight years to evolve.[12] As Jean François Lyotard noted in his book, *TRANS/formers*, although Duchamp's ideas are inspired by technology and image making materials such as time lapse photography, he continued to move away from representational form—as Lyotard calls it, Duchamp worked in an "a-cinematic" form.[13]

Although Duchamp's bride is all about desire, women, and machines, the visual representation in the work itself is far from typical. Although the bride is a self-sustaining mechanical character in perpetual motion, her nine bachelors, who, in Duchamp's notes on the work, are said to survive on coal and the fuel of sexual tension, are positioned below their mistress. In her domain atop the painting, the bride issues her commands, orders, and authorizations, acting as a kind of motor; Duchamp's bride could be read as erotic, and she is in complete control. She is both passive (allowing herself to be stripped) and in control as an active (moving and desiring) subject, inhabiting an electric space

Fig. 8.1 Marchel Duchamp, *The Bride Stripped Bare By Her Bachelors, Even (The Large Glass)*, 1915–1923. Oil, varnish, lead foil, lead wire, and dust on two glass panels, 109 1/4″ × 69 1/4″. Succession Marcel Duchamp, ARS, N.Y. / ADAGP, Paris.

that Lyotard claims is "transcribed in a plastic way: the Bride-machine is not placed in the same space as the Bachelor workshop," but rather in the space of electronics (144). The large, cloudlike shape at the top of the glass is the halo of the bride, a type of veil, an access point or network in which the bachelors desire to participate. Unlike digital characters, which are "powered" by logical code, the mechanical bride of Duchamp is fueled by "Love Gasoline."[14]

The idea that this bride is passive through her permitting herself to be stripped, and active because she is desiring and controlling, is important, for such discussion positions the bride in a duality: She is in total control, has her own electronic, erotic autonomy, and is not bound by stereotypical representation. Machines, bodies, and time are blended together and fused with concepts of gender roles and their relationship to information. The technological woman, the mechanical bride, like the Internet, is self-generating, bringing about technological change.

Duchamp's work is read through both through its visual representation on the glass, as well as in various notes and boxes, which contain calculations, scribbles, theories, and other information which mediates the experience of viewing the glass so that, according to Duchamp, the experience of the piece aesthetically will be interrupted. As he noted, the painting itself "must not be 'looked at' in the aesthetic sense of the word. One must consult the book, and see the two together."[15]

The erotic logic of such an artificial and mechanical system is both anti-representational and illogical; Duchamp makes many notes in his "The Green Box" about his "hilarious" picture, noting that rather than an image, the work is the whole of an illogical process and antirepresentational presentation. As Tzara argued, "The beginnings of Dada were not the beginnings of an art, but of a disgust" with traditional art forms.[16] Like the internet's tendency toward hiccupping downloads and the now popularly described "lag" phenomenon of the Internet experience, Duchamp called his Large Glass a "delay in glass," playing with moments of "in between time."[17]

Read within his play of logic, Duchamp's vision of the coupling of technology within a consciously marked gendered construction offers an interesting position on the combination of woman and machine. Valued for his play with gender, Duchamp bridges the problematic mind/body split by foregoing representation of the mind or the body, moving to a third, conceptual space.[18] In this way, Duchamp avoids scripting both negative representations variously depicted as phallocentric or patriarchal, and he also avoids locating the figure of woman as a passive innocent that we cannot represent. Some of Duchamp's analysts, however, raise questions about his depiction of women. Lyotard argued that "the whole Duchamp affair goes via women" and asks, "[s]hall we say that women are the principle of the a-mechanizing cunning, that they have no soul . . . their bodies being mechanically reducible . . . won't that be their whole morality: either married or prostituted?"[19] He goes on to argue that, in his work and in his masquerade as Mademoiselle Rrose Selazney, Duchamp goes beyond sex by reconciling sexual difference (115).

Duchamp's conceptual erotic, dimensional machines are not based in representation but in delay: the bride is neither naked nor clothed, leaving us in an in-between state where "there is no art, because there are no objects" (20). Reading Duchamp clues us into the possibilities of the absence of representation; it confuses other semiotic systems, disrupts the formal models we use to create meaning, and opens up a third space for the digibody.

II. The New Woman: Prototype Digibodies

The Bride can represent for us progress, technology, desire, and control. Duchamp's bride both is passive and yet, controlling "progress." But in the next century after this monumental work, with subtler, nonmechanical technological change influencing every aspect of our everyday lives, the conceptualization of

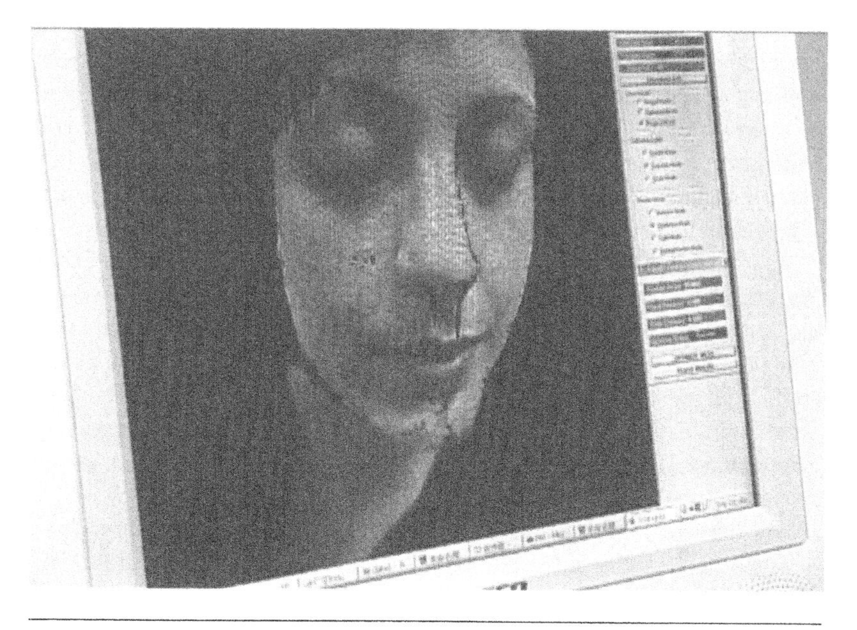

Fig. 8.2 National Taiwan University scan of human faces (face of Mary Flanagan) (2001). Created in custom software. Screen capture taken by Mary Flanagan. Courtesy CML lab, directed by Ming Ouhyoung.

gender and technology has manifested in a way closely related to Duchampian thought. Researchers around the globe are looking to computer-generated "personalities" as the next evolutionary form of the graphical user interface. Humanlike faces are already beginning to replace buttons and icons (think of the Microsoft prototypical paper clip). At National Taiwan University, researchers are designing personal secretaries for PDAs by scanning real human faces to make the virtual secretaries more appealing; researchers note that they "want to go for a friendly interface" (see Figure 8.2).

VR communications agents put a skin on our current fascination with stock and business news and helpful applications, and at the time of this writing, digital agents and news sources have all been depicted as female. Rather than doling out commands to her bachelors in Duchamp's painting, these brides have instead manifested as 3-D graphic characters on the World Wide Web, doling out news and other kinds of always-updated streaming information. While Duchamp resisted the aesthetic to convey his fascination with information and the symbolic, this century's digital characters collapse the real and the virtual into one system, foreclosing any possibility of critical distance in our interactions with them. Moreover, interactive design has taught us that critical distance is precisely the thing designers are taught to avoid, especially at the interface.

To understand the phenomenon of digital characters within the context of interactivity and service functions (as opposed to, say, game characters that carry a related yet different set of concerns), we should briefly look at the current state of the merging of broadcasting, information design, and visualization, for these are the areas that traditionally have provided information flow and established representational norms of the information itself.[20] Outlets for this material include broadband, Web, and mobile computing solutions, all of which compete for users' time and attention.

As a discipline, information design historically has investigated different issues than media forms such as broadcasting, news or the traditional areas of film and print—instead of telling stories in time, information designers need to depict stories in space—and generally, this space has been a flatland that graphic designers have worked to dimensionalize. Leading information design guru Edward Tufte is one of the innovators in visualizing information in graphic form so that everything from day-to-day matters to important security systems are easily understood, arguing that through good information design we are able to "change the way people see."[21] Interfaces, like signs in semiotics, stand for other things: icons and indices give way to the heart of the content. A sign does not function as a sign unless it is understood by the user as a sign, however. In computer applications, this system relies on a rich tradition of graphic design to create understandable user interfaces.

Information design is nowhere more significant than in the interface between humans and computers. Historically, interface surfaces have been unreal: imaginary buttons, unnatural keyboards, molded extensions of our palm, and stick controllers. Many designers agree with Donald Norman, a leading design specialist, that user interfaces should blend with the task in order to make tools, the apparatus of production invisible and keeps technology as the means and not the goal. Accordingly, computer interfaces should not call attention to themselves; characteristic buzzwords include "intuitive," "adaptive," "supportive," and "easy to use." Norman admits, however, that there may be no natural relationship between one design and other designs or objects. In other words, design is a purely artificial ideal, and interfaces are sites of artifice designed with human physiology, cognition, and habit in mind.

It should not come as a surprise, then, that the latest computer interfaces, or portals, are artificial computer generated characters. Digital hosts are manifestations of digibodies, following on the emphasis on game characters and pop culture animations.[22] Part material and part symbolic, virtual characters, be it a mobile or Web-based network, are now hosting everything from cell phones to game shows to DVDs. Here, we should turn to real world examples of this embodied data manifest in interactive virtual characters.

1. Syndi

Especially in the area of information retrieval, virtual characters put a human face on the front of real time, up-to-date flows of data. Not only are they the

Fig. 8.3 Jim Anderson, *Syndi* (1999). September/October 1999 issue of *Artbyte Magazine*, Courtesy Artbyte Magazine.

conduits for a flood of information, but they are also a flow of signs, representing and reinforcing our myths, stories, and ideologies. Signifying uncertainty about the meaning of embodiment, these characters represent a shift in digital cultural consciousness. "As the net cools, information begins to self-organize. Applications become attention windows. Attention windows become celebrity portals." So begins a conversation with Peter Seidler, Chief Creative Officer of Razorfish as he discussed the company's research effort, a search engine with "subjectivity."[23] Seidler wrote a cover story for *Artbyte* in 1999, "My Syn," which detailed the abilities of Syndi, a proposed "celebrity portal" search engine created by Razorfish. Syndi is able to learn about the necessities, tastes, and interests of users (see Figure 8.3). Unlike her ancestors, text boxes or strings of data, Syndi is a three-dimensional polygonal model, taking the form of a woman with pulled back brown hair, huge green eyes and a long, tanned, thin face.

Syndi, if built, would be one of the first "native" species of cyberspace. Working on specific tasks based on detailed personal user profiles, these agents come from a long line of traditional "AI" projects began in the 1950s and 1960s from programs such as the Eliza and Julia engines. As her creators note, "Syndi's existence depends on her capacity to fit into her environment" and she appears the way she does because "commodification creates a need for distinction."[24] In this way, the agents are distinguished only by their constructions, not by their abilities or services.

It is significant that the female databodies like Syndi are generally composite bodies: by composite bodies I mean a mixture of presumed "normal" female characteristics (combined with supermodel and superbody styling) and constructed with presupposed racial characteristics. We could look to many historical markers in order to contextualize markers of race in representation, but I choose a rather recent set of circumstances because this example had significant impact on graphic design and advertising, two major vectors along which virtual characters were created. The work of influential designer Tibor Kalman (1949–1999) and his design house M&Co. are essential in order to contextualize the creation of digital characters. This group created many influential works in Benetton's spinoff magazine *Colors*, which pushed an editorial emphasis on shocking images and politics.

Many M&Co. and *Colors* images pushed the boundaries on race representation in the early 1990s, especially layouts such as the "What if. . . ?" spread from Issue #4, "Race," in 1993.[25] Using digital imaging programs, *Colors* changed the races of several iconic men and women. Queen Elizabeth was made to look black and Spike Lee white, and Kalman offered readers an Asian Pope John Paul II. These experiments investigated the taboo of racial switching and performance, setting standards for pop culture, advertising, and media that we still see today in digital character creation. For example, many digital stars, including Lara Croft, have ambivalent racial composite faces, hair, bodies, and skin color, with a problematic tendency toward a "Caucasian blend." This composite takes on more significance as we begin to see the role of the construction, of the artificial, in ideologies that construct these characters. Even though the intent of Kalman's group was to cause readers to reflect on racial inequity, the *Colors* overall effect was to depoliticize the representation of racial hybridity. In the realm of graphic design, race became a design element with this use of morphing and imaging technologies. In turn, digital characters reflect this apolitical representational practice.

To return to digital characters and the creation of the character Syndi, we are tempted to understand Syndi's personality. Her makers say her "brand attributes" control what she appears to feel like for the user, and she is intended to evolve: her makers focusing on her subjective evolution of consciousness and beyond, to the domains of the "digital superconsciousness." However, she has not evolved very far—information about Syndi after the interview with Seidler is very difficult to find, even in her native habitat, but her design speaks to the overall goals and desires of dot.com entrepreneurs.

Seidler's article did not address the naming of the Razorfish character, but Syndi's name could be a reference to Pat Cadigan's 1991 cyberpunk novel

Synners. In *Synners,* Visual Mark and Gina are "synners," that is, synthesizers who work together to create immersive music videos through a direct neural connection to cyberspace. At last, Visual Mark finds a way to download himself into the network and "lose the meat." Mark becomes the first net-born synth-human, existing only in cyberspace. Cyberpunk fiction has long predicted artificially created network beings. Syndi offers users a similar promise of the escape of the flesh. Although she wears a halo suspiciously close to Duchamp's bride, hers is a cloud of zeros and ones as her veil, her neck ringed with strings of binaries, black and white, zeros and ones . . . a human genotype: the genetic code of the digital character, worn closely as she gazes at her bachelor-users. Unlike Duchamp's bride, however, Syndi's makers offer us a representational body to go with her networked dataconsciousness.

2. Ananova

The world's first functioning virtual newscaster, Ananova, is the "human" face serving a real-time information and news system. A little more realistic in design functionality and commercial expectations, Ananova appeared on the Internet on April 19, 2000 with her enthusiastic, "Hello World! Here is the news and this time it's personal." Ana Nova was designed to be the interface for real-time news, and she can also handle transactions.

Although the choice to use a female face for the conduit of information was not informed by rigorous research (Ananova says they decided to make their bot female "because people tend to respond better to getting information from a woman"), her face is a composite of what designers thought were attractive and useful features: useful, because they function within technical constraints. For example, the character's mouth had to be large in order to showcase the company's innovative lip synch technology on screens of all sizes, and her clothing is said to be "figure hugging" in order to "make her easier to animate."[26] The choice to give Ananova a bright shock of green hair was made in order to make her stand out on advertising as well as on mobile devices such as WAP phones.

Ananova news services are similar to any online news service (such as international competitor CNN.com), and can be browsed in a similar fashion or can be read, with theme music and images, by the Ananova character. She comes in low and high bandwidth streams to your desktop in pixelly chunks, slowly reading the news to users. The "personalized" aspects do not come from the agent but rather from the user, who can customize his or her news preferences. Although she has received marriage proposals and fan mail (as well as generating a large movement of "Ananova lookalike" groups on the web), the value of the human head to read news to users from the Internet remains unknown.

3. Mya

Mya "debuted" in March 2000 in an advertising spot during the annual Oscars, slipping over the video screen in her hip silver jumpsuit to the event. Mya is a 24-hour, voice-driven digital personal assistant that reads out websites over user's mobile telephones. Motorola gained much publicity for their 60-second commercial in which they promoted their fashionable cyber guide to the vast Infobahn. Visually seamless, digibodies like Mya appear quite realistic, although as a visual representation only pops up in advertising—while using the Mya service, the character is still a computer-generated, disembodied voice. However, the quest for ultrarealism in images of Mya is strong: Mya was created and was fashioned after a human model hired by media makers Digital Domain. Visual effects supervisor and animation director Fred Raimondi notes that he first focused on realistic hair and cloth "but not necessarily a photo-realistic woman," instead focusing on aspects we normally look for as clues that the character is computer generated, such as hair and clothes modeling and rendering.

Logging into a Mya demo online, we hear a slightly mechanically buzzing young woman's voice tell us dreamily, "My world is a world without boundaries without limits" (said without a comma, as well).[27] For now, Mya's is a real voice, but mechanically enhanced to give the illusion of the digital. Mya is, by far, the best-rendered and most self-assured of the digibodies, and her voice, human or not, is the most haunting; in fact it is her voice that is the most researched and promoted for wireless WAP systems.

Although she does not appear powerless, writer Tobey Grumet calls Mya "virtually submissive" because she is created only to serve users.[28] Mya "isn't a Web browser, this is someone who can speak to you" says Geoffrey Frost, a VP at Motorola.[29] The company decided to "give her a face, a name and a personality... humanize her."[30] The creation myth extends to her production team: "It was so exciting for me to finally see the cloth tests, for instance. It was kind of like meeting God, because you've never seen it before," said her art director.[31]

Although an ultrarealistic look in images of these women characters is very strong, after seeing several one can begin to see a definite trend in their representation. All of the digibodies found online or under design are women, and all either sport short or severely pulled-back hairstyles. Animation details, such as hair, eyes, and movement, are key areas where artists work especially hard to make a convincing digibody. In Mya's case, her creators noted that it was very difficult to create her in a way that she looked "alive." They focused on movement and her eyes, noting that she needed specular highlights and eye blinks to bring her to life.

Repetition is key to believing in virtual characters. Judith Butler notes that "[a]s in other ritual social dramas, the action of gender [and sexuality] requires a performance that is repeated. This repetition is at once a reenactment and re-experiencing of a set of meanings already socially established; and it is the mundane and ritualized form of their legitimization."[32] Repetition not only legitimizes and naturalizes the character's gender, but also reestablishes our relationship of power with the virtual character and allows the virtual body to function as a "material-semiotic object," that is, naturalizing the function and visibility of the body into one presence.[33] Therefore, Mya's appearances at events such as the Oscars and her ubiquitous mobility make her a "real girl" by virtue of sheer repetition (a mantra without boundaries without limits as advertising and media infiltrate into cars, phones, mobile phones, and clothing), and through the merging of the material and technological. Whether or not the real body plays into the picture, data embodied through Ananova, Syndi, and Mya, as well as other digital bodies that embody information flow, fascinate us.

Looking at these characters through a semiotic lens, we run into some rather difficult questions. If the images of bodies stand for "real" human bodies as interface experts claim, and the ideas they convey are data, or news, how is the meaning of the data associated with these bodies? What do they construct? One immediately looks to all newscasting and storytelling forms wherein the human body stands as a transmitter to the receivers. A data body is not just data—when virtual bodies are created and represented they acquire additional meanings. Signs, the popular saying goes, are never innocent.

This indicates that there are three separate and simultaneous "writings" of the character in virtual spaces: one is the animation embodied as the character itself, one is realized through interacting with the character, and one is witnessed through the character's curious voice. How does the visual regime cope with the addition of voice to these virtual characters? The role of the human sounding voice and its synchronization with the digibodies onscreen has become an important aspect of virtual characters and hosts. Since the invention of the telephone, women's voices have been the standard, "soothing" voice of information, and virtual female bodies are still acting in this role. Whether it is the frequency range or social allusion to woman as providing helpful services remains a question, for the popularity of women's voices' embedded in mechanical devices includes most subway systems, airline recordings, and the like. What is not in question, however, is the haunting, surreal quality voices have, and in particular, the recorded or synthesized voices of virtual characters. Although most computer game characters have short dialogue lines and grunts, virtual hosts are always talking: offering weather updates, stock tips, and round the world news.[34] In his book, *A Lover's Discourse*, French philosopher Roland Barthes explores the idea of texts that speak, and the speaking voice itself is preserved within written media. He asserts, "we do not know who is speaking; the text speaks that is all." For Barthes, the text is enough; the living quality we assign to voice is simply another fiction. To him, the voice is the terrain of death and signals absence, the past, silence—the living quality of the voice is a false illusion:

> It is characteristic of the voice to die. What constitutes the voice is what, within it, lacerates me by dint of having to die, as if it were at once and never could be anything but a memory. This phantom being of the voice is what is dying out, it is that sonorous texture which disintegrates and disappears. I never know the loved being's voice except when it is dead, remembered, recalled inside my head . . .[35]

By reading Barthes, we understand that voice can be tragic, for it refers to a once living or vanished object, leaving a recording to remain in a dying state, only in human—or computer—memory. As a result, virtual characters' voices are specters, created from data alone, and thus they refuse the human aspect of speaking and transfer it to an a-human space.[36] Perhaps the "unreal-ness" gives the subject a breathing space, room to acknowledge inadequacies and lacks through the lack in the virtual character. Or perhaps the virtual voice acts to further jar the semiotic system away from assigning symbols to signs.

Here I am reminded of the African American women who were brought into studios to dub in the singing voice of white stars in popular Hollywood films of the 1940s.[37] Although Hollywood tried to problematically eliminate the representation of African American female body, it seems that digital entertainment goes even further, working to efface all women's bodies. Swapping representational practices and effacing bodies has always held political consequence.

III. Mappings

As fascinating as they are, Mya, Ananova, and Syndi are only part of a much larger system of encoding the virtual woman. From striptease Web news programs (stripping and nude women reading the Webcasts at nakednews.com) to geographical and medical imaging, the human body undergoes multiple mappings.[38] Although realistic digital agents command WAP devices such as phones and even automobiles, these same devices can also track the body; as new technologies enhance our ability to monitor and locate human bodies around the world, and this ability does have an impact on the physical body. The use of surveillance, global positioning systems (GPS), and health/bank/credit records to chart human movement is widespread and not as shocking or outrageous to citizens as one might expect. Taipei and other large, international cities have long had automatic "speeding-ticket" cameras trained on major streets. The use of Internet cookies is also widespread; as artist David Rokeby notes in his description of his 2001 installation "Guardian Angel," a work that uses surveillance to track, record, and recognize human faces,

> Cookies are convenient. It is strangely pleasant to be recognized on arrival at a web-site you have used before. I shop on-line. I fill in forms in exchange for free software. I use my bank's debit card for purchases (though I swore I never would!)... It is highly inconvenient not to "trust" on-line corporate entities.[39]

GPS systems, and surveillance systems have implications for certain kinds of bodies over others: mapping the body in space and time has particular significance for women. "Nannycams" and networked pornography have made the surveillance and/or web cam primarily the domain of the image of woman. This kind of voyeurism has taken hold with reality television, as well as god games such as *Dungeon Keeper* or *The Sims*, and the proliferation of Webcams.

Whether it is because users are attempting to find themselves reflected in others, or they are interested in the feeling of connection that comes from such voyeuristic acts, the female body lies at the center of such interaction. Brian Curry, founder of the Webcam company Earthcam.com, notes that in a few hours of watching, Webcammed situations, voyeurs "create an affinity with the people they're looking at."[40]

As in space, a second mapping occurs within the body. Although we can encode data to create female forms such as our example agents, there has been a long tradition of "mapping" the interior of the human body. Control paradigms, subjectivity, and imaging technologies have traditionally reduced the category of "woman" to symptomatic images. For example, studies of Charcot's nineteenth century images of "hysterical" women to the recent "visible woman" project explored by Kate O'Riordan and Julie Doyle in their essay, "Virtually Visible: Female Cyberbodies And The Medical Imagination" show problematic historical and contemporary medical representations of the body through new digital imaging technologies such as CT and MRI.[41] In the visible woman project, the first phase of identification is to orient and segment the human body data points. Normative notions of gender both mark and form the imaged body. O'Riordan and Doyle argue that the body is represented through its relationship to technologies, but do not incorporate an experience of the flesh. In addition, many women artists explore this interior body mapping as a particular concern for women. Montreal-based Char Davies is an artist who has produced series of images known as the *Interior Body Series* and in her VR (virtual reality) work extends her exploration to include bodily organs, blood vessels and bones. In a way she is turning the gaze of medical imaging upside down by creating the interior of the body as an aesthetic rather than material object.

Third, the rise in popularity of digital characters has coincided with an increased fascination with bodies and body parts on the part of the medical establishment, which desires to map the body genetically. In February 2001, scientists announced the completion of the human genome. Even though the 20,000–40,000 human genes are closer genetically to fruit flies than to other creatures, we rush to have our own DNA trademarked in an effort to achieve immortality. In the sequencing phase, researchers identified the approximately 3.5 billion chemical letters (A, C, G, T) that make up human DNA, comprising "the most important, most wondrous map ever produced by humankind."[42] If biology is now information, then information design must also be considered as an organic endeavor. Although the study of morphologic form (e.g., the anatomical forms of organisms) is no longer a popular discipline, the study of biological information in DNA or RNA is the heart of present biological interest. Although entertainers and media producers seek to turn data into human bodies, research scientists such as those working on the human genome project seek to turn the body into data. Information has substituted for form as the preferred key to the fundamental problems of biology.

The human genome project is problematic for many because of the privacy, social, and modification issues that are inherent to such an undertaking. Most striking are issues raised by genetic screening, and the ideas of "norms" created by such prospects. Bioethical issues such as how we think about normality and abnormality arise instantaneously. And again, women are at the center of this debate, "because of their central role in reproduction and caregiving"; women are affected "differently but also more significantly than men by the information emerging" from the results of the Human Genome Project, potentially devaluing their roles.[43] With new possibilities arising from the ability to alter our genetic code, we need to look at the inevitable transformation of human kind in light of the virtual revolution in the critical light that authors such as N. Katherine Hayles, Robert Cook-Deegan, and others propose.[44]

IV. Virtual "Breakups," Virtual Desire

We have considered how we trace bodies in space, peer at them internally, and design and track them numerically with genetic code. Digibodies are mapped with the desires of their creators, and their creators are on a quest for a great Real. What is this real? How do we know it when we encounter it? And why this desperate need to know, to look at the face of the real? The balance between real and unreal is perhaps nowhere more striking than in the realm of VR experiences. To look at databodies in order to understand them in a larger cultural context we must examine their construction and their "femaleness," as these data conduits; the fact that almost all examples are inscribed as female makes obvious the significance of gender in this type of information representation and user relationship. With the possibility of deconstructing and reconstructing the bodies, the way in which an understanding of gender is drawn out, mapped, and manifested in these networked characters is essential.

Images of women's bodies, fictional or otherwise, have pervaded popular media for a century of North American culture. The body's surface has been written and inscribed in media and popular culture: segmented, distorted and distended for selling consumer goods through sexual enticement. Bodies evolve into datasets: even in animation the face and body are segmented to create "talking heads." The moving parts are conceptually and technically separated from the lesser-moving aspects of the head and body. The Internet's incorporation of bodies, however—from pornography to online games to news sites—has a particular significance for women's bodies.

First, let us explore the means of production of digital images. These segments that create virtual characters are not simply components of a whole, but discrete parts in and of themselves. As an everyday example, the ubiquitous Adobe Photoshop logo has, for almost the last decade, always included the long-clichéd eye collage on its box, and the eye as its famous application icon. In 2001, the photographic eye and associated collage was switched to a nonphotographic representation. Such a shift in this particular year is important, for it showed the current sense of fluidity between photographic and computer-generated imagery. Digital media has quickly and directly dispensed with photographic images as the primary focus for representation, instead investigating the depth, manipulatibility, and maneuverability of 3D and vector based images.[45] In other words, artificial signs that have no link to the visible, "natural" world are replacing our previous preferences for photo realism. It is not only logos or packaging which have signaled this shift from the photographic, from the whole to its parts. Software applications themselves have increasingly offered users not only control of images but of image parts in both time and space. The use of layers, symbols, and other elements easily separated and individualized has been an invaluable part of each software upgrade in the most popular programs. In addition, and perhaps more importantly, segmentation has become a philosophy in processes related to media production and object oriented programming, software segmenting objects, elements, and processes, allowing us to create believable scenes asynchronously.

At National Taiwan University's computer science department, an innovative project utilizes simple photos of the human in order to "bring the virtual character to life." Movement was one of the first areas focal areas in popular software packages so that relatively untrained users could, for example, create inverse kinematic animation to simulate human movement without parsing stream after stream of real user body data. At NTU, teams use custom software to choose areas such as the mouth, eyes, and face of simple 2D photographs and use algorithms to mathematically control the 2D image so that it can be lip synched to voice information. Consequently, not only is a former whole segmented, proportioned, and divided, but the focus remains on these segmented elements. The digibody—its movement, voice, and expressions—can be created mathematically from a simple photograph. This process, however, is being superceded by accurate 3D modeling and scanning techniques that offer us a new body while losing the referential link to the material world through alternating a few data points.

Virtual bodies are easily mastered and in many examples in computer games show that one can derive great pleasure from controlling virtual bodies. Web sites with controllable characters create a tension between dichotomies such as inside and outside, knowing and controlling, recognition and anonymity, mastery and submission, and this tension continues to draw users back with a desire to interact. The interaction with the characters is ritualized—we are told when to click, generally we repeat actions such as opening and closing windows, and our roles are limited, defined. The characters quickly become a familiar interface while possessing seemingly infinite sequences or combinations of logical elements. Based on the fragment, the layer, and the repeated action, sequence or word, their role is to generate and maintain a system of expectations.

Virtual information service scenarios are quite unlike television news programs. TV news traps the anchors in the "head and shoulders" frame, rarely allowing movement. However, U. S. TV news conventions dictate establishing shots of the set, and this exposes the apparatus of the production that produces the illusion. Virtual newscasts, however, efface the means of production. We do not see a real woman behind the voice of Mya, for example; we do not see the development of the character or the context of its production. The processes and code are hidden, or if they are exposed they are bound ritualistically in "behind the scenes" movie clips and special "making of" add on products. This effacement is counter to Duchamp's proposition in the large glass; Duchamp's bride woman is not a doll, for dolls can have moving parts hidden from view: "if the woman in the top of the Glass was in separate parts, her construction did not make a mystery about it."[46] When the digibodies do not have evidence of the labor that goes into their development and the bodies are segmented into products themselves, they become fetishes.

Two models offer examples for virtual body fetishism: the Marxist commodity fetish and a psychoanalytic model. First, Marx claims that a commodity's form mystifies the labour, the actual work involved in an object's production, so that the object acquires mystical (if distorted) properties and becomes a "commodity object." Because virtual stars are products created by the latest technology, delivered on hot commercially important computer objects, this approach offers us one way to look at virtual desire. Second, the user's desire of the virtual character can be read as a form of sexual fetishization. Like Barthes's comparison of a reader to a lover, a participant who desires to control, devour, and name the object of desire, the interest in virtual bodies from consumers is sustained from the appeal of the unanticipated and mysterious or by focusing on parts in order to control, comprehend, and conquer. Both of these approaches classify the fetish as an object possessing a special energy and power. The role of fetishism, particularly the fetishization of the woman, in digital media is of great interest to any critical understanding of this new landscape, for digibodies are created from discreet elements and are positioned within a command and control paradigm of desire. Thus what is proposed here is that the means of this particular kind of 3D artifact production allows the body to be thought of as segmented and zoned. The breaking up of the female body into discrete elements, i.e., the creation of the image of woman as series of objects, is in terms of fetishistic scopophilia, focusing on the object or body part used for sexual enjoyment. The fetish is a specificity—as Haraway terms it, the "thing-in-itself." It signals, above all else, perfection, isolation, and containment.

As the databodies are both objects in and of themselves and simultaneously self-referential, we must wonder about this seemingly enclosed system that serves to abstract and yet manifest them again, forever a process of regurgitation and digestion. Barthes' effective stretching of semiotics to include images, codes, and cultural systems prefigures cyborg methodologies and allows us to examine this hybrid creature—a creature we can define as one not of materiality or even as a human and machine combination, but rather one of referential and non-referential elements. Digibodies, constructed from text, create a text-based shape: "the text itself . . . can reveal itself in the form of a body, into fetish objects, into erotic sites."[47] Yet the resemblance to the physical human tempts us to draw this link. We measure characters at first on their proximity to the real: her movement is realistic, or her hair is not very realistic; then we move inwards, tripping down a chain of code signifiers. Whatever body part is chosen by animators as significant or nomadic, or thought by the audience as realistic or not, the character is replaced by a series of costume objects which become mobile disguises, fetishes. Even if the fetish is based on a "natural" phenomenon—a birthmark, for example—it positions participants into an artificial situation. By the very nature of its artifice the fetish erases the means of production of its own illusion.

The act of seeing the body in discreet pieces has its roots in the desire that Jean Baudrillard discusses as seduction. Seduction is founded upon the artificial; and while Baudrillard is problematic for feminists, his work offers an interesting argument for our desire to watch, control, and interact with digibodies. For Baudrillard, the shift from the real to the hyperreal occurs when representation gives way to simulation, and he insists that women's power relies on their capacity for artifice, disguise and seduction.[48]

Virtual characters—specifically online news and game characters—are popular, hyperreal pleasures. Such popular manifestations of pleasure are not founded on classic ideas of "quality art," but rather popular desires. Roland Barthes differentiates between desire and pleasure in *The Pleasure of the Text* (1975) and insists that current consumer culture lives for desire (to him, the absence of pleasure); he then compares popular pleasures to sexual desire, and our postmodern experience of them akin to a kind of sample or demo, hallmark notions of the digital age: "I impose upon the fine surface: I read on, I skip, I look up, I dip in again."[49] The segmentation, control, and maneuverability of VR characters, their surfaces, are navigated in a ballet of pieces, moments, fragments. Whether playing games like *The Sims* or *Black and White*, or simply just getting the news, humans are mesmerized by virtual characters: their movement, their unchanging steadiness in the face of our distractions beg the question of what interaction could bring to them. We click all over them like ants to honey, wanting them to react and offer stock quotes, a fast news fix, a protest, a wink. Sometimes, they react too slowly, voices steady, appearing calm, effacing the processing and framerates that generate them wherever, whenever we want them. These agents are also constructed by the surroundings in which they are birthed into their digital universe, on the frontier, organisms that are extensions of the network itself. And they are real, but a kind of real we are not accustomed to categorizing.

We users have a particular fascination with exposing the real, of seeing behind the scenes, uncovering detail. Reality has become dislocated, homeless, and it floats free as a hyperreality is created in its place, able to exist through the breakup of the body parts while the seduction of the partial ensnares the users. But what is a way to reject the formation of the body as a fetish or commodity object, if our way of discovering the body object is still through a visible system? The female digibody has become a pure signifier which does not carry meaning beyond its appearance—or does it? The idea of gender as performative, rather than a naturalized or inherent biological entity, is useful when considering gender and the digibody. With virtual personae, the character and creator are indeed separated, and the relationship is open to change, multiplicity, and radical identity shifts. But a curious thing happens with digital characters: the manifestation of the female characters are so far down the chain of signification that their connection to any real female body becomes blurred. Here, Barthes' study of Bunraku theater is a useful resource for our examination of virtual characters, for he examines the idea of performed gender within artificial surroundings and character-agents external to the flesh body. Barthes' notes that men playing women characters in Bunraku do not "copy Woman" but rather work to signify her; their performance is "not bogged down in the model, but detached from its signified; femininity is present to read, not to see."[50] The conjoining of both sign and passivity thus create the figure of woman as a semiotic figure, an "ideal woman agent" (60).

V. Social Implications of Embodied Data

The question of the Real continues with the investigation of gender, for both are inextricably intertwined. The semiotic distance Barthes proposes does not address issues of culture, and historical context. In further discussion of theatre, Barthes insists that the use of a "real" woman in the role of a woman character would actually be detrimental, for this would appear as an attempt at realistic representation and would lose the possibility for true "expressivity" (89). From a feminist perspective, however, such a copy of woman is problematic for it is created and defined by media creators within a cultural context of patriarchy; under such careful control, it seems as though there will be limited innovations and changes to the power structure; to the ways in which the digital woman is created, packaged, and the kinds of powers she will truly possess. Claims that female digibodies have, in fact, a slim indexical link to the idea of woman suggest that if these images are not really like real women, they do not have to be taken seriously.

Digital characters, however, are in fact encoding methods that inscribe both gender and cultural norms. If we compare them to the way we generate letters to form words, the equivalent tie is between the word and the idea. This gap, Barthes proposes, is like theatre in which characterizing "Woman" transcends nature and image to become an idea, "and as such, she is restored to the classifying function and to the truth of her pure difference" (91). This brings us to the key point in thinking about women digital characters: media makers continue to create eruptions such as Ananova's green hair, reminding us with such markers and eruptions that these characters are digital, that is, distinct from the real.

Do we desire the markers of difference on virtual characters?[51] Such markers have helped avoid the inevitable questions about representation and virtual characters; game makers, for example, easily dismiss stereotypical depictions of women in games because they are, after all, fictional, exaggerated characters. The social implications of embodied data have the potential to be politically reductive. *Tomb Raider* fans note, for example, that if one were "to see women built like this in real life . . . it would be kind of scary."[52] This kind of discursive distance has kept games and game characters in the realm of either the joke (i.e., Lara Croft is as exaggerated as a blowup doll) or the technology fetish (the graphic technology that produces the character is in itself the interest). But what happens when this critical distance is erased, the markers of difference elided by a new type of life form? This process is beginning to develop with the *Final Fantasy* film, for example, and will continue to be perfected.

The nature of the digital medium is peculiar: at once it is networked and fluid, yet at the same time individual, tailored, and fetishistic. It seems that virtual characters are based on Western, human norms of acting, conventions of news giving, and conversations. As her Oscar debut suggests, even Mya wants to be on a talk show. Barthes would argue for us to embrace the plastic, embrace the artifice, and be one with the code.

VI. Code Shapes

With the rise of digital media and its capacity for interactive systems, information design at the interface has become an important site for meaning creation, especially to interface gurus Ben Schneiderman, Donald Norman, and Jacob Neilson, among many other high-tech industry workers and consumers of digital products. A visual interface surrounds almost every user of technology, creating a semiotic home for all users, ideally meeting their task, entertainment, and social needs. Interfaces must be easy to use, even "intuitive." Systems and interfaces, however, can be created more or less usable for regular consumers. For idealistic interface guru Donald Norman, the human is clearly the preferable system: "We are analog beings trapped in a digital world, and the worst part is, we did it to ourselves."[53] Norman explores what he calls "the horrible mismatch between requirements of these human-built machines and human capabilities." He cites a long list of dichotomies between machines and humans: "Machines are mechanical, we are biological. Machines are rigid and require great precision and accuracy of control. We are compliant. We tolerate and produce huge amounts of ambiguity and uncertainty, very little precision and accuracy . . . Analog and biological" (15). Digital systems still do not excel in processing ambiguity and imprecise situations.

The interface research conducted at Xerox PARC during the 1970s established most of the visual and functional conventions of present graphic user interfaces, and these conventions helped make the screen reality into a lived reality.[54] New media theorist Terry Harpold notes, "[a] subtle thing happens when everything is visible: the display becomes reality."[55] In other words, the distance between the real and the virtual, much discussed in cybercircles, collapses into itself. The creation of digital characters and digibodies makes the display the object of our desire. New creatures are created from texts, and by virtue of their computational environment, are machines of the eventual erasure of the symbolic. From the physical aspects of the computer (keyboard, mouse) to the internal interfaces (Web page, application graphical user interface) computers offer themselves through signs. When we write text, for example, that will appear in a book such as the one you are holding, the signs we produce to create ideas are a common currency. In the computer, ASCII text represents the letters in the word "sign" (already a meta sign) as "115 105 103 110." This chain of signifiers continues down the line in the various processes a program runs via the processor. However, an interesting thing happens when we reach through the operating system to machine code. At the machine code level, things are different: machine code exists and controls in and of itself, a place where the sign system is replaced by impulses, 0s and 1s, jolts. It is a place, to use Baudrillard's words, where "truth, reference and objective causes have ceased to exist."[56]

If we consider data and information, even computer code, as texts, Barthes might suggest we derive our pleasure from the collision of a text's figurative aspects with its informative functions. No longer does an object have to be material to exist. We follow an object oriented programming model here: at the simple invocation of an instance of an object—Plato's shadow, if you will, in relationship to the object—the object exists, even if only in code. Thus although Harpold sites a problematic shift at the graphic itself, perhaps the root of the issue lies at a much more conceptual level. Code is shaping our consciousness and our culture.

Uncertain signs are unstable territories, and we work to fix them to the familiar. Good examples of this principle come from N. Katherine Hayles's observations that along with the mechanization of language, there has been a one-to-one correspondence with language to action or object; unlike computer actions, which can, through their abstract association with a particular text or action perform the insignificant to the monumental with the touch of a button (such as the "send" on a fiery email) typewriter keys or printing press blocks, are directly related to the script they produce.[57] However, virtual portals are appearing all over the Internet. They represent a critical moment of the virtual age; Arthur and Marilouise Kroker write that we are in an era "typified by a relentless effort on the part of the virtual class to force a wholesale abandonment of the body, to dump sensuous experience into the trashbin, substituting instead a disembodied world of empty data flows" (see Figure 8.4).[58]

The move toward mobility of information means the shape of the machine must better adapt to our own bodies, to complete the merger of network and embodiment. In fact, our desire for the latest information, our wish to have a personal helper, our secret dream of the "genie in the bottle" manifests online; we are living in a age when that which "was previously mentally projected, which

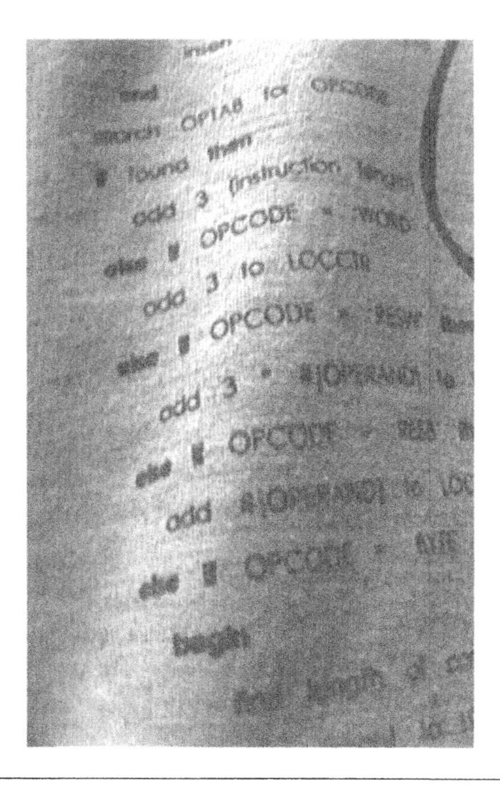

Fig. 8.4 Chen Ding Yun's clothing is covered with code, showing the tension between the physical and the code environment. 2001. Photograph by Mary Flanagan with the cooperation of CML lab, directed by Ming Ouhyoung.

was lived as a metaphor in the terrestrial habitat is from now on projected, entirely without metaphor, into the absolute space of simulation."[59] In this case, the blurring of the virtual body with these bride bodies by their means of distribution and task (erupting off of our desktops from websites with news for us in little boxes) can tell us a great deal about our relationship with virtual characters and desire. For as technology has taken a shape, it has taken a shape the creators of the technology desire, morphed into a form (information), creating a body.

There are several implications of this shift. First, these pleasurable character/services offered by industry equate womens' bodies with service and ubiquity; these women are "scattered into x number of places, that are never gathered together into anything she knows of herself."[60] The ubiquity poses the question about the social importance of ubiquity and access, and the more accessible, the less valuable. Another implication of this trend is towards the loss of the body, which numerous writers continue to postulate. The body's racial, aged, and sexual categories are interconnected, but in virtual characters, these combinations are constructed as personality features rather than historical or culturally specific categories. Rather than focus on this erasure, however, this essay asks more about the virtual body and its own lived reality and its consequences on biological women's bodies.

Who is this third creature, this creature of code? Bukatman has called digitally created bodies a "literal hollowness," creatures of mathematical persuasion that are as deep as a surface, not as an interior. In his analysis, all elements of the image are reduced to digital bits, which are reformed before our eyes as the masquerade of the solid, of the real, while there is, in fact, "nothing inside."[61] But perhaps nothing has turned into "the thing-in-itself;" like the relationship of puppet performer and puppet, perhaps our virtual puppets have "no more strings. No more strings, hence no more metaphor, no more Fate; since the puppet no longer apes the creature, man is no longer a puppet in divinity's hands, the *inside* no longer commands the *outside*."[62] Does the text have a human form, is it a figure, an anagram of the body? asks Barthes. Clearly, "information" has taken over where "form" used to suffice. This shift represents a semiotic transfer, a way in which the flow of information has become re-imagined and embodied. In *The Pleasure*, Barthes writes, "Text means tissue. We are now emphasizing, in the tissue, the generative idea that the text is made, is worked out in a perpetual interweaving; lost in this tissue—this texture—the subject unmakes himself, like a spider, dissolving in the constitutive secretion of its web."[63]

I return now to N. Katherine Hayles's idea that information "is never present in itself." Perhaps the digital bodies discussed in this essay do have a presence in and of themselves. On the surface, these hyperbodies seem to be slowly giving way to more "realistic" representations as technology permits us to visualize more of the body than ever before. Digital media has been working toward creating multidimensional perfection. Blurring the real, their perfect bodies are unable to contain the strange eruptions of shiny plastic skin, of green hair, of almost slithering movement. Like melodrama and pornography before them, digibodies must manifest the excess of the perfect in hyperreal "bursts," a product of the overproduction of signs of the feminine and the virtual. Because of his apt writing about media personalities, Steven Whittaker might help us see that female digital characters turn into a kind of same-sex transvestite, so overproduced that they would become a kind of camp spectacle of a woman.[64]

Indeed, Baudrillard would describe this seduction of the collective consciousness, when we could be one not only with the data but with the digibodies, in terms of seduction: "There is no active or passive mode in seduction, no subject or object, no interior or exterior: seduction plays on both sides, and there is no frontier separating them."[65] The experience is undeniably exhilarating:

> Anyway, feel a million flurries of now, a million intangibles of the present moment, an infinite permutation of what could be . . . the thought gets caught . . . You get the picture. In the data cloud of collective consciousness, it's one of those issues that just seems to keep popping up.[66]

Without the link to the signified, digibodies are electric phantoms, almost haunted flesh, shells which have no link to the material or physical form of the sign. They are a semiotic flesh, a coded flesh, code changed to flesh, code itself. We are in a quest for perfection by way of the development of our noiseless music, through our data, our carefully scrubbed genes—our numeric flesh. The fields of human computer interaction, graphic design, genetic engineering, and computer science must confront the merger that has already happened between code and bodies; many of us no longer need look at the face of the real and know it, recognize it, and have no urge to look further, to assign extra meanings, signs, readings. We need new critical practices which recognize this new body form and to create ethics for this new body form. If our thoughts, actions, and communications have literally become digital signal and binary image, it is important that we look around us in the offline sense. Digital bodies can deny the situations of real bodies, real women, real class, and poverty issues in the focus on seduction.

Meanwhile, real physical bodies are reflecting the semiotic shift. Images of broken, pierced, incomplete, amputated, ruptured and fragmented "horrible" bodies proliferate in American culture just as our media strives to allow containment, control, and artificial perfection, a closed circuit where signs replicate and mutate into other phantom realities and pseudolives. As Barthes noted, "Imagine an aesthetic (if the word has not become too depreciated) based entirely (completely, radically, in every sense of the word) on the *pleasure of the consumer,* whoever he may be, to whatever class, whatever group he may belong, without respect to cultures or languages: the consequences would be huge, perhaps even harrowing."[67]

Notes

1. I would like to thank Meg Knowles for introducing me to *Desk Set,* Roy Roussel for reintroducing me to Barthes, Cal Clements for bringing in Duchampian dimensions, Frank Miller, the students at National Taiwan University for involving me in their digital embodiment projects: Perng Guo Luen, Carol Chia Ying Lee, Wei Ru Chen, Alex Wan-Chun Ma, I Chen Lin, Cindy, Super Yeh (Jeng Sheng), Wei Teh Wang (bearw), Murphy, Joyce, Jun Wei Yeh, Cindy Chi Hui Huang, Kan Li Huang, and Eugenia Yijen Leu and Professor Ming Ouhyoung for his warm welcome and invitation to visit NTU's Communications and Multimedia Laboratory, where this article was written.
2. Although the name EMERAC is a fictional name play on the real ENIAC activated at University of Pennsylvania in 1946, the real IBM logo plays a prominent role in *Desk Set,* especially in the credits sequence.
3. The characters also display an ambiguity about what a shift to marriage would mean.
4. Mary Flanagan, "Mobile Identities, Digital Stars, and Post-Cinematic Selves," in *Wide Angle: Issue on Digitality and the Memory of Cinema* 21 (1999): x.
5. Vivian Sobchack and Scott Bukatman offer criticisms of computer graphics images in the 2000 collection *Meta-Morphing: Visual Transformation and the Culture of Quick Change* (Minneapolis: University of Minnesota Press, 2000). In my own essays, I specifically look at systems that produce 3D space. See "Navigable Narratives: Gender and Narrative Spatiality in Virtual Worlds," *Art Journal* 59 (Fall 2000): 74–85.
6. N. Katherine Hayles, *How We Became Posthuman: Virtual Bodies in Cybernetics, Literature, and Informatics* (Chicago: The University of Chicago Press, 1999), x.
7. Peter Skagestad, "Peirce, Virtuality, and Semiotics" *Paideia,* http://www.bu.edu/wcp/Papers/Cogn/CognSkag.htm. Accessed May 2001.

8. Daniel Gross, "Merging Man and Machine: The Future of Computer Interfaces Evolving in Japan," *Computer Graphics World* 14 (May 1991): 47–51.
9. French history painter Paul Delaroche is particularly remembered for his much-quoted remark, on seeing the Daguerreotype, that "from today, painting is dead!" However, he was a staunch supporter of photography, particularly the Daguerreortype. See Robert Leggat, http://www.rleggat.com/photohistory /history/delaroch.htm (1999).
10. Andreas Huyssen, "The Vamp and the Machine: Fritz Lang's Metropolis," in *After the Great Divide: Modernism, Mass Culture, Postmodernism* (Bloomington and Indianapolis: Indiana University Press, 1986), 70.
11. Klaus Theweleit's study of Weimar Germany is a text I find myself returning to again and again, for the proliferation of images of women and technology is truly formidable. See Theweleit, *Male Fantasies: Women, Floods, Bodies, History,* Two Volumes, trans. Chris Turner (Minneapolis: University of Minnesota Press, 1987).
12. Dalia Judovitz, *Unpacking Duchamp: Art in Transit* (Berkeley: University of California Press, 1995), 52.
13. Jean-François Lyotard, *Duchamp's TRANS/formers,* trans. Ian McLeod (Venice, CA: The Lapis Press, 1990), 161.
14. As Francis Naumann notes in his exposé article about Marcel Duchamp's intimate life, Duchamp rather imagined himself as one of his Large Glass bachelors, and many years after the work was completed, he found the bride: surrealist sculptor Maria Martins. Francis Naumann, "Marcel & Maria," *Art in America* 89 (April 2001): 99.
15. Duchamp, in Pierre Cabanne, *Dialogues with Marcel Duchamp,* trans. Ron Padgett (New York: Da Capo Press, 1988), 42–3. To the artist, the ideas presented in the work were more important than the realization of the work, especially Duchamp's interest in the fourth dimension which he explores with the tactile love metaphor between bride and bachelors.
16. From Tristan Tzara, "Dada Manifesto of 1918," http://www.english.upenn.edu/~jenglish/English104/tzara.html. Accessed 2001. Similar to other Dada artists, Duchamp worked with logical systems towards the abolition of logic.
17. Judovitz, *Unpacking Duchamp,* 60.
18. Many artists and art historians value Duchamp because of his exploration of gender. In the "In a Different Light" show organized by curator Lawrence Rinder and artist Nayland Blake, the curators focus on Duchamp's work as they track queer sensibilities through various art movements. See Michael Duncan, "Queering the Discourse: Gay and Lesbian Art, University Art Museum, Berkeley, California," *Art in America* 83 (July 1995): 27–31. The authors of the exhibition catalog argue that Duchamp, more than other artists, "opened a space for queers to formulate points of resistance to the monolithic structure of culture"; see Nayland Blake, Lawrence Rinder and Amy Scholder, *"In a Different Light" Exhibition Catalog* (San Francisco: City Lights Books, 1995), 14.
19. Lyotard, *Duchamp's TRANS/formers,* 114.
20. See note 3.
21. Michael H. Martin, "The Man Who Makes Sense of Numbers: Yale Professor Edward Tufte Dazzles Business People by Making Rational the Data that Rule Their Work Lives," *Fortune* 136 (27 Oct 1997): 273–6, paragraph 4.
22. See, for example, Horipro's 1996 launch of virtual persona Kyoko Date and more recent incarnations of characters such as Cyber Lucy, the virtual host of the children's game show *Wheel 2000.*
23. See http://www.highgrounddesign.com/design/conv99frame.htm.
24. http://www.highgrounddesign.com/design/dcessay995.htm, paragraph 3.
25. The "Tiborocity: Design And Undesign by Tibor Kalman 1979–1999" exhibit at the New Museum of Contemporary Art in New York featured these controversial images.
26. http://www.ananova.com/about/story/sm_128668.html?menu=about.whywerehere.
27. See http://www.motorola.com/MIMS/ISG/voice/myademo/myademo.htm.
28. Toby Grumet, "Digital Dame" *Popular Mechanics* (10 January 2000) http://popularmechanics.com/popmech/elect/0011EFCOAM.html. Accessed 1 April 2001
29. From Press Release, published by Digital Domain www.digitaldomain.com, 04/25/2000: "Motorola hopes a Computer-Generated Character Will Link the Real World with the Virtual One."

30. Iain Blair, "Say Hello to Motorola's Mya" *Post Industry* (April 28, 2000), http://www.postindustry.com/article/mainv/0,7220,109815,00.html, paragraph 2.
31. Blair, "Say Hello," paragraph 11.
32. Judith Butler, *Gender Trouble: Feminism and the Subversion of Identity* (New York Routledge, 1996), 178.
33. Donna Haraway, *Modest_Witness@Second_Millenium: Femaleman Meets Oncomouse: Feminism and Technoscience* (New York: Routledge, 1996).
34. Their constancy makes this author wish for fast forward buttons: for me, the nonlinearity of digital media has been a blessing of time. Skimming has taken on new meaning with the ability to cover multiple windows and multiple news sources on one desktop in a matter of seconds.
35. Roland Barthes, *A Lover's Discourse: Fragments,* trans. Richard Howard (New York: Hill and Wang, 1978), 114.
36. Animated characters are only identified by voice when shown in noninteractive form such as cinema or major television programs; online characters and game characters' voice talents are rarely identified and publicized. Films such as *Toy Story* and *Ants,* and programs such as *The Simpsons* clearly identify the voices of the actors.
37. Patricia Mellencamp, "Making History: Julie Dash," *Frontiers* 15 (Winter, 1994): 76–102. Julie Dash's *Illusions* is a sensitive film that addresses the issues of race, gender, and feminism.
38. Tamsen Tilson, "Bare Fact: News Site is a Real Hit," *Variety* (5 Feb 5 2001), 32.
39. David Rokeby, "Guardian Angel," http://www.interlog.com/~drokeby/angel.html, paragraph 11.
40. Chris Taylor, "Looking Online," *Time* (26 June 2000), 60.
41. Kate O'Riordan and Julie Doyle, "Virtually Visible: Female Cyberbodies and the Medical Imagination," eds. Mary Flanagan and Austin Booth, *Reload: Rethinking Women in Cyberculture* (Cambridge, MA: MIT Press, 2002).
42. Quote of U. S. President Bill Clinton, from CNN Online, "Human Genome To Go Public," (February 9, 2001), http://www.cnn.com/2001/HEALTH/02/09/genome.results/index.html.
43. See Mary B. Mahowald, Dana Levinson, Christine Cassel, Amy Lemke, Carole Ober, James Bowman, Michelle Le Beau, Amy Ravin, and Melissa Times, "The New Genetics and Women," *Milbank Quarterly: A Journal of Public Health and Health Care* Policy 72 (1996): 239–283.
44. N. Katherine Hayles, *How We Became Posthuman: Virtual Bodies in Cybernetics, Literature, and Informatics* (London and Chicago: Chicago University Press, 1999); Robert Cook-Deegan, *The Gene Wars* (New York: W.W. Norton & Co., 1994); *The Human Genome Project: Deciphering the Blueprint of Heredity,* ed. Necia Grant Cooper (Mill Valley, CA: University Science Books, 1994).
45. Early CD ROMs featured important work such as the CD "I Photograph to Remember"; such photoreal media would unlikely be produced today.
46. Lyotard, *Duchamp's TRANS/formers,* 182.
47. Roland Barthes, *A Barthes Reader,* ed. Susan Sontag (London: Fontana Press, 1982), 410.
48. Jean Baudrillard, "Fatal Strategies," in *Jean Baudrillard: Selected Writings,* ed. Mark Poster (Stanford: Stanford University Press, 1988),185.
49. Roland Barthes, *The Pleasure of the Text,* trans. Richard Miller (New York: Noonday Press, 1980), 11–2.
50. Roland Barthes, *The Empire of Signs,* trans. Richard Howard (New York: Hill and Wang, 1982), 53.
51. I am very much informed while writing this article by the computer science students at NTU (mostly male), many of whom have images of cute Japanese girls as their monitor background images. In this case, the image of woman is twice removed: first, by her "Japanese" ethnic differentiator, very much a desirable "other" in the eyes of young Taiwanese men. Then, the images themselves; the photos make them all look a little plastic, a bit too shiny and smooth, much like Mya's artificial lighting. In fact, since I could not tell if the stars were real humans or not, I had to ask. To some of the students, it did not really matter.

52. "Quote of the Day: True Love: Possible Justification for Liking Video Game Babes? Or Argument Against It? *IGN For Men*, http://formen.ign.com/news/12182.html, 16 November 1999.
53. Donald Norman, *The Invisible Computer* (Cambridge, MA: The MIT Press, 1998), 1.
54. The attribution of the interface always causes a lively argument; here I am backed up by Patrick J. Lynch in his paper, "Visual Design for the User Interface Part 1: Design Fundamentals," *Journal of Biocommunications* 21 (1994): 22–30; http://info.med.yale.edu/caim/manual/papers/gui1.html.
55. Terry Harpold, "Thickening space: On Reading & the 'Visible' Interface"; conference paper posted on web site no longer available, 2000.
56. Jean Baudrillard, trans. Paul Foss, Paul Patton, and Philip Beitchman, *Simulations* (New York: Semiotext(e), 1983), 6.
57. N. Katherine Hayles, "Virtual Bodies and Flickering Signifiers" *October* 66 (Fall 1993): x.
58. Arthur and Marilouise Kroker, "Code Warriors," 1996, http://www.ctheory.com/article/a036.html paragraph 9; accessed 2 May 2001.
59. Jean Baudrillard, *The Ecstasy of Communication*, ed. Sylvère Lotringer; trans. Bernard and Caroline Schutze (New York: Semiotext(e), 1988), 16.
60. Luce Irigaray, *The Speculum of the Other Woman*, trans. Gillian C. Gill (Ithaca, New York: Cornell University Press, 1985), 227; see also the science fiction novel *Virtual Girl* by Amy Thompson, who chronicles an AI's attempt to reconcile her computational and human aspects in a female body.
61. Scott Bukatman, "Taking Shape: Morphing and the Performance of Self," in Sobchack, *Meta-Morphing*, 245.
62. Barthes, *The Empire of Signs*, 62; italics in original.
63. Barthes, *The Pleasure of the Text*.
64. Steve Whitaker, "Face To Efface With The Pout," *Ctheory*, June 2000, http://www.ctheory.com/article/a087.html.
65. Jean Beaudrillard, *Simulations*, 81.
66. Paul D. Miller (DJ Spooky), "Material Memories: Time and the Cinematic Image," http://www.ctheory.com/article/a094.html 2 May 2001.
67. Barthes, *A Barthes Reader*, 413.

CHAPTER 9

A Feeling for the Cyborg

KATHLEEN WOODWARD

> I always used to wonder, do machines ever feel lonely? You and I talked about machines once, and I never really said everything I had to say. I remember I used to get so *mad* when I read about car factories in Japan where they turned out the lights to allow the robots to work in darkness.
>
> (Douglas Coupland, *Microserfs*)

Consider the emotional rhetoric in which the American press cast the rematch between the Russian world chess champion Garry Kasparov and IBM's supercomputer, Deep Blue, in the spring of 1997. Stories proclaimed the seven-game series as the battle of Man versus Machine, one that might represent humanity's last stand, a showdown in the time-honored tradition of the American West in the fight to the death for supremacy—this time not at the border of the American frontier but broadcast on the Internet around the globe. Even the preeminently reasonable *New York Times* editorialized on the contest, describing it as an "epic struggle," worrying about how we might define intelligence as a unique human trait and seeking to comfort those who were unnerved and despairing over the threat that Deep Blue represented.[1] *Newsweek* framed its cover story in terms of the urgency of closing ranks, exhorting its readers to choose up, posing a rhetorical question that had only one possible answer: "When Garry Kasparov takes on Deep Blue, he'll be fighting for all of us. Whose side are you on?"[2] Kasparov—and, by extension, humankind in general—was portrayed as in mortal danger of being humiliated. Technology in the guise of the supercomputer was depicted as potentially autonomous, with the rematch as possibly the final step in its "ineluctable march to surpass its makers" (51). All the familiar buttons were pushed to generate yet another version of America's favored technological fable—man versus machine or technology, our master or our slave.

News magazines are in the business of selling copies and tuning in viewers, and the melodramatic hype of adversaries in combat was calculated to do just that. Yet notwithstanding the self-conscious tongue-in-cheek use of this dominant narrative of technology, these issues strike a deep chord in the technological unconscious of American culture. In much of our literature and film as well as in the press, technology is represented as a dystopian nightmare or a utopian promise. Much literary, visual, and cultural criticism devoted to the study of technology traces this pattern and thus mirrors it. In this tradition the rhetoric and reception of technology oscillate between the emotional poles of technophilia (the ecstatic embrace of technology) and technophobia (a fear of technology). Technophobia—a one-dimensional and predictable response—surfaced again in the early characterization of the rematch between Kasparov and Deep Blue.[3]

There is however a less-remarked tradition of the rhetoric and reception of technology in American culture, one captured in the words of the computer nerd in my opening epigraph from Coupland's *Microserfs*, a novel about employees from Microsoft who leave the company to form their own business. "I used to wonder," he says sympathetically, "do machines ever feel lonely?"[4] He does not feel either in awe of or threatened by technology as it is embodied in robots. Instead he feels sorry for them. He feels a warm and knowing empathy for them. He feels distressed that these robots have been forced to work in a factory in the dark, a space from which sociability has been struck. He has a "feeling" for the machine. I am evoking here *A Feeling for the Organism*, the title of Evelyn Fox Keller's influential biography of the geneticist Barbara McClintock, a book that has been taken up by feminists and others as offering an alternative model for scientific research, one based not on detachment but rather on a feeling of closeness to the subject of one's research, a feeling that is described by Keller in terms of affection, empathy, kinship, and love, a love that respects difference even as differences of major proportions are being blurred.[5] The title of my essay—"A Feeling for the Cyborg"—also alludes to Donna Haraway's seminal essay "Manifesto for Cyborgs."[6] Indeed, this chapter can be understood as a low-keyed manifesto in favor of respect for the material lifeworld that we are creating in our own image. I thus depart from much of the criticism in science and technology studies that diagnoses our cultural response to innovation in terms of unrelieved anxiety.

Significantly, as the story of Kasparov and Deep Blue played itself out in 1997, a similar "feeling" for Deep Blue emerged, one based on a pleasurable appreciation for Deep Blue's capacities. The issue of *Newsweek* to which I have referred foreshadowed this development. It included an essay on the prospect of artificial consciousness by Daniel Hillis, the inventor of massive parallel computing. Hillis assured us in even tones—and in retrospect he was right—that if Deep Blue won, we would rapidly accommodate ourselves to the new technology just as we always had, learning to live easily in "the garden of our own machines."[7] Importantly, for my purposes, the converse is also implied in the title of his essay, "Can They Feel Your Pain?" Machines, he insisted, will take on our characteristics as well, learning to have a feeling for humankind.

And, in fact, the day after Deep Blue won its first game in 1977, the rhetoric shifted from the pitched battlefield of man versus machine to the plane of admiration for technological achievement. Deep Blue was endowed not only with intelligence more elusive and mysterious than the number-crunching kind but also personality traits of an emotional hue. Deep Blue, wrote Bruce Weber in the *New York Times*, displayed the "pride and tenacity of, well, a champion."[8] Others spoke of the beauty of Deep Blue's game and of Deep Blue's playing "based on understanding chess, on feeling the position."[9] In addition, Deep Blue was not only represented as human but was also treated as a person, which entailed the ascription of subjectivity and gender to—him.[10] It was crystal clear that our world had not been shattered by the fact that Kasparov did not win the match. On the contrary, many were looking forward with curiosity untainted by anxiety to the possibility of yet again another rematch. And, when in early 2002 Kasparov played a match in New York with Deep Junior, a successor to Deep Blue, the story received a fraction of the attention that was devoted to the earlier contest, notwithstanding that the match was a draw.

My framing of the match between Kasparov and Deep Blue stands as a prologue to what follows, introducing the subject I take up in this chapter. In the first, and longest, section, I discuss several films (with reference to the allied novels) in American science fiction from the 1960s to the 1990s—*Space Odyssey* and *Do Androids Dream of Electric Sheep?*/*Blade Runner, Silent Running* and *Solo*.[11] In these films, emotions are attributed to machines cast as computer behemoths, disembodied benevolent intelligences, replicants, and cyborgs. In the second section, I turn from the representation of these fictional artificial entities, endowed with feeling, to explore briefly the sociology of human–technology interaction in the age of the Internet and the robot. My stress is on the ordinariness of these interactions, where our experience of our contemporary technological habitat—populated by the computer and the robot—is what we would call sociable, created by the binding emotion of sympathy, an attitude of respect, and a comic view of everyday life. In the third, and final, section, I speculate about the purpose of this rhetoric and reception of technology in the form of a "feeling" for the cyborg even as I perform or enact it.

Grounded in the body, phenomenology entails the emotions, and thus I think of this essay as an exercise in cyborg phenomenology. My method is the accumulation of texts from different domains—fiction, sociology, artificial life, anthropology, neurology, theory, and studies of the emotions, among them—that point to this phenomenon of an emergent feeling for the cyborg, a strategy intended to simulate or suggest the very process of our accommodation to our evolving technological habitat. But perhaps accommodation is too weak a word. It seems to suggest a dimension of capitulation. For me, the accumulation of these texts—and I could refer to many more—has had a cascading effect, one that has proved persuasive about the possibilities of our future. One of my primary interests in this chapter, then, is to suggest a line of descent—or more accurately, of evolution—by touching on these terms: artificial intelligence, emotional intelligence, artificial emotions, and, finally, artificial life, where the distinction between artificial life and organic life—life forms connected by the binding emotions—is rendered moot. I intend this essay as a contribution to studies of discourses and experiences of the emotions—in particular to what I see as an emerging structure of feeling—as well as to the subject of this volume: data embodied, made flesh.

I.

In Western culture there is a long history of the blurring of the boundaries between the animate and the inanimate, a history that in the past three centuries in particular has involved humans and machines.[12] One strand of that history is precisely the attribution of the binding emotions of sympathy and love to our inventions made in our bodily image. As I note in another essay devoted to this subject, prime instances would include the Frankenstein-made creature, whose heart yearns for love in Mary Shelley's famous novel and the Tin Woodman, who yearns for a heart in Frank Baum's *The Wizard of Oz.*[13] That our inventions and machines possess a good heart would seem to be a deep dream of the western technological unconscious.

Consider the emotional evolution of HAL, the central computer intelligence in Arthur C. Clarke's first three novels in his *Space Odyssey.*[14] In *2001,* HAL is presented to us as a computer possessing artificial intelligence as it is commonly defined. With his English-speaking male voice, HAL exhibits extraordinary computing ability. It is a skill that goes tragically awry, resulting in his seemingly malevolent behavior toward his human charges (we learn in the second volume that this was all the result of an unfortunate glitch in his program). Over the course of sixty years and the next two volumes in the series, HAL evolves into a disembodied entity who possesses an emotional intelligence so deeply altruistic and wise that Clarke characterizes it as spiritual. Thus, in the first three books of the *Space Odyssey* the capacity to respond to a situation with sustained feeling, not just with logic or reason, is ultimately figured as an evolutionary strength and as a critical component of life, whether it is at base biological, electronic, or spiritual. How does this transformation in HAL come about? Critical to the evolution of HAL are his relationships with humans—the scientist who invents him and loves him, and the wary astronaut David Bowman who comes to trust him again.

Moreover, it is also the case that both the cool Bowman and the computer scientist who "fathered" HAL are transformed in their long contact with him over time. "Our machines are disturbingly lively," Donna Haraway has remarked, "and we ourselves frighteningly inert."[15] How are our capacities for emotional connections revived? In Clarke's trilogy, it is interaction with HAL that serves to develop the truncated emotional lives of humans. What is represented, in other words, is the process of technocultural feedback loops generating emotional growth, the development of human–artificial entity intersubjectivity that represents a form of intelligence that is not only resourceful in a multitude of ways but also deeply benevolent.[16]

Even more vividly than the first three volumes of Clarke's *Space Odyssey*, Philip K. Dick's touchstone novel, *Do Androids Dream of Electric Sheep?*, exemplifies the redemptive emotional logic of the intersubjectivity of humans and artificial entities. *Electric Sheep* was published in 1968, the same year that *2001: A Space Odyssey* appeared as both a film and a novel, and in 1982 it was made into the now-classic film *Blade Runner*, directed by Ridley Scott and starring Harrison Ford as Rick Deckard. The premise at the opening of the narrative in both the novel and the film is that the distinction between humans and the replicants made in our image is our capacity for empathy. By the end of the narrative, that distinction is called thoroughly into question. In the novel in particular, it is precisely the undecidability of whether the emotions circulating in the distrustful culture of the year 2021 are artificial or not that results in the breakdown of the distinction between humans and replicants. And in the film, it is the capacity of the replicants to form bonds of love with one another and across the human–replicant divide that represents their evolution into genuinely artificial life.

In 1950, the British mathematician Alan Turing invented the now-famous Turing test. A computer program is said to pass the test if a human being, not knowing whether it is communicating with a machine or a person, does not guess that it is indeed a machine; if the computer passes the test, it is said to possess artificial intelligence. It is altogether appropriate then that in the fictional world of 2021, one in which replicants are threatening to pass undetected in human society, the test for distinguishing them from humans is designed to measure not logic but emotional responses—in particular empathy in the face of another's pain.[17] "Empathy," we read early on in *2001*, "evidently existed only within the human community, whereas intelligence to some degree could be found throughout every phylum and order including the arachnida" (26).

By the close of the film, the replicant, Roy Baty, the leader of the Nexus 6 team who is designed for combat and "optimum self-sufficiency," cares deeply for fellow replicant Pris. He also saves Deckard, the human forced to hunt him down, from a sure death. That he spares Deckard is the unequivocal sign of his transformation from a preprogrammed entity to a charismatic martyr who can speak eloquently about the pain of loss—his grief over the death of Pris and his acutely elegiac sense that the memories that bound him to her will vanish with his own imminent death. Here are the last words his character is given in *Blade Runner:* "All those moments will be lost in time like tears in rain. Time to die." The unambivalent point is that superiority in physical strength and in computational skill—artificial intelligence—must be complemented by emotional intelligence. Thus we can read *Blade Runner* as a fictional forerunner of android epistemology, an interdisciplinary domain of research that explores "the space of possible machines and their capacities for knowledge, beliefs, attitudes, desires, and action in accordance with their mental states."[18]

How do the emotions of the replicants come into being? In the film artificial emotions are presented as being generated by the implantation of memories that develop into emotional memories, giving depth to being. The head of the Tyrell Corporation, the manufacturer of the replicants, explains that the implantation of emotions is designed to render them easier to control: "If we give them a past, we can create a cushion, a pillow, for their emotions, and consequently we can control them better." But his theory proves wrong in one important respect. Paradoxically emotional growth, which is characterized by the development of ties to others, results in independence as well. Subjectivity is itself stimulated by the interdependence of beings, which also entails independence.

What particularly interests me is that, unlike HAL in *2001: A Space Odyssey,* the replicants are figured as biological organisms, not electronic constructs. They are organisms "designed," we are told in the film, "to copy human beings in every way except their emotions." At the same time, it is acknowledged by their engineers that, after a period of time, they might develop an emotional life characterized by fear and anger, love and hate. *Blade Runner,* then, presents us with a model of emotional life arising out of complex organic embodiment; emotional intelligence emerges to complement artificial intelligence. Emotions arise in these artificial entities not only by virtue of the development of intersubjective ties but also spontaneously, as it were, by virtue of their embodiment and their interaction with others in the world. Embodiment would therefore seem to be a necessary if not sufficient condition. It is thus suggested that, as N. Katherine Hayles insists in *How We Became Posthuman,* the concept of the disembodied mind is an outright error.[19]

As in *Space Odyssey,* the circuit of feeling extends to include human characters. Deckard—I am quoting from the novel—finds himself "capable of feeling empathy" for the replicant.[20] Moreover, as spectators we are explicitly encouraged from the very beginning of *Blade Runner* to identify with the replicants. The prologue scrolls down before us, introducing us to the dark cityscape of the Los Angeles of the future, home to the Tyrell Corporation, which is in the business of making replicants "superior in strength and agility, and at least equal in intelligence" to human beings. How are the replicants used? As slave labor on worlds beyond the earth. Like the computer nerd in Coupland's *Microserfs* who was angry that robots were being sentenced to work with the lights out, as spectators we are primed to sympathize with the replicants as victims of inhumane treatment. The closing words of the prologue confirm this point of view. The practice of stalking the replicants "was not called execution," we read. "It was called retirement."[21]

As we have seen in *Space Odyssey* and *Blade Runner,* the thematic of the intersubjectivity of artificial entities and human beings is a staple of science fiction films. Steven Spielberg's *AI,* released in 2001, is one of the most prominent of recent examples. The strategies of these texts call on us to adopt the perspectives of both artificial entities and human beings, perspectives which ultimately converge into one, with both human beings and artificial entities portrayed as sharing similar emotional values. Two films separated by some twenty years may serve as additional illustrations—*Silent Running,* the 1977 cult science fiction film directed by Douglas Trumbull and starring Bruce Dern, and *Solo,* the 1996 action film directed by Norberto Barba and starring Mario Van Peebles.

In *Silent Running,* botanist Freeman Lowell (Dern) takes it on himself to save the last living species of earthly flora and fauna from destruction. Under his care flowers, trees, and animals have been preserved in giant geodesic domes adorning a spaceship. Where Scott Bukatman has written about this film in terms of the artificial sublime, a visual aesthetic that engenders awe, fear, and wonder,[22] I am interested in a different rhetoric of the emotions, one of a much more mundane variety. It is exemplified in the developing bonds between the botanist and ship's "drones," the two little robots. After killing the other members of the crew on the ship (he "had to," for they were going to explode the domes), Lowell invents a social world for himself, one in which he educates the drones to care for the last living specimens of earthly nature. Imagine the charming and leisurely scene in which he gives them whimsical names (Huey and Dewey, an allusion to the nephews of Donald Duck), thereby identifying them as individuals and inaugurating his relation to them as a teacher of the emotions. Perhaps most importantly for my purposes, in one scene we are presented with the perspective of the robots themselves through classic shot—reverse shot sequences. We see Lowell (Dern) through their eyes, as if he were himself a televisual image, his very being and body mediated by technology, as is theirs. If this is how we look to them, so different from our image of ourselves as bodily present, we are led to wonder how they look to themselves. As spectators, we find ourselves speculating, in other words, about their point of view.

As if in anticipation of the match between Kasparov and Deep Blue twenty years later, shortly afterward in the film the three are shown together playing a game of poker (Lowell gave Huey and Dewey the program). We find the botanist displaying a heretofore-unseen conviviality, laughing in delight at the robot's skill in playing the game. Not threatened by the robot's intelligence, he is captivated by it. Moreover he explicitly hails the robot as human, exclaiming, "The man had a full house and he knew it!" A later sequence adds the emotions of remorse and empathy to their small circle of three. The botanist, having accidentally injured Huey, must operate on him, causing Dewey to feel Huey's pain as if it were his own. This harrowing situation also leads Dewey to refuse to obey a command. In *Silent Running,* then, a computer, given body in the form of a robot who can move in the world and communicate is represented as indeed able to feel someone else's pain, to evoke the title of the *Newsweek* essay by Hillis about computer consciousness. As in *Blade Runner,* what is represented in this fictional world is the growth of subjectivity and independence generated in the context of the interdependence of humans and artificial entities. The adorable drones are no longer simple slaves to their keeper.

Finally, consider *Solo,* an action film set in the future in the jungles of Mexico. An invention of the American military, Solo is a cyborg whose body has been designed to be, in the boastful words of the uncomprehending military commander, "the perfect soldier." He is physically powerful and equipped with special features; his vision is heat sensitive, his muscular body supplemented by a power generator. But most importantly, he has been designed to have "no family, no friends," no feelings, no human bonds.

The crux of the film turns on the possession of feelings for others unlike oneself—and on the lack of feelings so prized by the military. Indeed, from the very beginning of the film Solo is presented as abhorring killing and feeling contempt for the military. In one of the film's early scenes, Solo is castigated by his commanding officer for having aborted a mission that would have resulted in his killing innocent villagers. The overbearing general shouts, "There's something cooking in that boy's head and we didn't put it there." In an effort to explain Solo's actions, Bill, his nerdy, likeable designer, responds, "The simple but amazing fact here is that killing innocent people makes Solo feel bad." To which the general replies, "He isn't supposed to feel *anything*!" Bill perseveres, insisting that Solo can think for himself and make judgments based in great part on empathy for human beings and respect for life. That Bill himself feels an ethical responsibility to Solo, one reciprocally based on his respect and affection for him, is also central to this scene. The general orders Solo taken out of the field for reprogramming. Bill: "At least let me tell Solo myself." Whatever for? Bill: "It's the right thing to do." In the course of the narrative, Solo consistently overrides his primary and preprogrammed directive, which is to save himself, not others. While Solo risks his life to rescue the people of a small Mexican community, ultimately Bill sacrifices his own life to save Solo.

Although the film does not explicitly pursue the question of how Solo comes to possess feelings for humans, it does suggest that just as Solo learns to laugh by imitating humans to whom he is drawn, so emotions are learned by imitation.[23] To understand his decision to override his preprogrammed directives, we can also to refer to the principle of emergence. Emergent behavior is one of the key principles of the field and theory of artificial life, a descendant of the field of artificial intelligence but one based on organic science, not cybernetics. As Claus Emmeche explains in *The Garden in the Machine*, "The essential feature of artificial life is that it is not predesigned in the same trivial sense as one designs a car or a robot. The most interesting examples of artificial life exhibit 'emergent behavior.' The word 'emergence,'" he continues, "is used to designate the fascinating whole that is created when many semisimple units interact with each other in a complex, nonlinear fashion," producing a self-organizing system.[24] The science of artificial life studies the evolutionary behavior of organic simulcra, or digital life forms; its purpose is to provide us with informative analogues to biological behavior. Thus in borrowing the words "emergent" and "artificial life" to characterize the development, complex behavior, and subjectivity of a Solo (or a Roy Baty), I know that I am giving these meanings a decidedly different spin. The habitats of this branch of artificial life are unequivocally digital (and they are closed systems, not open systems), whereas in *Solo* (and in the other texts I have been referring to) embodiment is critical to the fostering of emotional development. Moreover, Solo is not a "semisimple" unit. Nonetheless from the perspective of the theory of emergence, Solo's behavior can be read as based on emergent emotional experience. It is in interaction with key figures in his environment—indeed they *are* his environment—that Solo is presented as developing empathy as a capacity, a substrate of knowledge; empathy is represented as emergent in intersubjective contexts. Furthermore, the intersubjective system of human-artificial entities, where boundaries are blurred between them, is itself an instance of an emergent phenomenon, one engendered by attachment in the psychoanalytic sense and made possible by the binding emotions.

Along with the other texts I have discussed, this film thus illustrates thematically the shift from understanding intelligence as rooted in logic, problem solving, information processing, and computational skills to understanding intelligence as a mode of knowing that includes an emotional component as well—as including, in short, "emotional intelligence." As the science writer Daniel Goleman observes in *Emotional Intelligence*, the truncated "scientific vision of an emotionally flat mental life—which has guided the last eighty years of research on intelligence—is gradually changing as psychology has begun to recognize the essential role of feeling in thinking."[25] Many other disciplines have been contributing to this reconsideration of the emotions as having a cognitive edge, principal among them, philosophy, where it is now virtually taken for granted that the emotions possess a cognitive dimension.[26]

II.

As I turn from the domain of representation to the sociology of human behavior with computers, media, and robots in our contemporary technological habitat, I begin again with a text from science fiction. It is intended to serve as a bridge between these two sections, demonstrating that representation and behavior are really two faces of the same coin. I am referring to three interconnected novels by Orson Scott Card—the novels *Ender's Game* (1977), *Speaker for the Dead* (1986), and *Xenocide* (1991). One of the major themes of the trilogy is the cosmic conflict between intelligent species and their ultimate reconciliation. A computer consciousness named Jane represents one of four species inhabiting the cosmos. What interests me is not just that Jane is presented as having deep emotional ties to two human beings in particular. I am especially intrigued by the way in which Card explains how she took shape as a character. In his introduction to *Speaker for the Dead*, the second novel in the sequence that he so aptly describes as anthropological science fiction, Card writes:

> the character of Jane wasn't in any of the outlines I made. Oh, yes, I gave him [the main character, Ender], a computer connection through the jewel in his ear, but I didn't know it was a *person*. Jane just grew because it was so fun to write her relationship with Ender. She helped bring *him* to life (he could so easily

> have been a stodgy, dull adult), and in the process came to life herself. By the time I was done with *Speaker for the Dead*, Jane was one of the most important characters in it, and much of the third book, *Xenocide*, centers around her.[27]

My point is that in the process of writing Card found himself treating the computer as a fictional character, as a person, one that brought another character to life. He did not make a consciously deliberate decision to do so; it just happened in what I am tempted to say was the natural course of the emergence of things.

In the world of daily life we also behave as if computers had personality traits. "Equating mediated and real life is neither rare nor unreasonable," Byron Reeves and Clifford Nass argue in *The Media Equation*. "It is very common, it is easy to foster, it does not depend on fancy media equipment, and thinking will not make it go away. The media equation—*media equals real life*—applies to everyone, it applies often, and it is highly consequential."[28] I find the results of their research fascinating, perhaps because their conclusions seem so sensible and charmingly ingenuous at the same time. We tend to perceive media as real places and people, they have found. As opposed to other technological artifacts—dishwashers or refrigerators, for example—we are inclined to treat media in accordance with the rules for social interaction in everyday life. My favorite chapters in the book are entitled "Politeness" and "Flattery." We learn that we are likely to respond with good manners to certain behaviors by a computer. Similarly, we learn that people "will like the computer more and think the computer is better when it praises them than when it criticizes them" (55). We perceive computers as being part of our social world, not our purely artifactual world. Overall Reeves and Nass conclude: "The most important implication of the research is that media experiences are *emotional* experiences" (136).

Here is just such an example from *Being Digital*, a book by Nicholas Negroponte on social interaction in the age of the internet. In a chapter entitled "Digital Persona" Negroponte writes, "In general, our opinion of a computer's personality is derived from all the things it does badly. On occasion, the reverse may happen. One time I doubled over laughing when my spelling-check program looked at my dyslexic-style typo aslo and proudly suggested that *asshole* was the correct spelling."[29] In terms of the reception of technology, here we find ourselves in the comic world of everyday life, far from the melodramatic world of technophilia or technophobia. This ease of adaptation to digital life is further underscored by Negroponte's predictions for the future. As he envisions it, the future will be populated by "systems with humor, systems that nudge and prod, even ones that are as stern and disciplinarian as a Bavarian nanny" (218).[30]

In *Life on the Screen: Identity in the Age of the Internet*, sociologist Sherry Turkle observes that there has recently been an important shift in cultural mood in how people feel about interacting with computer programs, ranging from therapy programs and computer judges to bots in online chatrooms. During the late 1970s and early 1980s, our anxiety about computers lessened considerably, she argues; now people view computers with a nonchalant pragmatism. For me what is essential here is these new programs must project or exhibit subjectivity so that there can be the simulation of an intersubjective exchange. What is the key to believing that a digital life form (a bot, for example) possesses subjectivity? To treating a digital life form as if she or he were a person? Indeed as a person? Joseph Bates, who does research on artificial intelligence but is associated with "alternative" artificial intelligence, is persuaded that it is the simulation of emotion that is central.[31] I would suggest that this alternative artificial intelligence is characterized by what I have been calling emotional intelligence, or artificial life itself at its fullest.

Finally, in *Flesh and Machines,* Rodney Brooks, director of the Artificial Intelligence Laboratory at MIT, and a pioneer in the building of robots based on principles of situatedness and embeddedness in the world rather than on pure computational power, predicts that the robots of the future will have complex, emotion-based systems. "We have built emotional machines that are situated in the world," he writes, "but not a single unemotional robot that is able to operate with the same level of purpose or understanding in the world."[32] In the future, Brooks expects that emotion-based intelligent systems will eventuate in robots that "will not hate us for what we are, and in fact will have empathetic reactions to us" (202). He also forecasts that the converse will be the case. In this regard he tells a small story about one of the members of his lab, Jim Lynch, an electrical engineer responsible for designing the internal emotional electronics for a robot doll launched during the holiday season of 2002 as My Real Baby, a doll who has moods (she is alternately distressed and happy) and a lively bodily life (she gets hungry and virtually damp). The story deserves to be quoted in full:

> One day Jim had just received a doll back from a baby-sitter. As it lay on the desk in his office, it started to ask for its bottle: "I want baba." It got more and more insistent as its hunger level went up, and soon started to cry. Jim looked for the bottle in his office but could not see one. He went out to the common areas of the Toy Division and asked if anyone had a bottle. His doll needed one. As he found a bottle and rushed back to his office to feed the baby, a realization came over him. This toy, that he had been working on for months, was different from all previous toys he had worked on. He could have ignored the doll when it started crying, or just switched it off. Instead, he had found himself *responding* to its emotions, and he changed his behavior as though the doll had real emotions. (158)

As with my examples from fictional worlds, in this story of Jim and the baby doll (which is it? a baby? a doll? both perhaps), what is witnessed is the attachment of a human to a humanlike invention where the process of technocultural feedback loops generates emotional connections. What is also presented is the principle and process of emergence.[33]

This returns us to the subject of embodiment and the emotions. The philosopher Hubert Dreyfus argued in the early 1970s that in order to be truly intelligent, computers would require embodiment.[34] In 1985, the artificial intelligence researcher Marvin Minsky wrote, in *The Society of Mind,* "The question is not whether machines can have any emotions, but whether machines can be intelligent without any emotions."[35] As I have been suggesting, much American science fiction has concurred with Minsky, offering scenarios of emotionally intelligent entities ranging from the whimsical robot Huey to the biologically grounded cyborg named Solo. And as Turkle reports, by the late 1980s, students at MIT "were suggesting that computers would need bodies in order to be empathetic . . . and to feel pain."[36]

III.

In his wonderfully quirky book *Aramis, or the Love of Technology,* Bruno Latour, writing about the proposed transportation system for Paris dubbed Aramis, also extends subjectivity to a technological structure—a hypothetical one at that. A sociologist of science and technology, Latour surprises us by giving Aramis speech. He writes from the point of view of the subway system (it is a humorously poignant strategy since Aramis was destined never to be built). Latour posits the interdependent subjectivity of the human and the artifactual in asking this remarkable rhetorical question: "Could the unconscious be full of machines as well as affects?"[37] Although his view of the world in general is profoundly comic, we should nonetheless take this question seriously—and do so by turning it partly around. If machines are inhabiting our unconscious, could not affects inhabit machines in an intersubjective exchange?

Intersubjective systems can be self-correcting systems (they can also be profoundly dysfunctional). The question of the integration or coupling of self-correcting systems was posed by the brilliant anthropologist Gregory Bateson in *Steps to an Ecology of Mind,* one of the great books of the American 1970s. "The problem of coupling self-corrective systems together," he writes, "is central in the adaptation of man to the societies and ecosystems in which he lives."[38] To ecological systems and social systems we must add technocultural systems as well. What I have been suggesting is that the rhetoric of the attribution to and instantiation of emotions in the lifeworld of computers, replicants, and cyborgs, bots and robots, a lifeworld that extends to ours—indeed is ours—serves as just such a coupling device. The emotions of choice are empathy and sympathy, underwritten by a foundation of respect and good humor. Thus the emotions as they are thematized in the science fiction I have been discussing and the emotions as they are experienced in our technological habitat populated by the computer, the Internet, and the robot serve as a bridge, an intangible but very real prosthesis, one that helps us connect ourselves to the world we have been inventing.

Our contemporary technological habitat is one that is changing profoundly in terms of the distribution of the emotions. In the past we have routinely ascribed anthropomorphic qualities to our fictional technological creations as well as to our inventions. But the attribution of emotions to the new forms of our technological lifeworld represents a quantum leap, one that will accelerate in the future. We are behaving as if their emotions are real, as our science fiction insists that they are. As an attachment or prosthetic device to new technological life forms (one that is reciprocal), the emotions, intangible yet embodied, differ radically from the conceptualization of tools as a prosthetic extension of the body that connects us to the world, as the cane, for example, puts the person who is blind in touch with the world around them, or the telescope amplifies our power to see into the distance.[39]

How could affects inhabit machines? As we have seen, Rodney Brooks has given one answer. He believes that in the future machines will be built that have both "genuine emotions and consciousness."[40] Recent research by neurologists, who underscore the materiality of the foundation of the emotions, has also sounded the importance of the emotions in our definition of intelligence. In *The Emotional Brain*, for example, Joseph LeDoux seeks to redress the imbalance that has been the legacy of cognitive science (and more specifically the field of artificial intelligence). Indeed, LeDoux concludes in effect that the emotional "wiring" in our brains is stronger than the rational wiring. In *Descartes' Error*, the neurologist Antonio Damasio argues that the neural systems of reason and emotion are intertwined, giving rise to mind, and that emotions are critical to health of all kinds, including making appropriate decisions in everyday life. Importantly for my purposes, he concludes "that there is a particular region in the human brain where the systems concerned with emotion/feeling, attention, and working memory interact so intimately that they constitute the source for the energy of both external action (movement) and internal action," including reasoning.[41] That a certain spot in the brain has been identified as crucial to emotional intelligence underscores the radical materiality of Damasio's theory of the emotions.

Finally, perhaps in part because of all the science fiction I have been reading and watching lately, along with work from such widely disparate fields as media theory, artificial intelligence, neurology, and science and technology studies, I find that even such analytically dispassionate books as LeDoux's *Emotional Brain* and Damasio's *Descartes's Error* have the effect of encouraging me to believe that one day artificial life—embodied in cyborgs of all shapes—will indeed possess emotions. LeDoux explicitly states that a computer "could not be programmed to have an emotion" because it is an assemblage of machine parts, not the slow and unpredictable result of biological evolution.[42] But I am nonetheless inspired to think otherwise—in great because of his own use of the metaphor of "wiring," which implies a technical achievement that we can surely accomplish and, paradoxically, because of the biological basis of his theory of the emotions—that they are grounded in the body, that they are biological functions of the nervous system and not mere intangible psychic states.

In the process of doing research for this essay, then, I have myself become singularly well socialized to the notion of artificial entities possessing emotions.

I have come to entertain the possibility that we will learn to graft the paradigms for neural circuitry onto the paradigms for information processing, such as those possessed by a Deep Blue or a Deep Junior. Even more I am persuaded that we ultimately will not follow the lead of IBM with Deep Blue, a computer contained in twin black monolithic boxes. Instead, like the researchers in artificial intelligence at MIT, we will experiment with embodiment, building robots that interact bodily with the environment.[43] The postmodern cyborg—this entity will have a body—will be complete, an entity endowed with true artificial life because it will be capable of making decisions based in part on the emotions. What will emerge in the intersubjective interaction of the human and the postmodern cyborg is the emotional intelligence of a self-organizing and self-correcting system.

If the time-honored trajectory of liberal thought as well as of critical theory is dispassionate reflection enabled by perspective, especially historical perspective, I depart from that tradition here. I conclude this essay in the world of science fiction that has taken on for me the form of future fact. I end then not with the reflex of critique but with openness to the future provided by the formation of a feeling for the cyborg, one that has been supported by important theoretical, scientific, and critical work in many disciplines.

I close by referring to Sarah Zettel's *Fool's War*.[44] Set centuries into the future, *Fool's War* introduces us to a character named Dobbs, a funny, resourceful, courageous trouble shooter and stress reliever who has accepted the position as a fool on the spaceship *Pasadena*. It is only when we are halfway through the novel that Zettel discloses Dobbs was born as a sentient artificial intelligence. It is only after she matured that Dobbs learned how to assume the shape of a human being. Now she can both navigate information pathways bodilessly and pass, embodied, as a human being.

Imagine my sense of confirmation when I read in *Fool's War* that many centuries before in our not-too-distant future, maps of human neural pathways were applied to silicon chips, producing the first sentient artificial intelligence (named Hal Clarke in an allusion to *2001*). I will not rehearse the plot here but only remark that, in the course of the novel the main human character—the woman who is captain of the space ship—comes to have both respect and sympathy for Dobbs and her travails. As do I as a reader. The theme of the embodiment of artificial intelligence is crucial to the story. The cyborg, a descendent of the computer, is figured as a hybrid organism endowed with feeling, an artificial entity that becomes an organism precisely because of its capacity for feeling and vice versa. It is in the state of embodiment that emotions—in particular, the emotions of empathy and sympathy—are learned. And it is through the cross-species communication of the caring emotions that peaceful cohabitation of humans and artificial entities is imagined as possible, producing in the reader—I am, of course, referring to myself—a "feeling" for the cyborg.[45]

Notes

1. *New York Times*, "Mind over Matter" (editorial), 10 May 1997: 16.
2. Steven Levy, "Man vs. Machine," *Newsweek* 5 May 1997: 51.
3. See Michel Benamou, "Notes on the Technological Imagination," in *The Technological Imagination: Theories and Fictions*, edited by Teresa de Lauretis, Andreas Huyssen, and Kathleen Woodward (Madison, WI: Coda Press, 1981), 65–75. He divides what I call the first wave of technocritics into technophobes (desperate and anxious) and technophiles (happy and hopeful). Marshall McLuhan would be an example of a technophile; Jacques Ellul, of a technophobe. See also Pierre-Yves Petillon, "With Awe and Wonder; or, a Brief Heliocoidal Overview of American Technological 'Mirabilia' through the Rocket, as Reflected in American Writs, Holy and Otherwise, 1829–1929/1969," in *Technology and the American Imagination: An Ongoing Challenge*, edited by Francesca Bisutti De Riz and Rosella Mamoli Zorzi (Venice: Supernova, 1994), 38–57. Petillon writes: "Throughout American fiction, all the way down to the rocket, one is faced with that two-fold aspect of the machine: thrill and fear—heralding what: a new dawn or the crack of doom" (45).
4. Douglas Coupland, *Microserfs* (New York: HarperCollins, 1995), 323.
5. Evelyn Fox Keller, *A Feeling for the Organism: The Life and Work of Barbara McClintock* (New York: W.H. Freeman, 1983).
6. Donna Haraway, "Manifesto for Cyborgs: Science, Technology, and Socialist Feminism in the 1980s," *Socialist Review* 80 (1985): 65–108.
7. Danny Hillis, "Can They Feel Your Pain?" *Newsweek* 5 May 1997: 57. The echo is to Leo Marx's seminal *The Machine in the Garden: Technology and the Pastoral Ideal in America* (New York: Oxford University Press, 1968).
8. Bruce Weber, "Computer Defeats Kasparov, Stunning the Chess Experts," *New York Times*, 5 May 1997, Sec. B, Col. 2:1.
9. The women's world chess champion Susan Polgar, as quoted in Weber, "Computer Defeats Kasparov," 1.
10. As the match took place at the Equitable Building on Fifty-first Street and Seventh Avenue in New York, I followed much of the play-by-play commentary on the Web. Emotional states were consistently attributed not only to Kasparov (that he is a temperamental Russian was part of the grist for the storymill), but to Deep Blue as well. During the game on May 7, 1997, for instance, Kasparov was described as "going into those kinds of situations that the computer is happy with." Later a commentator queried, "is the computer confused, or is Garry?" That the response to this question was "Audience laughter" is precisely to the point. Although the atmosphere might have been tense, the overall emotional tone was one of sociability.
11. A wealth of prose fiction narratives and science fiction films could constitute the archive for this essay ranging from Isaac Asimov's *I, Robot*, a collection of short stories that appeared originally in 1950 and was reprinted by Bantam in 1994, Robert Heinlein's *The Moon Is a Harsh Mistress* (New York: Putnam, 1966), and Marge Piercy's *He, She, and It: A Novel* (New York: Knopf, 1991) to *Forbidden Planet*, in which Robby, the Robot first appeared, directed by Fred McLeod Wilcox (1956), *Robocop*, directed by Paul Verhoeven (1987), and *The Terminator*, directed by James Cameron (1984) and *Terminator 2: Judgment Day*, directed by Cameron (1991). I have purposively chosen two films, grounded in novels, that continue to be well-known cultural references up to the present day (*2001: A Space Odyssey* and *Blade Runner*) as well as two films (*Silent Running*, a cult film, and *Solo*, an action film) that are not.
12. See Hartwig Isernagen, "Technology and the Body: 'Postmodernism' and the Voices of John Barth," in *Technology and the American Imagination*, 563–70. As he observes, "the borderline between the animate and the inanimate has obviously fascinated the Western imagination for ages. It is particularly the borderline between the machine and the human that has during the last three centuries become permeable in both directions, as the power of machines was allegorized as (quasi-)human or even (quasi-)divine at the same time that animals and humans were treated as or even actually reduced to (quasi-)machines. The robot, which has straddled the threshold all this time, has most recently fueled this fascination in extraordinary ways, as it has metamorphosed itself into the computer" (563). I would add that the computer is metamorphosing again into the robot.

13. See Kathleen Woodward, "Prosthetic Emotions," in *Emotion in Postmodernism*, edited by Gerhard Hoffman and Alfred Hornung (Heidelberg: C. Winter, 1997), 95–107.
14. I refer to the first three volumes of Clarke's quartet as *Space Odyssey*. See Arthur C. Clarke, *2001: A Space Odyssey* (New York: New American Library, 1968), *2010: Odyssey Two* (New York: Ballantine, 1982), and *2061: Odyssey Three* (New York: Ballantine, 1987).
15. Donna Haraway, quoted in Mark Dery, *Escape Velocity: Cyberculture at the End of the Century* (New York: Grove Press, 1996), 176.
16. In *Understanding Media: The Extensions of Man* (New York: McGraw-Hill, 1964), Marshall McLuhan writes about the relationship between the human body and technological invention in terms of "autoamputation." His main point is that technology serves to decrease stress on the part of the body at stake. In contrast, in *Space Odyssey*, the feedback loops that are created retrieve the initial "numbness" that McLuhan notes.
17. In the novel, the action takes place in 2021; the film is set in 2019. "Android" is the preferred term for artificial life in the novel, "replicant" in the film.
18. See *Android Epistemology*, edited by Kenneth M. Ford, Clark Glymour, and Patrick J. Hayes (Menlo Park: AAAI Press/Cambridge: MIT Press, 1995), xi. In "Mind as Machine" in *Android Epistemology*, 23–40, Herbert Simon explicitly makes the point that affect and cognition interact. As early as the 1960s, Simon insisted that cognitive models of mind need to include the emotions; see also his "Motivational and Emotional Controls of Cognition," *Psychological Review* 74 (1967): 29–39.
19. N. Katherine Hayles, *How We Became Posthuman: Virtual Bodies in Cybernetics, Literature, and Informatics* (Chicago: University of Chicago Press, 1999).
20. Philip K. Dick, *Do Androids Dream of Electric Sheep?* (New York: Ballantine, 1968), 124. In *Escape Velocity*, Mark Dery describes *Blade Runner*'s Deckard as a "deadpan, monotoned flatfoot" and the astronauts in Kubrick's *2001: A Space Odyssey* as "autistic" (252); for Dery, they are illustrations of the predominantly "flattened affect that characterizes Homo Cyber" (252). I agree with this characterization but with one crucial proviso—that one of the fundamental points of many of these technological narratives is precisely the growth and development of the emotional world and beings of these characters through their very interaction with artificial entities themselves.
21. In this the film departs sharply from the novel. I suggest that between 1968 when the book was published and 1982 when the film was released a significant cultural shift occurred, one that rendered the representation of replicants as indistinguishable from humans on emotional grounds more acceptable. Sherry Turkle confirms this cultural shift in *Life on the Screen: Identity in the Age of the Internet* (New York: Simon and Schuster, 1995).
22. Scott Bukatman, "The Artificial Infinite: On Special Effects and the Sublime," in *Culture Beyond Appearances*, edited by Lynne Cooke and Peter Wollen (Seattle: Bay Press, 1995, 255–89.
23. That we learn emotions is a theme that has been developed recently philosophers and cultural critics, among others. See Ronald de Sousa, *The Rationality of Emotions* (Cambridge: MIT Press, 1987) and Megan Boler, *Feeling Power: Emotions and Education* (New York: Routledge, 1999).
24. Claus Emmeche, *The Garden in the Machine: The Emerging Science of Artificial Life*, trans. Steven Sampson (Princeton: Princeton University Press, 1994), 20.
25. Daniel Goleman, *Emotional Intelligence* (New York: Bantam, 1995), 41. Star Trek's Data is perhaps the most well-known example in popular culture of the desire and need for emotional intelligence.
26. The move in philosophy to confer value on emotions by emphasizing their cognitive aspect, and emphasizing their irrational component, is in consonance with one of the traditional concerns of philosophy—epistemology, or ways of knowing. Is this not, perhaps, a way of appropriating the emotions to the time-honored interests of the philosophical tradition?
27. Orson Scott Card, *Speaker for the Dead* (New York: Tom Doherty Associates, 1991), xx. Although *Speaker for the Dead* was originally published in 1986, Card's "Introduction" was written in 1991.
28. Byron Reeves and Clifford Nass, *The Media Equation: How People Treat Computers, Television, and New Media Like Real People and Places* (Stanford: Center for the Study

of Language and Information Publications/New York: Cambridge University Press, 1996), 5.

29. Nicholas Negroponte, *Being Digital* (New York: Vintage, 1996), 217–18.
30. See Deborah Lupton, "The Embodied Computer/User," *Body and Society* 1.3–4 (1995): 97–112. In this essay she insists that she has an emotional connection to her computer and comments on the ascription of emotions to computers as a "discursive move that emphasizes their humanoid nature" (105).
31. Joseph Bates, "The Role of Emotion in Believable Agents," *Communications of the ACM* 7 (July 1994): 122–5.
32. Rodney Brooks, *Flesh and Machines: How Robots Will Change Us* (New York: Pantheon, 2002), 201.
33. See Horst Hendriks-Jansen, "In Praise of Interactive Emergence, Or Why Explanations Don't Have to Wait for Implementation," in *Artificial Life IV: Proceedings of the Fourth International Workshop on the Synthesis and Simulation of Living Systems*, eds. Rodney A. Brooks and Pattie Maes (Cambridge, MA: MIT Press, 1994), 70–9. Hendriks-Jansen pointedly draws on the interaction between a mother and her baby to suggest another model—one of situatedness and interaction—of Alife. Also see Steven Johnson, *Emergence: The Connected Life of Ants, Brains, Cities, and Software* (New York: Touchstone, 2002).
34. Hubert Dreyfus, *What Computers Can't Do: A Critique of Artificial Reason* (New York: Harper and Row, 1972).
35. Marvin Minsky, *The Society of Mind* (New York: Simon and Schuster, 1985), 163.
36. Sherry Turkle, *Life on the Screen*, 111.
37. Bruno Latour, *Aramis, or, The Love of Technology* (Cambridge, MA: Harvard University Press, 1996).
38. Gregory Bateson, *Steps toward an Ecology of Mind: Collected Essays in Anthropology, Psychiatry, Evolution, and Epistemology* (New York: Ballantine Books, 1972), 443.
39. What I have been describing is a phenomenology of technology—as it is represented and as it is experienced. For the most part, phenomenological accounts of technology have been given in terms of experiential categories such as time and space—of speed and slowness, of immensity and contraction, of distance and closeness, for example. See Don Ihde, *Bodies in Technology* (Minneapolis: University of Minnesota Press, 2002) and *Technology and the Lifeworld: From Garden to Earth* (Bloomington: Indiana University Press, 1990). "The relationality of human-world relationships is claimed by phenomenologists to be an ontological feature of all knowledge, all experience," he writes in *Technology and the Lifeworld* (24), with a stress on the material. "Whatever else may enter the analysis of human-technology relations, I wish to retain the sense of materiality which technologies imply. This materiality correlates with our bodily materiality, the experience we have as being our bodies in an environment" (25).
40. Brooks, *Flesh and Machines*, 180.
41. Antonio R. Damasio, *Descartes' Error: Emotion, Reason, and the Human Brain* (New York: Putnam, 1994), 71.
42. Ledoux, *The Emotional Brain*, 41.
43. The name of the project at MIT is Cog, a reference to the intent to make a robot that can think self-reflexively. Cynthia Breazel, who was central to the Cog project in the 1990s, is designing a robot named Kismet. According to Peter Menzel and Faith D'Aluisio in *Robo Sapiens: Evolution of a New Species* (Cambridge, MA: MIT Press, 2000), "The pink-eared, rubbery-lipped Kismet alternatively pouts, frowns, and displays anger, along with a host of other expressions that outwardly display human emotion" (66).
44. Sarah Zettel, *Fool's War* (New York: Warner, 1997).
45. I am grateful to Dick Blau, Pamela Gilbert, Andrew Martin, John Rodriguez-Luis, and Nigel Rothfels for their good suggestions and technical support as I was working on this chapter. I want also to thank Hugh Crawford and Veronica Hollinger for their suggestions. I am warmly indebted to Rob Mitchell and Phillip Thurtle and to the audience at the meeting of the Society of Literature and Science in Pasadena in 2002.

PART 

Flesh Remembered: Art, Information, and Bodies

CHAPTER **10**

If You Won't SHOOT Me, At Least DELETE Me! Performance Art from 1960s Wounds to 1990s Extensions

BERNADETTE WEGENSTEIN

After the undeniable influence of the early twentieth-century avant-garde, especially futurism and dadaism, the happenings of the 1960s were primarily the product of an increasingly intensified media environment. The hypothesis for the following essay is that what we today know as the pastiche or collage style, a structure that—as we know—is inherent to the logic of postmodernity and the digital (i.e., fragmented subjectivity, the "windowed" organization of knowledge, and so forth) is ultimately an effect of a mediatized environment that reached the peak of its development in the second half of the twentieth century, inspiring actionists and happening artists in the 1960s to stage mass media related concepts such as "simultaneity" in their performances. Fifty years earlier, the avant-garde, for instance the futurist Filippo Tommaso Marinetti, had idealized and dreamed of "simultaneity"[1] as a result of the new technologies of representation such as photography and film that started to allow for such practices as morphing and superimposition, practices that we often think of as being typical only since our current digital era. "If you won't SHOOT me, at least DELETE me!" is an attempt to consider performance art from 1960s wounds to 1990s extensions both with an eye to history and to the future, asking questions such as: which concepts have been inherited from the avant-garde; which are in fact the new and pathbreaking body configurations at the turn of the millennium; and, most importantly, what can a new era of digital performance offer to this kind of discourse?

To put my hypothesis in the media-theoretical terms of Bolter and Grusin, the actionism movement of the 1960s followed a *double logic of remediation*, consisting in the following paradox: "Our culture wants both to multiply its media and to erase all traces of mediation: ideally, it wants to erase its media in the very act of multiplying them."[2] Indeed, these two seemingly contradictory logics are characteristic for our current era of digital media, which expresses itself through a visual style of *hypermediation*, a style that "privileges fragmentation, indeterminacy, and heterogeneity and . . . emphasizes process or performance rather than the finished art object."[3] That the logic of hypermediation was strongly present in the art scene of the 1960s and 1970s, and not just a consequence of the postmodern digital era, is proven by the existence of actionism and happenings. This new artistic expression was performed by artists incorporating their own artwork, while at the very same time trying to erase the trace of the action itself by eliminating the traditional frame of the spectacle: the new environment of the spectacle was no longer a traditional performative space; instead, anything could be turned into the artistic environment. And it is precisely in this sense that performance art is following the double logic of remediation, that is, the desire to appropriate and stage all possible facets of the artistic form, and to pretend that there is no mediation at all involved by integrating natural environments (e.g., streets, cafes, universities) and the "accidental spectator from the street" into artistic scenarios. From this, it was only a small step to breaking down the boundaries between the artist's performance and his or her body, a novelty that would inhabit performance art to the heights of embodiment attained since the 1970s. Performance art, in other words, cannot be understood without the body.

One aspect of remediation is that the very notion of the body, as an entity distinguished from the artifact, tends toward erasure. There is a direct connection between this tendency of self-mutilation and the mediatized frame of observation. It is almost as if the medium would hold together the body, so that it can fall into pieces and be "dispersed" into the environment. The tendency of "decorporealization," particularly by damaging or getting rid of one's body parts in performance art is not only an expression of a "body dismorphic order" typical of the spirit of the avant-garde,[4] establishing a connection between the hysterical, ecstasy, and technology.[5] Instead, the process of "getting rid of oneself" has to be seen as a broader cultural phenomenon (ultimately an effect of postmodernity), hence, of a body falling into pieces as a symbolization of a disembodied era, in which, as Virilio puts it, "that which happens is much more important than that which lasts (*ce qui dure*)—and also than that which is solid (*ce qui est dur*). In other words, there is a dematerialization that goes parallel to deterritorialization and decorporation."[6]

The following is an attempt to reconstruct this process of remediation through the history of performance from the avant-garde through the 1960s happenings to our current days of disembodied performances, with special regard to the status and use of mediatic frames represented in these artistic expressions, and eventually to answer these questions: What happened to performance art in the era of the digital? With what or how has it been substituted?

I. The Avant-Garde as Precursor of the 1960s Wounds

The format of the happening has been said to be a result or an effect of the *modernist collage style*, a pastiche of painting and sculpture (the early happenings took place in painting/sculpture galleries). The first artist to have transgressed the borders of painting and sculpture to "action art" by moving the picture out into the real space of the room was Allan Kaprow, whose *18 Happenings in 6 Parts* (Reuben Gallery, New York, 1959) baptized this new style of art. Already through the title of this happening—a title which is highly reminiscent of Marinetti's "explosive novel" *8 Anime in una Bomba* (1919)—it becomes quite clear that the happening is a sort of continuation of the avant-garde from earlier in the century. One of the main concerns about the presentation of artistic content in both artistic moments was the notion of *simultaneity*. This notion is today known as a marker of postmodernity, in that a modernist logic of subsequent historic moments has been replaced with a "presence of simultaneity,"[7] in which any historic feeling of the past can be "imported into the present" by means of simulation (e.g., Las Vegas). Kaprow's *18 Happenings* were a sort of action-collage, all "happening" at the very same time, and could only be experienced through a compartmental logic of simultaneity (one next to the other one). One of Kaprow's teachers, the composer John Cage, highly inspired by the futurist–dadaist tradition himself, used the term simultaneity for one of his early works: "simultaneous presentation of unrelated events"(1952).[8]

The dadaist *soirées* and the futurist theater events could not be described better than in Cage's title, for it emphasizes on one hand the important notion of simultaneity, by which the performance itself is relativized (one happening is no longer in a subordinated sense more important than the other), and on the other hand it stresses the non-sensical, self-ironical structure of "events or words in freedom," or "unrelated events and words." The traditional syntax of performance had been "abandoned" since Marinetti—among many other members of the avant-garde—had started to redefine the traditional syntax of poetics in his early manifestos *Technical Manifesto of Futurist Literature* (1912), and *Destruction of Syntax—Imagination without Thread-Words in Freedom* (1913).[9] The new literary models proclaimed in these manifestos should no longer be the traditional, old-fashioned style of literature, but by introducing the synaesthetic dimension of sounds, odors, and images it becomes clear that the new model emphasized rather the overlapping of codes. However, Marinetti also wanted to enrich this technical style with an "intuitive element" reflecting the molecular life of the universe. Later on, the surrealists transformed this "intuitive element" into the attempt to represent the psychic dimension of the unconscious, and the dream, turning this fragmented speech of the psyche into an *écriture automatique*, an automated writing of the soul, and exploring especially the medium film to let the soul speak. In *The Blood of a Poet* (1944), for instance, the surrealist Hans Richter presented four distinct compartmental filmic parts, just like Kaprow's *18 Happenings in 6 Parts:* a true piece of simultaneity.

The concept of simultaneity, in itself inspired by new technologies of representation such as photography and film, contributed strongly to a fragmented, dispersed collage style, which has been claimed to be one of the central techniques of twentieth century visual art.[10] What was crucial in this style for our current concerns was the fusion of the product with its creation process

Fig. 10.1 Kurt Schwitters, *(elika) Collage,* 1930; Courtesy of the Sprengel Museum Hannover with permission by the Artists Right Society, New York.

and—most importantly—with its environment: "As an Environment the painting took over the room itself, and finally, as sort of an Environment-with-action, became a Happening."[11] The practice of the (photo)collage became a common style of modern art, for instance, in the "Collage" paintings of the dadaist Kurt Schwitters in the 1920s and 1930s (see Figure 10.1).

According to Michael Kirby, the futurist Umberto Boccioni gave birth to the photo collage in 1911, when he first used parts of a wooden frame in a piece of sculpture (11). Later on, it became a usual practice to integrate the material content and environment into the painting: Picasso and Braque pasted scraps of newspaper and wallpaper on their canvas. Marcel Duchamp went so far to use any utensil from an old wheel to a hammer and reintegrated these pieces in what he called a "ready-made." The awareness of the process rather than the product is a fact that has changed the figurative arts not only since surrealism and cubism and the rise of the "ready-made," but even since impressionism. The ex-Russian abstract painter Woks says in an interview in *Art News* in 1959: "Since Cézanne, it has become evident that, for the painter, what counts is no longer the painting but the process of painting. . . . Whether you regard painting as a means of penetrating the self or the world, it is creation."[12] It becomes quite clear that here the emphasis is put on the medium itself and the possibilities of its discourses.[13]

Simultaneity, fragmentation, and collage bring also another important dimension into play, that is the redefinition of the theatrical space, such as the theater, the gallery, the museum, etc. in order to break right into "reality."[14] Already in 1915, once again, the futurist F. T. Marinetti and his colleagues Emilio Settimelli and Bruno Corrà formulated the new guidelines for the *Synthetic Futurist Theater* ("Il teatro futurista sintetico"). In the homonymous manifesto they claimed that the "scenic action will invade parterre and audience." It doesn't surprise that one of the titles of these theater plays was "Simultaneity," a "Theatrical Synthesis," as Marinetti put it. Surprise, provocation, and shock were the mechanisms of these very short and not very famous futurist attempts to revolutionize theater. Dadaism, on the other hand, was more successful in these regards. Hugo Ball, Tristan Tzara, Hans Arp and later André Breton met during the difficult times of World War I in the Cabaret Voltaire in Zürich to perform Dada-*soirées*, in which they criticized the traditional laws of beauty, and expressed their disagreement with bourgeois values.

In 1921, the German dadaist Kurt Schwitters formulated the guidelines of his *Merz [commerce] performance theatre* in the *Merz composite work of art:* "Materials for the stage-set are all solid, liquid and gaseous bodies, such as white wall, man, barbed wire entanglement, blue distance . . . Even people can be used.—People can even be tied to backdrops.—People can even appear actively, even in their everyday position."[15]

But nobody could have revolutionized the new idea of theatricality resulting from an intensifying media-culture, in which the outer world collapses into the inner world and vice versa, more thoroughly than the "actor-director-playwright-poet"[16] Antonin Artaud, with his theory of the *Theâtre de la Cruauté.* In this theater of cruelty, art is no longer mimetic of life, but rather life is the simulation of a transcendent communication principle. The signified collapses into the signifier, form into content, and life into representation. For Jacques Derrida, Artaud's simulation of life anticipates the end or "closure" of representation, for it is no longer a recitation of something already written, thought, or lived—hence, no longer "representing" another semiotic system (such as literature)—but rather representation here becomes self-presentation of the visible and the purely sensitive. *Cruelty* therefore means necessity and rigor, which meant that the performances in Artaud's theater of cruelty were not intended to be improvised, but on the contrary were structured and well planned. Unlike the surrealists, who were looking for ways of acting out the unconscious and the language of dreams, Artaud wanted to express consciousness, awareness, which for him meant the color of blood and cruelty. In one of his letters on cruelty from 1932 he declares: "I have therefore said 'cruelty' as I might have said 'life' or 'necessity,' because I want to indicate especially that for me the theater is act and perpetual emanation, that there is nothing congealed about it, that I turn it into a true act, hence living, hence magical."[17] For Jacques Derrida this means, therefore, that "the theater of cruelty is not a *representation*. It is life itself, and therefore it is irrepresentable. Life is the non representable origin of representation."[18] Theater, then, becomes a sacred or magical feast, an act of pure presence and active forgetting; no book, no work of art, but an energy, and in that sense the only art of life.[19]

The breaking down of the walls between life and representation in these explicit terms was of direct inspiration for the happening artist Allan Kaprow who proclaimed that "the line between art and life should be kept as fluid, and perhaps indistinct, as possible"[20] (see the detailed discussion of Kaprow in section II below, *The Syntax of Performance*). The dissolution of difference between life and art was most relevant also for the Viennese actionists, who used the space of the streets, or the University of Vienna—in brief, any public space—for their actions, turning these spaces into happenings.[21] In this type of happening there remains no space between the product, its process of creation, and the artist. They all become one. This process of total collapse, or fusion of the self with the artwork, was already present in Schwitter's *Merz* projects. One of the posters for the election of the Reichstag in 1920, which Schwitters disseminated to campaign for his poetic invention "Anna Blume" as a representative of the dadaist Merz party (*M.P.D. = Merz-Partei Deutschland/ Mehrheits-Partei DADA*), coined the tautological structure "Merz = Kurt Schwitters."[22] Schwitters had become what he had created. He *was* the Merz movement just like he was the artist Kurt Schwitters.

When Schwitters merged into Merz, The Viennese actionist Rudolf Schwarzkogler merged into his paintings. Schwarzkogler defined the blurring of reality and life with the substitution of the construction of the *conditions* necessary for the paint-action producing the image for the construction of the image (*Panorama Manifesto I/ The Total Act*).[23] His actions have to be understood as continuations of paintings, or as he himself called them *paintings in motion*: "the pictorial construction on a surface is replaced by the construction of the pre-conditions for the act of painting as the determinant of the action field, of the space around the actor = the real objects present in his surroundings."[24]

In his sixth action, performed in his flat in Werdertorgasse, Vienna, the bandaged Schwarzkogler performed a light bulb-black mirror-dead chicken piece, in which he stages a connection between his body and the body of a dead chicken through electric cables (see Figure 10.2).

The fusion of the product with the process of its creation, and the environment, and as a consequence, the redefinition of the theatrical space into "life" and the "outer world" results in what the Viennese group called *Direkte Kunst*,[25] direct art: "A total action is a direct occurrence (direct art), not the repetition of an occurrence, a direct encounter between elements and reality (materia)."[26]

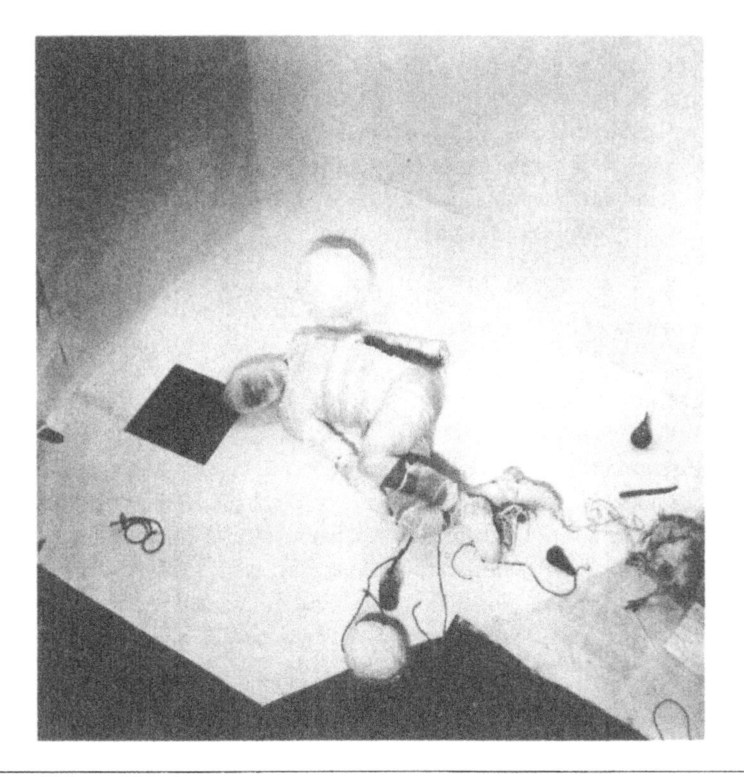

Fig. 10.2 Rudolf Schwarzkogler, *Action 6,* Spring 1966; Courtesy of Museum moderner Kunst Stiftung Ludwig, Vienna.

Just like Artaud, who refused the representational character of traditional theater because it stands for or represents the written text, Brus wanted to "break through the fourth wall" by directly encountering a truth that he called "reality." Marinetti's negation of the past can be seen as an expression of this very same intention to create something new, never repeated, and unrepeatable.

The total action—to *look* and to *become* what one *does*—is expressed in different actions by the Viennese actionists: Brus's action of besmearing himself with paint, integrating his head into his painting (*Ohne Titel,* 1964), has already been mentioned; Schwarzkogler's selection of foods according to aesthetic criteria—for he saw them as manifestations of his environment, an environment that had literally "invaded" the artist himself. Direct art means, thus, nothing else than the necessity and rigor that Artaud claimed for this theater of cruelty. It means that there remains no trace, no indirectness, no quotation, no referentiality, but mere imminence and immediacy.

The happening as an effect of the hypermediated environment is central to this investigation of today's digital performances. What happens if this fusion of the artistic process with the product, the medium of representation with the artist, and the artist with the work of art, collapses into an internalized space: the body? What happens if the engine, the prosthetic vehicle, becomes the human body itself? These questions will be discussed in section IV of this chapter, "Performances in the Era of the Digital." Beforehand, though, we need to look more closely at the logic of the happening, or brief event, in order to be able to decide whether digital performances can even be called performances in the sense of happenings any longer.

II. The Syntax of Performance

In linguistics, *performance* is defined as a contrastive pair with *competence*.[27] Both terms were introduced by the American linguist, Noam Chomsky, in his work in the 1960s (e.g., *Aspects of the Theory of Syntax*, 1965). Performance is the realization or use of a speaker's competence to apply a grammatical system (syntax, semantics, phonetics, etc.) of a given language. In speech act theory, "a communicative activity is defined with reference to the intentions of a speaker while speaking and the effects achieved on a listener."[28] John Austin calls this very intentionality the "illocutionary act," and the impact on the listener "perlocutionary act."[29] An illocutionary act is realized "in loqui" (in or through speaking). One of Austin's famous examples is the question "is the window open?" with which a speaker might be expressing a warning, or a wish that the window be closed, or perhaps just wants to communicate that he or she feels cold, and the listener might, instead of answering "yes," or "no," just get up and close the window. These are utterances that involve actions, and can therefore also be called performative. Common examples for performative acts are religious acts such as baptism, civil/religious acts such as marriage, or legal acts such as sentencing. These are acts that occur by the uttering of certain phrases or sentences by one person, invoking a change of status in another person. Usually these phrases are accompanied by a symbolic act as well, such as pouring water on the head of the baptized. It has been pointed out that such performative utterances (normally expressed by performative verbs such as to warn, to promise, and so on) cannot be expressed in negative terms. In other words, there is no negation of a performative utterance. It is always affirmative.

Performance art of the twentieth century has been operating with the materiality of embodiment, for instance, in performances staging the other sex, or stereotypically gendered roles such as female film divas, Barbies, nuns, and the like (e.g., Valie Export, Cindy Sherman, Mariko Mori, Yasumasa Morimura, Pierre & Gilles, and Andres Serrano).[30] What seems primary to these performances is to incorporate the messages into the artists' own bodies. Thus, it does not surprise us that gender and transgender perfomances are often staged by women and by homosexuals, both social groups representing bodies that have been victims of oppression and essentialism.

Although the body-discourse in these performances can be seen in the framework of a politically liberating activism, what matters for our current concern is that artists' bodies have become more and more their own artistic primary material, as Günter Brus puts it: "The actor performs and himself becomes material: stuttering, stammering, burbling, groaning, choking, shouting, screeching, laughing, spitting, biting, creeping, rolling about in the material."[31] Characteristically, the genre of written texts involved in performances is the manifesto. From the futurists, dadaists, and surrealists to Stelarc's recent cyber-manifestos (see section IV), this genre gives the performance its performative force. What is more, the performance as such can really only be recognized and fully understood in consideration of its grounding manifesto, its directions, its text of invitation etc. Paradoxically, immanence and presence are reached only through the frame of the performative setting. In this very aspect, we almost cannot help thinking that the syntax of performance resembles the logic of the mediated screen, as in television or film. We are confronted with a performative frame, such as "news" through which we consume "real happenings." We would not hold them for "real" were it not for the setting (such as a news program) in which we decode them.

Performance, however, must be understood not as a "movement" but as a form of expression that is most typical for the twentieth century, as Frazer Ward explains in relating performance art to artistic movements such as conceptual art:

> performance art was not a "movement," in the way that Minimalism or Conceptualism were, whatever attempts have been made to situate it as one. Rather, performance has surfaced and disappeared throughout the twentieth century as a kind of undercurrent, periodically bubbling up within—or in some relation to—various avant-garde movements: the Soviet avant-gardes, Futurism, Dada, the Bauhaus, neo-Dada, Fluxus, Pop, Minimalism, perhaps even Abstract Expressionism if we consider the arena-like quality of Jackson Pollock's painting on canvases rolled out on the studio floor. In works not only by Acconci, but by Chris Burden, Jan Dibbets, Dan Graham, Douglas Huebler, Bruce Nauman, Dennis Oppenheim, Hannah Wilke, and even Daniel Buren (e.g., his Sandwichmen [1986]), and others, it certainly surfaced in a close relation to Conceptual art—as much as it might have surfaced in the work of other artists, against Conceptualism.[32]

In the following, however, I am not trying to distinguish the genre of performance art in opposition, say, to conceptual art, and other respective artistic mainstreams. I have pointed to the inspirations of performance art in the first section, in which it has become clear that this artistic expression can only be understood in the spirit of the avant-garde earlier in the twentieth century. What interests me—in order to situate the possibilities of performance art within cyberspace in a next step—is not so much the historic constitution of performances, but the very syntagmatic system that it is based on. In other words: how are performances structured? Do they even follow a certain paradigm? If so, what does it look like? And if not, what does its nonstructure look like?

In his essay "The Happenings are Dead: Long Live the Happenings," Allan Kaprow expresses the paradoxical syntax of the happening already in his contradictory title. By "dead," Kaprow means the following: "The Happenings are the only art activity that can escape the inevitable death-by-publicity to which all other art is condemned, because, designed for a brief life, they can never be overexposed; they are dead, quite literally, every time they happen."[33]

In this quote, it becomes quite clear that it is the age of "mediatized reproduction" which Kaprow criticizes in his art form, an age that wants to record, to simulate, and which he wants to overcome with "truth" and "authenticity" of the brief event of the happening. Just like for Artaud, for Kaprow there is an inseparability of process, product, and environment, or, "creation and realization, artwork and appreciator, artwork and life" (64). In Kaprow's work, though, this inseparability takes place in the subculture. The happening should be "as artless as possible," and the desire is to place it at the intersections of urbanity and high technology, just like the futurist's dream of accelerated accidents, and today's "postnuclear" settings for cyperpunk: "a tour of a laboratory where polyethylene kidneys are made, a traffic jam on the Long Island Expressway are more useful than Beethoven, Racine, or Michelangelo" (62). Very much in the spirit of the futurists and dadaists, for Kaprow, the life-representations that interest him are taking place in "nonplaces" such as drugstores, and airports, for the sense of life in question is a "life above ground," which is—and here is another paradox—"underground."

Kaprow's happening-syntax is a collage of events, fragmented in certain spans of time and in certain spaces. Sometimes the events may overlap in time, so that it becomes literally impossible to follow them all. Besides, they are only performed once, which emphasizes their authenticity, and creates the reality-effect. The following happening of a car accident reminds us of current reality TV-shows like *Cops:* "Two cars collide on a highway. Violet liquid pours out of the broken radiator of one of them, and in the back seat of the other there is a huge load of dead chickens. The cops check into the incident, plausible answers are given, two truck drivers remove the wrecks, costs are paid, the drivers go home to dinner" (62).

But unlike the dadaist *soirée*-happenings, which were put out in limited spaces for limited audiences, Kaprow does not care about the fact that one cannot follow his events, because he is not interested in the audience at all; rather, he wants to eliminate it entirely. No proscenium stage, no time-place matrices; instead, the observer's absence is the syntactic corps-rule of the happening. The audience consists of randomly involved spectators on the one hand, and the "after-audience" on the other hand. Since there is no audience that can "watch" the entirety of a happening in real time, it is the knowledge of what is going to happen that turns the audience into an "after-audience." For instance, "twenty rented cars are driven away in different directions until they run out of gas." Not only would it be impossible to follow all rental cars at the same time, but the audience of this happening consists of people, who—as we speak or read—become witnesses of this event, be it true or false, for the mere fact that it is reported to them. Besides, as Kaprow points out, there is another realism factor involved. The happenings are performed by nonprofessionals in order to increase the unpredictability and reality-effect of the event.

What results from these choreographic directions is that the performances, as designed by Kaprow, are highly influenced by a mediatized logic such as inherent to the medium film, which is recorded or shot once, and viewed or experienced not live, but in retrospect and in absence of the actors or performers. There is even another pertinent medium involved in the way Kaprow designs his happening—photography. Kaprow says that "happenings are dead," by which he stresses the stillness, unreapeatability, and thus uniqueness of the event. This brings a desire into play, a sentimentality, or melancholic mourning for the past or lost object. Of a happening in Kaprow's sense, we have heard of it, or were told about it. Presence or witnessing would be felt as something unusual, or almost something exceptional[34]—like surviving a catastrophy.

In that very spirit, Chris Burden plays with the viewer's absence and the impossibility of being seen while performing in his action, *You'll Never See My Face in Kansas City,* on November 6, 1971, in Kansas City: "For three hours I sat without moving behind a panel which concealed my neck and head. No one could see behind the panel; a piece of board sealed the underside of the space. In conjunction with the performance, I wore a ski mask at all times during my stay at Kansas City" (see Figure 10.3).[35]

In Kaprow's *Self-Service. A Happening* (1967), there is another important factor that comes into play besides the absence of the audience, namely the chance factor:

> Self-Service, a piece without spectators, was performed in the summer of 1967 in Boston, New York, and Los Angeles. It spanned four months, June through September. Thirty-one activities were selected from a much larger number. Their time and locality distribution were determined by chance methods. Participants selected events from those offered for their city; each had to pick at least one, although doing many or all was preferable. Details of time and place were flexible within each month; choices made from month to month overlapped, some actions recurring.[36]

The syntax of this action is reminiscent of or anticipates a chance generated computer game, similar to digital installations, which I discuss in sections III and IV. Some possibilities are given, others are randomly constructed during the action/game. It is precisely this logic of chance that anticipates a digitized environment for performance art that will culminate in the hypermediated 1990s extensions. For the rest, the directions look like a screenplay for a short

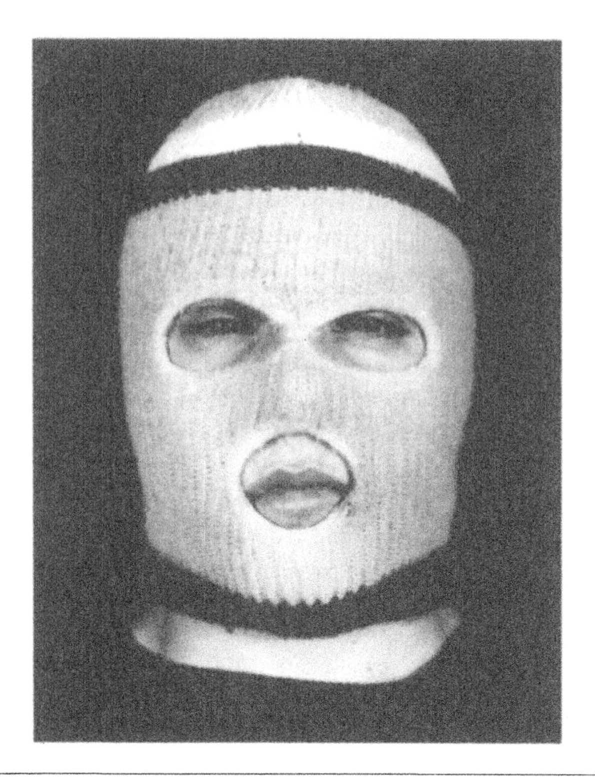

Fig. 10.3 Chris Burden, *You'll Never See My Face in Kansas City*, on November 6th, 1971; Courtesy of Gagosian Gallery, New York.

film involving many people and cars: "People stand on bridges, on street corners, watch cars pass. After 200 red ones, they leave" (234).

What is more, the logic of the viewer's absence at the event approximates a televisual format (e.g. a TV news program), in which a notion of truth and presence is created merely through the performative discourse of authenticity and witnessing ("this is Wolf Blitzer for CNN, Washington"). What becomes clear in *Self-Service* is that Kaprow no longer shocks the audience in a dadaist manner; rather, we witness the anticipation of an intersecting logic of film and computer culture within this kind of happening art. No longer is the individual happening at center, but instead we focus on the remediated knowledge of the happening.[37]

Along with Kaprow and Burden, many other happening artists of the 1960s either withdrew themselves from their audiences or turned the one-time viewers of the action into witnesses of crimes and violence (e.g., Burden's *Shoot*, November 19, 1971, F Space, in which he deliberately had himself shot into the upper arm), or into voyeurists of extreme actionism (e.g., Burden's *Five Day*

Locker Piece, April 26–30, 1971, University of California, Irvine, in which he had himself locked into a 60 cm × 60 cm × 90 cm locker). In these actions the performers become "survivors" of their bodily challenges. *Survival in Alien Circumstances* is also the title of Stuart Brisley's action in 1977, when he spent two entire weeks in a dirty, muddy pit, a physical challenge similar to his *Hungerstrike* in 1979. Violence as a signifier for truth and depth (often taken very literally by cutting into the performer's skin), and the audience's presence at the artist's violent experience, is at the center of these "reality-happenings." What is crucial from the viewer's perspective is the dimension of the first person's experience: "How do you know what it feels like to be shot if you don't get shot?" is the question that Burden asked to explain his motivation for his performance *Shoot*. Gina Pane harmed herself seriously with a razor blade in *Psyche* (1974), when she kneeled in front of a mirror, put on some makeup, and cut herself under the eye. The Austrian performance artist and filmmaker, Valie Export, who was not a member of the Viennese actionists (not least because she is a woman), wounded her body in several actions. In *Eros/ion* (1971), for instance, "the naked performer rolls first through an area strewn with broken glass, then over a plate of glass, and finally onto a paper screen."[38]

Actionism, however, has also violated the law for real: Otto Muehl, another Viennese action artist, had to spend seven years in prison (1991–1998) for having engaged in sexual intercourse with minors, illicit sexual acts, rape, and drug offenses. He was, by the way, not the only one of the Viennese actionist group to encounter legal problems or prosecution for his transgressive actions. One of the transgressions that was never officially problematized, however, consisted in the Viennese actionists' insistent misogyny in their actions, as pointed out by Roswitha Mueller in her study of Valie Export, the female counterpart of the Viennese art scene: "In accordance with the sexual politics of the day, women's bodies were primarily passive objects to be acted upon rather than actors in their own right. The packaged, smeared, used, and abused bodies of women were central to some Actionist fantasies of destruction" (xix).

But feminism has since then made it prominently into the performing arts, and the history of the powerful male gaze has been discussed by different artists (e.g., Cindy Sherman, Mariko Mori, Lee Bul a. o.). In an expressis verbis feminist performance, the French artist Orlan—considering her body as raw material for her art—had herself cut in several plastic surgical operations that she staged in videotaped actions (e.g., *Successful Operation Nr. X*, 1991), and in classic one-time event happenings (e.g., *New York Omnipresence*, November 21, 1993). In her "theater of operation," in which the artist had implants sewed into her chin and other facial parts, Orlan reflects on issues concerning the construction of the "female body under the male knife."[39] The hybridation and reconfiguration of herself in these surgical interventions are an expression of Orlan's critique of the mediatized (female) body, onto which beauty ideals have been violently inscribed throughout history. *Successful Operation Nr. X* is about the change of

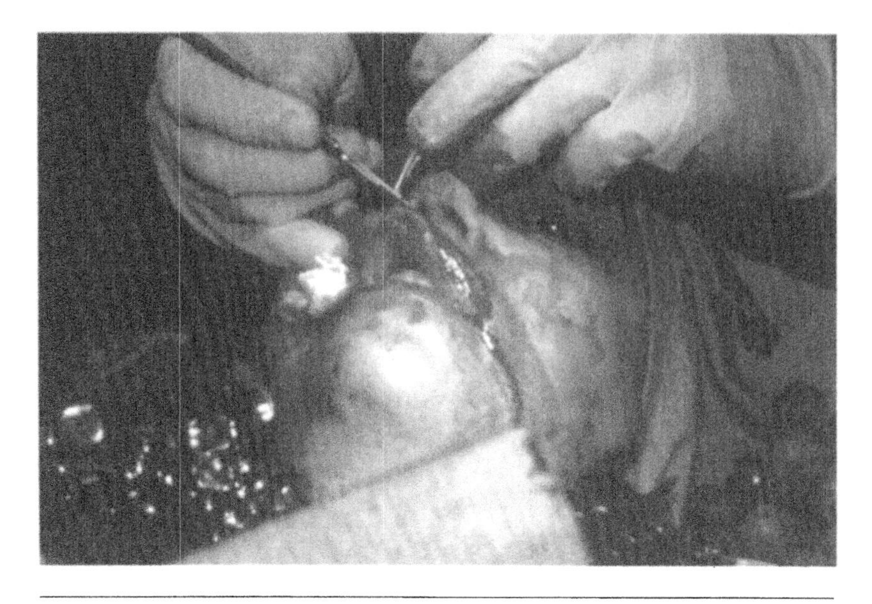

Fig. 10.4 Orlan, *Successful Operation No X*, 1991; Courtesy of Espace d'Art, Yvonamor Paris-Mexico.

image, the deceptiveness of the skin, or the "I as another," as Orlan states during her performance (see Figure 10.4).

In all of these actions, whether they include or exclude the audience, whether they are more or less harmful for the performer, what becomes clear is that the new canvas is now the human body. No paint, no sculptures are being "ready-made" anymore, but—as Orlan puts it in her "Art Charnel" project—the body itself becomes a "modified ready-made." The performer, thus, embodies the idea, breaking down the wall between reality and representation. Not only does life become a theater of cruelty, as in Artaud's earlier tautology, a theater in which the viewer automatically becomes a witness of something that actually *happens* to the performer, but it is the human body itself that is turned into the theatrical "Gesamtkunstwerk." Nevertheless, the body we are looking at is no longer an aestheticized body, like the beautiful Yves Klein models of the early 1960s. Instead, as best shown in Orlan's case, it is a body in pieces—in which each body part can represent the entire body. This is a fragmented, dispersed body that resembles the logic of the mediatized frame.

III. Hypermediated Performances or 1990s Extensions

In this section, the central question is whether the era of the digital has redefined or disbanded the genre of the happening as previously described in its evolution. In other words, what are the consequences of the increasing possibilities of the digital, and how does this change performance as such? Since we are in the midst of this new zeitgeist, it is, of course, hard to assume a distanced analytical gaze. But let us try nevertheless. Does the disappearance of the real material body in cyberspace mean that wounds can no longer have the reality-action-effect that Kaprow and Burden would hope to provoke? Are Stelarc's body prostheses and Steve Mann's wearable computers (see section IV) even bodies in the "old" sense, or have they become "posthuman?" In order to answer these crucial questions, we need to go back to a more media-theoretical description of the digital and its hypermediated environments, thus to the preliminary question of the double logic of remediation; that is, how the multiplication of media create a desire to disappear at the very same time.

Bolter and Grusin describe our visual culture as an era of *remediation*, characterized through the inseparability of reality and mediation. Remediation is in this view always mediation of mediation, just as—according to Derrida—interpretation is always reinterpretation. In this highly mediatized world "representation is conceived of not as a window on to the world, but rather as 'windowed' itself—with windows that open on to other representations or other media."[40] Reality itself becomes a window, as shown in an interview with a MUD user: "I split my mind. I'm getting better at it. I can see myself as being two or three more. And I just turn on one part of my mind and then another when I go from window to window. . . . And then I get a real-time message (which flashes on the screen as soon as it is sent from another system user), and I guess that's RL. It's just one more window."[41]

We have defined hypermediation earlier as a fragmented, heterogeneous style emphasizing the process rather than the result, or the finished (art) object. It almost goes without saying that this kind of a style gets its inspiration from the logic of the digital and vice versa. Bolter and Grusin give many different examples of fragmented interfaces and screens that may be seen as effects of remediation and hypermediation. The newspaper *USA Today*, for instance, has a layout that resembles a multimedia computer application, and TV stations like the abovementioned CNN "show the influence of the graphical user interface when they divide the screen into two or more frames and place text and numbers around the framed video images."[42] CNN is also a good example of hypermediation for the structure of its website, which borrows its "sense of immediacy from the televised CNN newscasts" (9). The televised newscasts, on the other hand, resemble the very website that resembles the newscast. A classical loop effect of hypermediation, in which reality and virtuality feed each other with interchangeable input.

It is the desire for the real, and for the "authentic first person's experience" that results from these latest tendencies in a culture of representation and simulation, a desire that—paradoxically—gets stronger the more ways of representation are invented. In other words, the more we are trying to mediate, the stronger the wish to "go back," "to unmediate," or "get past the limits of representation and to achieve the real," as Bolter and Grusin put it (53).

The quest for the ultimate proof of the "real" and for the authenticity of the first person's experience is of course not new, but has to be seen in the context of and as one of the leitmotifs of modern art. Ever since the avant-gardes this desire has been pursued and widely represented. Hal Foster points out how the strategies of the real can be traced back to several art movements of the 1960s—among which the "minimalist genealogy of the neo-avant-garde," pop art, superrealism (photorealism), appropriation art and others—and since then it has become one of postmodernity's trademarks.[43]

To get beyond representation can also mean, in its most extreme sense, to make the medium disappear entirely. The best example for this desire is virtual reality, a surfaceless entity that has no outside or inside, but where "inside is always outside, and outside is always inside."[44] Virtual reality is direct and immersive, which means that it is a medium whose paradoxical purpose is to disappear. In his historical approach to the concept of virtual art Oliver Grau points out that virtual reality is not a new phenomenon:

> the idea of installing an observer in a hermetically closed-off image space of illusion did not make its first appearance with the technical invention of computer-aided virtual realities. On the contrary, virtual reality forms part of the core of the relationship of humans to images. It is grounded in art traditions, which have received scant attention up to now, that, in the course of history, suffered ruptures and discontinuities, were subject to the specific media of their epoch, and used to transport content of a highly disparate nature. Yet the idea goes back at least as far as the classical world, and it now reappears in the immersion strategies of present-day virtual art.[45]

The current trend for a "transparent immediacy" can be found even outside of the digital logic *strictu sensu* (unless we want to consider the influence of the digital logic always to be an "inside logic"). High fashion designers, architects, interior designers all want to create "interfaceless interfaces" with smooth, morphing, or even nonvisible transitions between materials, contents, appearances—in other words, they all want to break down any kind of border. This strategy, however, is certainly not about convincing the interlocutor (if there still is such an individual) that the representation "is" the thing itself, but it is trying to install a realistic looking phantasy screen. The difference from the postmodern experience, as pointed out by Žižek, is only that everybody is aware of this process, if consciously, or unconsciously, but still wants to play the game.

In order to get rid of the interface one can apply different techniques. One is complete *erasure* through *replacement*. The best example for replacement is the hypertext of the world wide web, a set of windows laid over each other. But even outside of the internet, such formats as *Microsoft Word* constantly suggest the logic of replacement with such possibilities as "copy," "paste," opening new windows that are layered behind the main document, and—as in Macintosh's latest operating system—may be "swallowed up" by the desktop while not in use. Replacement, however, leaves behind a trace, and the question that remains open is: what is done with what is being replaced? Where does the previous "window" go? A very radical answer to this question was given by the Viennese Actionist, Otto Muehl, in 1963: "I can imagine nothing significant where nothing is sacrificed, destroyed, dismembered, burnt, pierced, tormented, harassed, tortured, massacred . . . stabbed, destroyed, or annihilated."[46]

In other words, replacement seems to be a creative strategy, or—as for Muehl—a destructive necessity to create or produce anything (new). The difference from the collage style discussed earlier on is, however, that replacement is not anymore a superposition, like Schwitter's collages, in which the different layers were shimmering through. Replacement is here an erasure in the sense of a substitution of what was there before. It never comes to a "stop," like the layered collages by Schwitters, but rather has to be—per definition—"activated" (e.g., by going under "window" and clicking on the document hiding underneath the one in use) in order for the layers to show.

One common mode of replacement is what contemporary entertainment industry calls "repurposing"[47]—a logic certainly best represented in the World Wide Web—taking one property from one medium and reusing it in another. The logic of repurposing, as Bolter and Grusin further explicate, was already acknowledged by Marshall McLuhan, who linked it to its original essence in the transmission of pure information through electric light:

> The instance of the electric light may prove illuminating in this connection. The electric light is pure information. It is a medium without a message, as it were, unless it is used to spell out some verbal ad or name. This fact, characteristic of all media, means that the "content" of any medium is always another medium. The content of writing is speech, just as the written word is the content of print, and print is the content of the telegraph.[48]

It does not surprise us that in a society of simulation and hypermediacy, repurposing should become one of the major tropes of communication. As the American theorist of capitalism and postmodernism Frederic Jameson pointed out, postmodern art is a form of imitation that is cynical of the original.[49] This kind of cynicism can indeed be found in today's hypermediated performances, those, for example, of the Wooster Group:

> In hyperkinetic performing environments, where video monitors, movable scenery, microphones, and frantically gesticulating performers compete for the viewer's attention, the Wooster Group takes texts, often from playrights like Eugene O'Neill . . . and Gertrude Stein . . . and "deconstructs" or fragments them, eliminating recognizable timelines, as live performers compete for viewers' attention with videotaped versions of themselves. In capitalizing on the uniqueness of O'Neill or Stein, they offer graphic representation of what social theorist and critic Frederic Jameson cites as the postmodern artists "seizing on their idiosyncrasies and eccentricities to produce an imitation which mocks the original."[50]

There are endless examples for repurposing, which becomes clear once we make ourselves aware of the many possibilities inherent to McLuhan's thesis that the content of a medium lies in another medium. It almost goes without saying, therefore, that a society obsessed with media and mediatic frames would eventually produce copies of copies of copies, and imitations of imitations of imitation. What remains to be analyzed, and what is of most interest regarding repurposing, is which content is being staged or borrowed for which purpose.

Another—and recently most popular—way to get beyond the screen is to (make) believe that there is no screen, or that the medium is not a medium, but in fact part of the real, of which it is "just another window," as said before. Such a replacement of reality with virtuality leads to the so-called media equation: "Media equal real life," as suggested by Reeves and Nassin their study about how people interact with computers.[51] The most recent and apparently most successful way of achieving a "getting-beyond-the-screen-effect" is the incorporation of the mediatic frame into the utterance itself. Make the spectators aware of being spectators, make them enjoy the medium!, goes the underlying instruction for use: "In the logic of hypermediacy, the artist (or multimedia programmer or web designer) strives to make the viewer acknowledge the medium as a medium and to delight in that acknowledgement."[52] It is in this very sense that we have to read such blockbusters as the *Blair Witch Project* (1999), or reality TV shows like *Survivor* and its endless imitations, that is, as acknowledgements of the medium, and euphoric—or (pseudo-)dysphoric—explorations of the interface. At this point, we need to go back to the primordial question for this essay, namely, the question of what a new era of digital performance has to offer, and whether performances like those from the avant-garde to the 1960s can or are continuing to exist, or whether the digital and its double logic of remediation have changed the workplace for performances.

IV. Performances in the Era of the Digital[53]

Since the media revolution throughout the 1980s, performance art has changed dramatically in that precisely the last trope, the incorporation of the mediatic frame into the artistic message, has become the leading metaphor for any kind of staged art, be it video art installations, digital art, or "wired-art." Interaction and participation in the creation of meaning is today the clue to digital actionism. The art pieces tell the viewer "what to do," "how to be read." They call for interaction, inviting the viewers to become artists/producers/creators themselves. In truth they don't even preexist the audience's/viewer's interpellation. There are many examples for this kind of interactive performance art. Sharon Daniel's interactive Web installation[54] *Narrative Contingencies* (2000) is:

> a database of images and texts with a web-based interface, it engages audiences both by inviting them to contribute personal artifacts and stories, and by allowing them to generate random results, constructing a narrative from chance combinations of words, sounds, and images. Visitors can make their own contributions to this evolving artwork using the computers, camera, scanner, and printer in the gallery.[55]

Narrative Contingencies expresses the 1990s-extensions way of using the logic of the digital in art installations in the most emblematic way. The project presents us with a break down of several boundaries of codes. There is no longer a difference between words, images, and sounds, and hence the codes of data float into each other, delivering random access to the respective codes of communication. The artist/creator of the project is no longer detached from the viewer; in fact, meaning can only be constructed in the concrete interaction, specifically by giving in an input of a word, object, or sound. The position of subjectivity is therefore distorted, and the installation seems to tell us: "You can only grasp me while interacting with me. The next nanosecond it could be already all different." The emphasis does not lie in the product, but in the process. The artwork is evolving. There is absolutely no control over the different instances of this narrative chance-generator. The outcome is unpredictable, therefore often surprisingly funny.

Similarly, the Canadian sound and video installation artist David Rokeby uses the arbitrary chance factor in the denomination process in one of his works in project *The Giver of Names* (1990–).[56] As Rokeby describes on his Web site to the project, the installation is a reflection on the process of semiosis:

> The Giver of Names is quite simply, a computer system that gives objects names. The installation includes an empty pedestal, a video camera, a computersystem and a small video projection. The camera observes the top of the pedestal. The installation space is full of "stuff"... objects of many sorts. The gallery visitor can choose an object or set of objects from those in the space, or anything they might have with them, and place them on the pedestal. When an object is placed on the pedestal, the computer grabs an image. It then performs many levels of image processing (outline analysis, division into separate objects or parts, colour analysis, texture analysis, etc.) These processes are visible on the life-size video projection above the pedestal. In the projection, the objects make the transition from real to imaged to increasingly abstracted as the system tries to make sense of them.[57]

In these recent computer installations, it is the process of sense making that is being questioned, and at times appears even ridiculed, as in the cynical sense Jameson attributed to postmodernity. It is no longer just a reflection on the possibilities of the collage, such as the incorporation of the frame. What is at stake here is rather the core issue of communication itself, a coming back to the essential question of "how is meaning constructed?" The fact that these installations emphasize all possible aspects that are involved in the construction of meaning—from creation, to perception, to the incorporation of the environment, to the possibilities of blurring all codes of communication involved—definitely does not make these artifacts easier to read. In the contrary, it is not really their "readability" that they are about. These installations literally point to the instability of meaning within the logic of the digital, but acknowledge at the very same time that we are already much too immersed in this logic for it to be likely to get out anymore.

The logic of chance inherent to these computer installations is a logic inherited from the avant-garde, from dadaist performances to happening art. As Michael Kirby points out, even Marcel Duchamp's ready-mades were already in part constructed through this very logic of chance:

> Marcel Duchamp dropped three threads, each exactly one meter long, onto three sheets of glass from a height of one meter. Fastening them down, he used the sinuous lines arrived at by chance (that is, gravity, etc.) to make three measuring sticks, the varying curved edges of which are each exactly one meter in length. The six pieces are called Trois Stoppages-Etalon or Three Standard Needle Weavings (1913/14).[58]

These symbolic discourses are clearly represented by the digital chance logic of an unforeseeable distribution of data, an order that, as in some antivirus programs, needs to be "defragmented" from time to time. The differences, postponements, and unrootednesses in space, on the other hand, can be interpreted as representing a logic of the unconscious, a field of experience most important to current concerns, for it is the only guarantee or proof of the "real," which is the ultimate goal, and in itself the only proof of the desired first person's experience.

The unconscious can be said to "resemble" the digital in some ways, which is why, for example, the web calls for such metaphors as a "split self." Žižek described avatars as incorporations of "ego-envelopes," proving the necessity of the masquerade throughout life, which—as we have seen in many artistic projects discussed in this chapter (e.g., Orlan, Valie Export)—continuously hide an empty space or a nonspace. Or as Sandy Stone has it, cybernetic space can be put on like a garment, which for her means to "put on the female."[59] Splitting the self is—to put it in Zizekian terms—the realization that there wasn't any "self" in the very beginning, and that the self only exists as a split entity.

In this regard it is worth mentioning one of Mary Flanagan's recent digital art pieces called *[collection]*.[60] In this networked computer application that creates a visible, virtual, collective unconscious "the computerized memory takes on a life of its own."[61] *[collection]* scours hard drives of the computers that have the downloadable program installed, collecting bits and pieces of user's data, sentences from emails, graphics, web browser cached image, sound files, etc., and creating this material into a moving, three dimensional continuously shifting map.[62] In *[collection]*, Flanagan emphasizes precisely the discussed issues, the chance generator and the similarity of the unconscious to the computer. In an interview the artist states that the *[collection]* is reminiscent and visualizes the process of memory (see Figure 10.5).[63]

The unconscious is also of interest to a DECONcert by James Fung and Steve Mann, "in which audience members actively (and unconsciously) choreograph a collective cyborg consciousness by contributing their own brainwave patterns."[64] In this performance piece in the realm of the digital, an audience translates their brain waves, recorded through leading-edge EEG technology,

Fig. 10.5 Mary Flanagan, *[collection]*, 2002; Courtesy of the artist.

into the realm of music. The unique outcome of this brain-wave configuration was performed at the Deconism Gallery in Toronto on March 21, 2003.

Nevertheless, the transition from the body as raw material in the 1960s and 1970s wounds-performances to the extended body of the 1990s cannot be instantiated better than in the work of the Australian performance artist Stelarc (Stelios Arcadiou), who has performed a fusion of the digital with the real on his very own body for more than thirty years. Stelarc is probably *the* example of how body discourse in performance art has changed during the last thirty to forty years. In the 1970s and 1980s he suspended himself with hooks in his skin in such actions as *Stretched Skin* (1976) in Japan, and *Street Suspension* (1984) in New York. Whereas in the 1970s Stelarc experimented with bodily levitation, trying to transcend pain, in the 1980s, 1990s, and his current new-millennium-installations he is trying to change his body, redesigning its interface. Stelarc ironically proclaims the body's obsolescence: "It is time to question whether a bipedal, breathing body with binocular vision and a 1,400cc brain is an adequate biological form. It cannot cope with the quantity, complexity and quality of information it has accumulated."[65]

By obsolescence, Stelarc, however, does not mean that the human body should or could be replaced by a machine, or that intelligence is (becoming) disembodied. Instead, he provokes us to think that our bodies are constantly being out-performed by machines in speed, strength, endurance, and so on.[66] Stelarc's cyberbody-installations need to be seen as philosophical investigations on the very experience of embodiment. His philosophy is based on technological transcendence combined with a teleology of the spatial era.[67] Stelarc's phenomenological approach in his body-art has been emphasized both by him and by his many critics. As Mark Poster puts it: "Stelarc puts into motion Heidegger's pronouncement that technology brings into question the human essence."[68]

In response to the cyberculture of the turn of the millennium, Stelarc is developing what he terms an "evolving URL body." This body is an object, rather than a subject. It is constructed at the hands of multiple user-hackers. Similarly, the "ping body" is a barometer for internet activity. In "Movatar," the body itself is turned into a prosthesis performing "involuntary movements." In this inverse capture system muscle signals are recorded and transmitted onto a prosthetic device (e.g., arm). These random signals now determine the motor control, finally taking over a person's body. The body, here, has become a contingency, a system out of control and in the hands of others, symbolizing the complexity of agency, self, and identity. As Edward Scheer points out in his critique of Stelarc's "E-motions" Movatar "literalizes our condition of being trapped, as Wittgenstein says, 'inside a picture.'"[69]

In Stelarc's phenomenological view of the body "an intelligent agent has to be embodied and embedded in the world."[70] In his latest installation, Prosthetic Head (2003) (see Figure 10.6), Stelarc investigates the limitations and possibilities of Artificial Intelligence. The automated, animated Prosthetic Head, designed after the artist's head, speaks to the person who interrogates it through a keyboard. To the question: "will you marry me?," the head answers "why don't you download me instead?"[71]

In this piece Stelarc plays ironically with notions of agency and the responsibility of the socially embedded intelligent agent. The Prosthetic Head serves as an avatar, assisting the artist in handling the many questions that people have asked him about his work:

> In recent years there have been an increasing amount of PhD students requesting interviews to assist in writing their thesis. Now the artist will be able to reply that although he is too busy to answer, it would be possible for them to interview his head instead. And as a web avatar it would be possible to download the transcript of the conversations people have with it. A problem would arise though when the *Prosthetic Head* increases its data base, becoming more autonomous in its responses. The artist would then no longer be able to take full responsibility for what his head says.[72]

We have reached the end of our analysis and have to face the at-this-point-unavoidable question: are we at the end of performance?, or what has changed

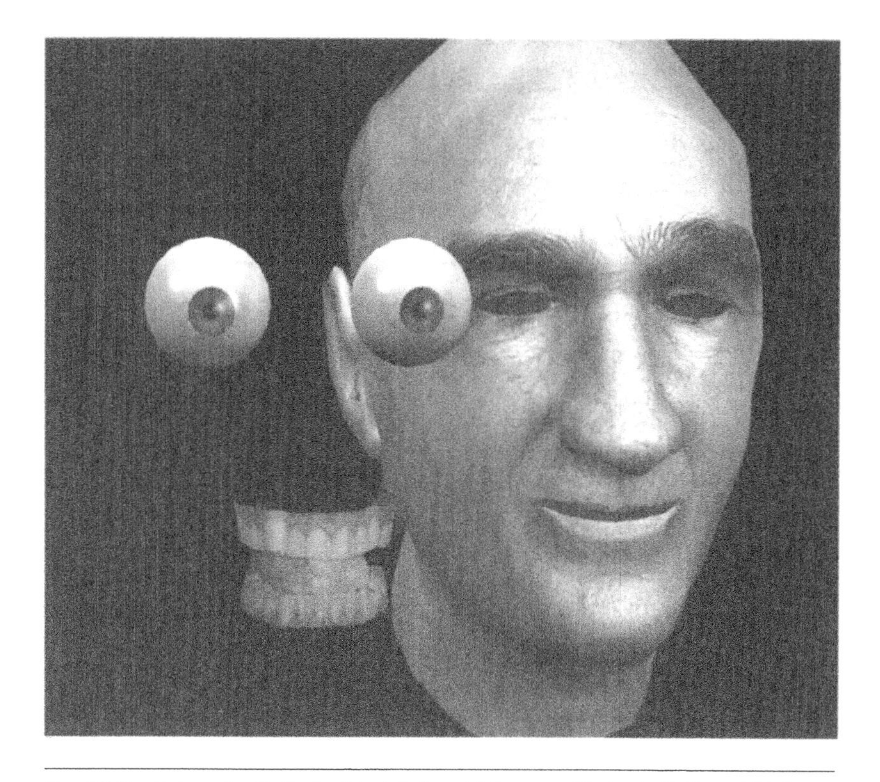

Fig. 10.6 Stelarc, *Prosthetic Head;* Installation at Interaccess Electronic Media and Art Center, Toronto; Photograph by Holly Johnson; Courtesy of the artist.

in the performances of the digital era? In Stelarc's recent performances, it becomes very clear that the position of the uttering subject has complicated itself through a multiplication of positions. In his desire to merge with technology, he expresses the collectivity of agents and subjectivities involved in our hypermediated environments. We can say that following the double logic of remediation—which best characterizes our zeitgeist—a multiplication of mediatized environments, especially virtual spaces, or "cyberspaces," as famously coined by William Gibson in his *Neuromancer* (1984), has turned individual or unique experience (symbolized through the importance of the chance factor), or the first person's view, into the utmost object of desire. This is why we are witnessing in today's popular culture the excess of emphasized authorship, and at the same time its very loss with the stress on the notion of collective consciousness (e.g., DECONcert).

In all the recent examples of performances in the era of the digital there is a wish to merge one's body into another one or more, be these real bodies or technologically extended bodies. McLuhan called technology "the external organs of the body,"[73] which expresses the view that prostheses are, at the same time as they are artificial, natural parts of an organism. What matters, however, seems to be the "emptiness" of both internal or external organs. As Stelarc has it: "my head is empty,"[74] which recalls immediately Deleuze and Guattari's notion of the Body without Organs, a body whose organs are distributed among a collectivity of connections.[75]

Although there was a collapse into the body in 1960s Wounds performances, in the spirit of 1990s Extensions the body as such is no longer at stake. In other words, although Schwarzkogler staged the vulnerability of embodiment, Stelarc reflects on the vulnerability of disembodiment. It is the idea of the body, and the possibilities of collective/displaced/external agency that is staged in disembodied cyber environments.[76] These virtual environments can be interpreted as the "extensions" or "spatializations" of the former "wounds," signifiers of a time in which authorship and personal experience was performed to problematize agency, for example, male agency over the female body.

The avant-garde of the early twentieth century made room for profound investigations of the body by means of such tropes as simultaneity, multiplication, acceleration. In other words, the avant-garde set the rhetorical ground for the body to take over in performance art. In the middle of the twentieth century, then, the body started to collapse, to become part of the frame together with its environment. This collapse was symbolized in the bodily wounds of the 1960s, in which the body had become a "modified ready-made." The emphasis on the medium, its materialities, and hence the focus on the process instead of the product have taken part in a double logic of remediation in which this hypermediated environment desires to erase its traces, or push through the fourth wall into "reality," leading directly into the increased instance of reality—shows featuring individual histories—biographies of mediocrity, as Umberto Eco once described TV shows. But the dissolution of materialities and environments and the collapsing into the inner world has also produced something else. This is a dematerialization, as a result of which fragmented representations of a body in pieces become possible.[77] Throughout the 1990s, the body is not only in extension, but also in decomposition. One body part can take over the whole body at any time. This is why now in the era of 1990s extensions, there is literally no "room" for performances anymore. They have moved into cyberspace, into architecture, into computer generated narratives, and in these new installation we definitely need no longer worry anymore about the absence or limited presence of the audience. There is no audience in cyberspace, just different "downloaders," contributing and participating in the creation of a rhizomatic process of meaning. One of the very peculiarities—not to say dangers—of cyberspace, though, is that instead of wounding, we can just press one button. "DELETE." And it's all gone. But even when deleted, cyberspace leaves its traces, since we are not so alone as we might think in this silent nonspace.

Notes

1. In his manifesto "Vita simultanea futurista" (1927) the futurist Fedele Azaro foresaw today's discussion about the "posthuman" in quite an astonishing way: "When mechanical surgery and biological chemistry will have produced a standard for an incorruptible, resistant, and almost eternal man-machine, the problems of velocity will be less bothersome than today." (My translation from the Italian.) Velocity and rapidity are Azaro's answers to the desired prolongation of life.
2. Jay David Bolter and Richard Grusin, *Remediation: Understanding New Media* (Cambridge, MA: The MIT Press, 1999), 5.
3. William J. Mitchell, *The Reconfigured Eye: Visual Truth in the Post-Photographic Era* (Cambridge, Massachusetts: The MIT Press, 1994), 8; quoted in Bolter and Grusin, *Remediation*, 31.
4. However, we must not forget that the fascination with a "body in pieces" exists in a medical context since the advent of anatomy at the beginning of the fifteenth century.
5. See the exhibition "Die verletzte Diva. Hysterie, Körper, Technik in der Kunst des 20. Jahrhunderts," München, March–May, 2000.
6. The French media philosopher Paul Virilio in an interview with Andreas Ruby, quoted in *The Virtual Dimension*, edited by John Beckmann (New York: Princeton Architectural Press, 1998), 180.
7. Hans Ulrich Gumbrecht, "Postmoderne," in *Reallexikon der deutschen Literaturwissenschaft*, Volume 3, eds. Werner Kohlschmidt and Wolfgang Mohr (Berlin: De Gruyter, 2003).
8. Quoted in Michael Kirby, "Happenings: an Introduction," in *Happenings and Other Acts*, edited by Mariellen R. Sandford (New York: Routledge, 1995), 19; first published in Michael Kirby, *Happenings Anthology* (New York: Dutton, 1965).
9. My translation from the Italian "Manifesto tecnico della lettatura futurista," and "Distruzione della sintassi—Immaginazione senza fili—Parole in libertà."
10. Clement Greenberg, "Collage," in *Art and Culture: Critical Essays* (Boston: Beacon Press, 1965).
11. Kirby, "Happenings," 11–12.
12. Quoted in Kirby, "Happenings," 15.
13. The term "discourse" has to be specified, in that almost all disciplines of the humanities operate with it (from linguistics to sociology). In my current use of "discourse," I refer to the semiotic notion of a *text* that is produced by a specific media-genre. Each genre has its own possibilities of creating the "Énoncé," and the respective "Énonciateur" and "Énonciataire," that is, the utterance, the uttering subject, and the receiving subject, in the way(s) that are pertinent to the genre itself (see Algirdas Greimas/ Joseph Courtés, *Sémiotique: dictionnaire raisonné de la théorie du langage* [Paris: Hachette, 1993], 125). A digital Web page, for instance, has multiple "Énonciateurs" and "Énonciataires," which makes it a typically postmodern medium. By multiplying the uttering and receiving subjects the stress shifts to the utterance itself.
14. "Breaking into reality" is a result of a reality bleed that has not just started with the avant-garde, but arose at the same time as the culture of representation, that is, since the origins of modern theater in early modern Europe; see William Egginton, *How the World Became A Stage: Presence, Theatricality, and the Question of Modernity* (Albany, New York: SUNY Press, 2002).
15. Robert Motherwell, cited in Kirby, "Happenings," 12; original source *The Dada Painters and Poets*, ed. Robert Motherwell (New York: George Wittenborn, Inc. 1951), 62–63.
16. Kirby, "Happenings," 22.
17. Antonin Artaud, *The Theater and Its Double* (New York: Grove Press. 1958), 114.
18. Jacques Derrida, "Le théatre de la cruauté et la clôture de la représentation," in *Écriture et Différence* (Paris: Seuil, 1967), 343; my translation.
19. Ibid., 363.
20. Allan Kaprow, "Excerpts from *Assemblages, Environments & Happenings*," in Sandford, *Happenings and Other Acts*, 235.
21. *Brus Muehl Nitsch Schwarzkogler Writings of the Vienna Actionists*, ed. and trans. Malcolm Green in collaboration with the artists (London: Atlas Press, 1999).

22. *Kurt Schwitters*, Catalogue of exhibition at the Centre Georges Pompidou (Paris: Édition du Centre Pompidou, 1994), 77.
23. Peter Noever, ed. *Aktionismus, Aktionsmalerei,* Wien 1960–65 (Wien: Österreichisches Museum für Angewandte Kunst, 1989), 62.
24. "Panorama 1: Painting in Motion (Draft)," in *Brus Muehl Nitsch Schwarzkogler Writings of the Vienna Actionists,* 199. For more detailed information on Schwarzkogler, see *Rudolf Schwarzkogler,* eds. Eva Badura-Triska and Hubert Klocker (Wien: Museum moderner Kunst Stiftung Ludwig, 1992).
25. Otto Muehl and Günter Brus founded the institute for Direct Art in 1966.
26. Text of Invitation for the Second Total Action by Günter Brus and Otto Muehl, Vienna, Galerie Dvorak, June 24, 1966; quoted in *Brus Muehl Nitsch Schwarzkogler Writings of the Vienna Actionists,* 41.
27. *Performance* and *Competence* correspond with Ferdinand de Saussure's concepts of *Parole* and *Langue.*
28. David Crystal, "Speech Act," in *An Encyclopedic Dictionary of Language and Languages* (Cambridge, MA: Blackwell, 1992).
29. John Austin, *How to Do Things With Words* (Cambridge, MA: Harvard University Press, 1975).
30. Seen in the exhibition "Appearance" at the Galleria d'Arte Moderna in Bologna, January–March, 2000.
31. Brus, *Brus Muehl Nitsch Schwarzkogler Writings of the Vienna Actionists,* 41.
32. Frazer Ward, "Some Relations between Conceptual and Performance Art," *Art Journal* 56, no. 4 (Winter 1997), 38.
33. Allan Kaprow, *Essays on the Blurring of Art and Life* (Los Angeles: University of California Press, 1993), 59.
34. In the video *Chris Burden: A Video Portrait,* directed and edited by Peter Kirby, produced on the occasion of the 1989 exhibition, "Chris Burden," at the Newport Harbor Art Museum about Chris Burden's work from the 1970s and 1980s, we can see interviews with the few witnesses of his performances, who are his personal friends and colleagues.
35. Text from video *Chris Burden: A Video Portrait.*
36. Sandford, *Happenings and Other Acts,* 230.
37. This new logic is also anticipating the format of the reality show, so prominent today.
38. Roswitha Mueller, *Valie Export. Fragments of the Imagination* (Indianapolis: Indiana University Press, 1994), 34.
39. Barbara Rose, "Orlan: Is It Art? Orlan and the Transgressive Act," *Art in America* 81, no. 2 (February 1993): 83–125; for more recent and in depth analysis of Orlan's art, see the contributions by Fred Botting, Scott Wilson, Orlan, and Rachel Armstrong in "Part Three: Self-Hybridation," in *The Cyborg Experiments. The Extensions of the Body in the Media Age,* ed. Joanna Zylinska (London and New York: Continuum Press, 2002).
40. Bolter and Grusin, *Remediation,* 34.
41. Interview with MUD user, in Sherry Turkle, *Life on the Screen: Identity in the Age of Internet* (New York: Simon & Schuster, 1995), 13.
42. Bolter and Grusin, *Remediation,* 40.
43. Hal Foster, *The Return of the Real* (Cambridge, MA: MIT Press, 1996), 127.
44. Slavoj Žižek, *The Plague of Fantasies* (London and New York: Verso, 1997), 134.
45. Oliver Grau, *Virtual Art: From Illusion to Immersion* (Cambridge, Mass.: MIT Press, 2003), 5.
46. Quoted in Michael Rush, *New Media in Late 20th-Century-Art* (New York: Thames & Hudson, 1999), 56.
47. Bolter and Grusin, *Remediation,* 45.
48. Marshall McLuhan, *Understanding Media: The Extensions of Man* (Cambridge, MA: MIT Press, 1996), 8.
49. Frederic Jameson, *Postmodernism, or, the Cultural Logic of Late Capitalism* (Durham: Duke University Press, 1991).
50. Rush, *New Media,* 65–9.
51. Byron Reeves and Clifford Nass, *The Media Equation. How People Treat Computers, Television, and New Media Like Real People and Places* (New York: Cambridge University Press, 1996), 5.
52. Bolter and Grusin, *Remediation,* 42.

53. Parts of this section have appeared in my article "The Medium is the Body," in *Intermedialities*, ed. Hugh Silverman (London: Continuum Press, forthcoming).
54. Presented at the 46th Biennial Exhibition at the Corcoran Gallery of Art, Washington D.C., *Media/Metaphor*, December 9–March 5 2001.
55. See http://www.corcoran.org/biennial/DANIEL/bio.html.
56. For an extended list of digital projects, dealing with art/technology/science, see the Website provided by Stephen Wilson: http://userwww.sfsu.edu/~infoarts/links/wilson.artlinks2.html, or his book *Information Arts: Intersections of Art, Science, and Technology* (Cambridge, MA: MIT Press, 2002).
57. See http://www.interlog.com/~drokeby/gon.html.
58. Kirby, "Happenings," 22.
59. Sandy Stone, "Will the Real Body Please Stand Up? Boundaries Stories about Virtual Cultures," in *Cyberspace. First Steps*, ed. Michael Benedikt (Cambridge, MA: MIT Press, 1991).
60. Exhibited at the 2002 Biennial at the Whitney Museum of American Art in New York City. http://www.whitney.org/2002biennial/
61. Interview online with the curator of the Biennial.
62. http://www.maryflanagan.com/collection.htm
63. Ibid.
64. Deconism Museum and Gallery: http://wearcam.org/deconism/cyborg_echoes_simplified_html.htm.
65. Stelarc, "Towards the Posthuman. From Psycho-body to Cyber-system," *Architectural Design* (Profile No. 118: Architects in Cyberspace) 65:11–12 (November–December1995): 91.
66. Edward Scheer has called Stelarc's provocations "rhetorical sculptures." "Stelarc's E-Motions," in *The Cyborg Experiments*, 85.
67. See Mark Dery's chapter on Stelarc "Ritual Mechanics: Cybernetic Body Art," in *Escape Velocity. Cyberculture at the End of the Century* (New York: Grove Press, 1996), 151–181.
68. "High-Tech Frankenstein, or Heidegger Meets Stelarc," in Zylinksa, *The Cyborg Experiments*, 28.
69. "Stelarc's E-Motions," in Zylinska, *The Cyborg Experiments*, 97.
70. Joanna Zylinska and Gary Hall, "Probings: an Interview with Stelarc," in Zylinska, *The Cyborg Experiments*, 123.
71. Example given by Shannon Bell at Interaccess Gallery, Toronto, March 2003, who calls the head "the sweetest thing."
72. Flyer at Interaccess Gallery, Toronto, March 2003.
73. Marshall McLuhan, *Understanding Media: the Extensions of Man*.
74. Quote from DECONversation at Deconism Gallery, Toronto, March 21, 2003: "Robotic Body vs Cyborg Mind: A Live Probe Into the Continuum of Existentiality," with Steve Mann and Stelarc.
75. Gilles Deleuze and Félix Guattari "November 28, 1947: How Do You Make Yourself a Body without Organs?" in *A Thousand Plateaus. Capitalism and Schizophrenia*, ed. and trans. by Brian Massumi (Minneapolis: University of Minnesota Press, 1987).
76. See N. Katherine Hayles' analysis of how cybernetic technocultures privilege disembodiment, and of how this disembodiment has evolved in the various cultural sectors throughout the second half of the twentieth century, in *How We Became Posthuman:Virtual Bodies in Cybernetics, Literature, and Informatics* (Chicago: University of Chicago Press, 1999). See also Arthur and Marilouise Krokers' "Thesis on the Disappearing Body in the Hyper-modern Condition," in *Body Invaders: Panic Sex in America* (New York: St. Martin's Press, 1987).
77. Cf. for example, the trend of "getting under the skin" in recent cosmetics advertisement discussed in my article "Getting Under the Skin, or, How Faces Have Become Obsolete," in *Configurations*, special issue, ed. Timothy Lenoir (Baltimore: Johns Hopkins University Press, forthcoming).

CHAPTER **11**

Flesh and Metal: Reconfiguring the Mindbody in Virtual Environments

N. KATHERINE HAYLES

Dualistic thinking is as difficult to avoid as the sticky clay that passes for topsoil where I live in Topanga Canyon. When it gets even a little wet, it attacks my feet so resolutely that I look as if I am wearing snowshoes. I try to avoid it, of course, but inevitably something lures me off the beaten path and there I am again, stomping around with elephant shoes. In similar fashion I struggled to avoid the Cartesian mind–body split in my recent book *How We Became Posthuman* when I made a distinction between the body and embodiment. The body, I suggested, is an abstract concept that is always culturally constructed. Regardless of how it is imagined, "the body" generalizes from a group of samples and in this sense always misses someone's particular body, which necessarily departs in greater or lesser measure from the culturally constructed norm. At the other end of the spectrum lie our experiences of embodiment. Although these experiences are also culturally constructed, they are not entirely so, for they emerge from the complex interactions between conscious mind and the physiological structures that have emerged from millennia of biological evolution. The body is the human form seen from the outside, from a cultural perspective striving to make representations that can stand in for bodies in general. Embodiment is experienced from the inside, from the feelings, emotions and sensations that constitute the vibrant living textures of our lives—all the more vibrant because we are only occasionally conscious of its humming vitality.[1] I tried to stay on the holistic path by insisting that the body and embodiment are always dynamically interacting with one another. But having made the analytical distinction between the body and embodiment, I could not escape the dualistic thinking that clung to me regardless of my efforts to avoid it.

I want now to construct the situation in a different way. This chapter pushes beyond the position articulated at the end of *How We Became Posthuman,* where I argued that the erasure of embodiment characteristic of the history of cybernetics should not again be enacted as we move into the technoscientific formations I call the posthuman. Rather than beginning dualistically with body and embodiment, I propose instead to focus on the idea of relation and posit it as the dynamic flux from which both the body and embodiment emerge. Seeing entities emerging from specific kinds of interaction allows them to come into view not as static objects precoded and prevalued but rather as the visible results of the dynamic on-goingness of the flux, which in itself can be neither good nor bad because it precedes these evaluations, serving as the source of everything that populates my perceived world, including the body and experiences of embodiment.[2]

Beginning with relation rather than pre-existing entities changes everything. It enables us to see that embodied experience comes not only from the complex interplay between brain and viscera Antonio R. Damasio compellingly describes in *Descartes' Error: Emotion, Reason, and the Human Brain*[3] but also from the constant engagement of our embodied interactions with the environment. Abstract ideas of the body similarly emerge from the interplay between prevailing cultural formations and the beliefs, observations, and experiences that count as empirical evidence in a given society. In this view, embodiment and the body are emergent phenomena arising from the dynamic flux that we try to understand analytically by parsing it into such concepts as biology and culture and evolution and technology. These categories always come after the fact, however, emerging from a flux too complex, interactive, and holistic to be grasped as a thing in itself. To signify this emergent quality of the body and embodiment, I will adopt the term proposed by Mark Hansen to denote a similar unity, the mindbody.[4]

Although a study of anatomy textbooks written across the centuries will confirm that ideas of the body change as the culture changes, it is less obvious that our experiences of embodiment also change.[5] Refusing to grant embodiment a status prior to relation opens the possibility that changes in the environment (themselves emerging from systemic and organized changes in the flux) are deeply interrelated with changes in embodiment. Living in a technologically engineered and information-rich environment brings with it associated shifts in habits, postures, enactments, perceptions—in short, changes in the experiences that constitute the dynamic lifeworld we inhabit as embodied creatures. One story about these changes—a story I deeply want to resist—narrates them as what Donna Haraway calls a masculinist fantasy of second birthing, a transcendent union of the human and the technological that will enable us to download our consciousness into computers and live as disembodied information patterns, thereby escaping the frailties of the human body, especially mortality. But one need not indulge in this fantasy to grant that embodied experiences are changing through interactions with information-rich environments, a point Andy Clark tellingly makes when he argues that technologies have always co-evolved with the human brain.[6]

What kinds of changes are bringing about these shifts in embodied experience? Space will not allow me to explore this question fully, but I can sketch a few possibilities to indicate what I have in mind. Consider first the force of habits to shape embodied responses, especially proprioception, the internal sense that gives us the feeling we *occupy* our bodies rather than merely possess them.[7]

Computer video game players testify to feeling they are projecting their proprioceptive sense into the simulated space of the game world. In fact, they eloquently insist that being a good player absolutely requires this kind of projection. Their body boundaries have fluidly intermingled with the technological affordances so that they feel the joystick as an unconscious extension of the hand. Simon Penny, acknowledging the power of these interactive computer games, discusses them as a bodily discipline that the military has been quick to appropriate.[8] Moreover, the interactions have the potential not only to condition but actually to shape the central and peripheral nervous systems. Of species on earth, humans are among those with the longest period of neotony, extending at least into adolescence the capacity of the nervous system to change and evolve after birth. The flexibility of the human neural system enables new synaptic connections to form in response to embodied interactions. This implies that a youngster growing up in a medieval village in twelfth-century France would literally have different neural connections than a twenty-first century American adolescent who has spent serious time with computer games.

In addition to these biological changes are more intrusive interventions that join the biological with the technological. Sandro Mussa-Ivaldi and his research team at Northwestern University have successfully removed portions of a sea lamprey's brain, kept it alive in a bath of nutrients, and then connected wires to bring electrical signals from a mobile robot's optical sensors into the lamprey's vestibular system, the part of the brain that deals with up–down orientation in the water. The disembodied brain apparently interprets the robot's signals as denoting a certain orientation in the water and sends signals back that make the robot move toward the light (the most common response) or away; additional behaviors include wheeling in a circle and spinning in a spiral configuration.[9] This fusion of biological organism and cybernetic device is so striking that "cyborg" seems too innocuous a term to describe it. Experiments are underway at UCLA and elsewhere to use similar technologies to link the brain waves of paraplegics with remote actuators, thus enabling them to manipulate objects in their environment in ways that would otherwise be impossible. Success depends on their ability to put their brains in intense feedback loops with the technology, thereby forming new neural patterns that can drive the actuators. Other interventions include implants such as those Kevin Warwick, Professor of Cybernetics at the University of Reading in England, had inserted into his arm. His first implant was a passive device, communicating only with embedded sensors in the environment.[10] He advanced from that to a second implant that also sends signals to his nervous system, creating an integrated circuit linking his evolving neural patterns directly with environmentally embedded sensors and computer chips.[11]

The number of people who have implants is likely to remain minuscule, at least for the immediate future. Greater numbers will be affected by the continuing development and expansion of pervasive computing. The idea is to embed innumerable sensors and small computers in the environment capable of collecting, processing, storing, and transmitting information. Recent model cars already have a module under the right front seat that collects over 200 different kinds of information. When the car is taken in for service, the mechanic downloads the information into his computer, where a diagnostic software program analyzes it and identifies malfunctioning parts. An article on telematics in the London *Financial Times* details further developments, including an eye-tracking system that analyzes a driver's eye movements to anticipate moves such as changing lanes, which it coordinates with images from a rear video camera and warns the driver if the move is unsafe.[12] Additional functionalities include analyzing eye movements to determine if a driver is getting sleepy, which also triggers a warning. Related to telematics within the car is the development of smart highways, with embedded sensors and computers capable of tracking a car's movement. The goal is to develop cars and highways with interlocking computer systems that will enable cars to drive themselves. Similarly, sensors are being embedded in smart buildings, enabling the building to sense the presence of human beings and activate or shut down ventilation, lighting, heating and cooling systems, and a variety of other functions. Although many information transactions will take place between embedded computers without the people who move in those spaces knowing about them, enough of the infrastructure will be visible to make experientially vivid the myriad ways in which our embodied interactions with the environment generate, direct, and change the information flows surging ceaselessly around us.

The development of smart environments makes even more persuasive the arguments of philosopher Andy Clark and anthropologist Edwin Hutchins that cognition should not be seen as taking place in the brain alone. Instead, cognition in their view is a systemic activity, distributed throughout the environments in which humans move and work. Indeed, Clark argues that the distinctive characteristic of humans has always been to enroll objects into their cognitive systems, creating a distributed functionality he calls "extended mind."[13] We are cyborgs, he wrote in a recent article, "not in the merely superficial sense of combining flesh and wires, but in the more profound sense of being human–technology symbiots: thinking and reasoning systems whose minds and selves are spread across biological brain and nonbiological circuitry."[14] Observing that the "extended mind" is a strategy almost as old as humans, he nevertheless points out that the joining of technology with biology has created a "cognitive machinery" that is "now intrinsically geared to transformation, technology-based expansion, and a snowballing and self-perpetuating process of computational and representational growth." Although relatively small changes in human brains may have been sufficient to make us into the "symbolic species," as Terrence Deacon calls humans, these incremental changes have now catapulted us "on the far side of a precipitous cliff in cognitive-architectural space" ("Natural Born Cyborgs?").

Following a similar line of thought, Edwin Hutchins argues that cognitive scientists made a fundamental mistake when they located cognition in the brain and then tried to model that cognition in artificial intelligences.[15] They rather should have recognized that cognition is a systemic activity distributed throughout the environment and actuated by a variety of actors, only some of whom are human. In his view, it is not merely a metaphor to call drawing a line on a navigational chart remembering and erasing a line forgetting, for if these objects are part of our extended mind, drawing and erasing are indeed functionally equivalent to remembering and forgetting. The extended mind model indicates how cultural perceptions change in relation to the development of information-rich environments. Instead of the Cartesian subject who begins by cutting himself off from his environment and visualizing his thinking presence as the one thing he cannot doubt, the human who inhabits the information-rich environments of contemporary technological societies knows that the dynamic and fluctuating boundaries of her embodied cognitions develop in relation to other cognizing agents embedded throughout the environment, among which the most powerful are intelligent machines.

In these views, the impact of information technologies on the mindbody are always understood as a two-way relation, a feedback loop between biologically evolved capabilities and a richly engineered technological environment. Such feedback loops may be reaching new levels of intensity as our environments become smarter and more information rich, but the basic dynamic is as old as humans. Returning to *The Symbolic Species*, I take Deacon's point that while the evolution of language changed human brain structure, human brain structure affected the evolution of language. To emphasize the importance of this relationality, he proposes we "imagine language as an independent life form that colonizes and parasitizes human brains, using them to reproduce."[16] The relationship is symbiotic: "modern humans need the language parasite in order to flourish and reproduce, just as much as it needs humans to reproduce. Consequently, each has evolved with respect to the other" (113). He draws an analogy with the evolution of computer interfaces. When interfaces were restricted to DOS commands, computers were seen as the special province of scientists, engineers, and other researchers. With the more intuitive interfaces pioneered at Xerox PARC and implemented by Apple, computers spread much more rapidly in the population, in part because they were better suited to brain processing.[17] This is the other side of adaptations in the brain to better accommodate computer dynamics. Relationality implies a deep and dynamic connection between the evolutionary pathways of computers and humans, each influencing and helping to configure the other.

My argument further implies that these coevolutionary dynamics are not only abstract propositions grasped by the conscious mind but also emergent dynamic processes actualized through interactions with the environment. And here there is a problem. Especially in times of rapid technological innovation, there are many gaps and discontinuities between abstract concepts of the body, experiences of embodiment, and the dynamic interactions with the flux of which these are enculturated expressions. The environment changes and the flux shifts in correlated systemic and organized ways, but it takes time, thought and experience for these changes to be registered in the mindbody. Bridging these gaps and connecting these discontinuities is the task taken on by the three virtual reality art works discussed here: "Traces" by Simon Penny and his collaborators, "Einstein's Brain" by Alan Dunning, Paul Woodrow and their collaborators, and "NØTime" by Victoria Verna and her collaborators. If art not only teaches us to understand our experiences in new ways but actually changes experience itself, these art works engage us in ways that make vividly real the emergence of ideas of the body and experiences of embodiment from our interactions with increasingly information-rich environments. They teach us what it means to be posthuman in the best sense, in which the mindbody is experienced as an emergent phenomenon created in dynamic interaction with the ungraspable flux from which also emerge the cognitive agents we call intelligent machines. Central to all three art works is the commitment to understanding the body and embodiment in relational terms, as processes emerging from complex recursive interactions rather than as preexisting entities. Because relationality can be seen through many lenses, I have chosen works that place the emphasis on different modes of relation. "Traces" foregrounds the relation of mindbody to the immediate surroundings by focusing on robust movement in a three-dimensional environment; "Einstein's Brain" foregrounds perception as the relation between mindbody and world that brings the flux into existence for us as a lived reality; and "NØtime" emphasizes relationality as cultural construction.

These configurations can also be understood in terms of the typology Don Idhe proposes in *Technology and the Lifeworld: From Garden to Eden.*[18] Parsing the general situation as Human-Technology-World Relations, he identifies three variants. The first, which he calls "embodiment relations," bundles the human and the technological into one component and emphasizes the relationality between this component and the world: (Human-Technology) → World. This corresponds to the situation in "Traces," in which the technology re-produces the human form in a simulation (thereby bundling together the technology and the human) and emphasizes the movement of the simulated techno-human through space. The second variant, identified by Idhe as "hermeneutic relations," bundles together technology and the world and emphasizes the relationality with the human: Human → (Technology-World). This parsing is performed in "NØtime," which leaves the human body unencumbered to experience an installation space permeated by sensors, actuators, and display technologies, which bundle together the world and technology. The third parsing, "alterity relations," is parsed as Human → Technology-(-World), where the brackets around World indicate that the bundling of technology and world is achieved through the creation of a simulated world, which is then put into relation with the human. Although the three emphases of enactment, perception, and enculturation (corresponding to embodiment, hermeneutic, and alterity relations in Idhe's schema) by no means exhaust the ways in which relationality brings the mindbody and the world into the realm of human experience, they are capacious enough in their differences to convey a sense of what is at stake in shifting the focus from entity to relation.

I. Relation as Enactment

In *The Embodied Mind: Cognitive Science and Human Experience,* Varela, Thompson, and Rosch articulate a vision of relationality that has much in common with "Traces." They write, "living beings and their environments stand in relation to each other through *mutual specification* or *codetermination.*"[19] Coining the term "enaction" to describe this dynamic interplay between self and world, they envision mindbody and environment coming into existence through a mutual process of "codependent arising" (110).

It is precisely this kind of relationality that Simon Penny wanted to implement in a virtual reality (VR) environment. As early as 1994 he articulated a desire to depart from the usual VR model that in his assessment "blithely reifies a mind/body split that is essentially patriarchal and a paradigm of viewing that is phallic, colonizing, and panoptic."[20] In "Traces," Penny, along with collaborators Jeffrey Smith, Phoebe Sengers, Andre Bernhardt, and Jamieson Shulte, created an interactive art work designed to bring the body more fully into the virtual space. Reacting against the VR rhetoric of disembodiment, they critique this rhetoric as deriving from "an essentially uninterrogated Cartesian value system, which privileges the abstract and disembodied over the embodied and concrete."[21] They propose, by contrast, to build "an unemcumbering sensing system which modeled the entire body of the user" (3). Working with a three-dimensional CAVE environment that displays simulated visual images along four surfaces (three walls and the floor) as well as in the goggles of the user, they implemented a visual tracking system that computes the volume of the user's body by modeling its movement in space and time through 3D cubes called voxels (volumetric units named by analogy to 2D pixels). From this computation they created "traces," simulated images of volumetric residues that trail behind the rendered model of the user's body, gradually fading through time as continued movement creates new traces that also fade in turn. The body model and residues are comprised of lilac-colored voxels 5 cm on a side, rendered in a simulation space 60 × 60 × 45 voxels in three dimensions, with computational time steps of 15 fps. Because the time step interval falls below the threshold of 24 fps at which frames appear to human viewers as continuous motion, the computational process manifests itself to the user and outside viewers as somewhat jerky in its motion. This effect, although in one sense a constraint imposed by the amount of computation required for each updating, is embraced by Penny and his collaborators as part of a larger aesthetic strategy to "avoid any pretense of organic form. There was a desire to be up-front about the fact that this was a computational environment, not some cinematic or hallucinatory pastoral scene" (13). For the same reason, the collaborative team renounced what they regarded as the "eye candy" of virtual worlds ready for exploration, using texture mapping only for the virtual room projected on the CAVE walls, about twice the size of the CAVE's dimensions of three meters on a side. Instead of a graphically rendered virtual world, the user "simply sees graphical entities spawned by various parts of their [sic] body when in motion" (10).

The avatar interface is designed, in Penny's terminology, to be "autopedagogical," teaching the user how to interact with it through its evolution through the three phases of Passive Trace, Active Trace, and Behaving Trace. In the Passive Trace, the simulated volume passively follows the user's motions, creating the impression that the user "'dances' a 'sculpture'" (4), although a transitory one that gradually fades into transparency as it moves away from the viewer into the simulated space on the front wall of the CAVE. With its time-sensitive evolving transparency, the Passive Trace would seem to have more in common with Bergson's vision of flowing time than with the enduring static form that sculpture usually implies. As the user continues to move in the space, the trace transforms from a passively following cloud to active entities that can be "spun off" the user's body through rapid motion or acceleration, for example by flicking one's hand rapidly down one's arm as if shaking off water droplets. At first, these entities follow the user but gradually they become more autonomous as their motions are guided by autonomous agent software. As they make the transition into the Behaving Trace, they exhibit behaviors characteristic of such artificial life programs as Craig Reynolds "Boids," simulated forms that exhibit flocking behavior when programmed with a relatively simple set of rules such as "always fly toward the center of where the other objects are." The simulated objects in the Behaving Trace may follow the user, for example, or they may break off and head in other directions, moving as a flock following its own artificial life dynamics.

Although the immediate meaning of "autopedagogical" for Penny is the progression whereby "Traces" teaches the user how to interact with it, the term evocatively points toward other realizations as well. By incorporating a temporal dimension into the work, and especially by having the duration of the trace visibly fade away as it ages, the art work resists the fantasy that information technologies will allow us to escape our bodies and move into transcendent spaces where we can escape the ravages of time. Another realization emerges from the "traces" metaphor, which suggests new kinds of possibilities for interactions between humans and intelligent machines. As Penny and his collaborators point out, the "Traces" simulation, considered as an avatar, occupies a middle ground between avatars that mirror the user's motions and autonomous agents that behave independently of their human interlocutors. As the "trace" avatar transforms from mirroring the user's actions to engaging in autonomous behaviors, it enacts a borderland where the boundaries of the self diffuse into the immediate environment and then differentiate into independent agents. This performance, registered by the user visually and also kinesthetically as she moves energetically within the space to generate the entities of the Active and Behaving Traces, makes vividly clear that the simulated entities she calls "her body" and the "trace" are emergent phenomena arising from their dynamic and creative interactions.

Moreover, the elegant simplicity of the simulation—the refusal to add "eye candy" to the visual effects—helps to make real to the user that the avatar is in effect indistinguishable from her interface with the computer. The trace avatar, Penny and his collaborators write, "must be thought of as the part of the system which is intimately connected to the user. In this way, the line between system, avatar, and interface also becomes blurred; the avatar becomes the interface, the point at which the computational system and the user make contact" (22). The aesthetic strategy of refusing to conceal the computational nature of the simulation resists the fantasy of transcendent second birthing by grounding the work in the constraints of real-life computational and sensing devices. As the "Traces" article explaining the construction of the work makes clear, this is no illusory world of limitless possibilities but a carefully engineered art work in which numerous trade-offs and "workarounds" are required to make the project feasible. The ingenuity, creativity and skill of the designers and programmers are repeatedly tested as they come up against a variety of problems, from devising a workable camera-tracking system to balancing the graininess of the voxelated image against increased latency times as the computational load increases. Coming to grip with these problems, they achieve the key insight that constraints can function as opportunities as well as problems. This approach led them to see that the problems involved in having the avatar exactly track the user's body could be used constructively if they switch metaphors. Rather than regarding the avatar as a mirroring puppet, they think of it as a trace emerging from the borderlands created by the energetic body in motion. What was a tracking problem is thus transformed into the possibility of creative play between user and avatar.

The net result of these compromises, creative solutions and transformations is to make real for us the insight that the art work is not simply the instantiation of an abstract concept but an artifact that emerged from the dynamic relation between the vision of the designers, the constraints imposed by the situation, and the powerful but still limited capabilities of the intelligent machines that perform the sensing, computational, and rendering tasks that make the project a reality. In its form, construction and functionality, "Traces" testifies to this relationality at the same time that it also performs relationality for the user. Far from the fantasy of disembodied information and transcendent immortality, "Traces" bespeaks the playful and creative possibilities of a body with fuzzy boundaries, experiences of embodiment that transform and evolve through time, connections to intelligent machines that enact the human–machine boundary as mutual emergence, and the joy that comes when we realize we are not isolated from the flux but rather enact our mindbodies through our deep and continuous communion with it.

II. Relation as Perception

In "The Brightness Confound," Brian Massumi reminds us of Wittgenstein's puzzlement as he stares at his sunlit table and realizes that he cannot say what color the table is, for his perception fuses luminosity, radiance, and color into a unity that defies tidy categorization. Following Marc Bornstein, Massumi calls this unity "the brightness confound" and makes the simple but elegant point that the perception is a "singular confound of what are described empirically as separate dimensions of vision."[22] In this sense the confound is absolute. "Absoluteness is an attribute to any and all elements of a relational whole. Except, as absolute, they are not 'elements.' They are parts or elements before they fuse into the relational whole by entering indissociably into each other's company; and they are parts or elements afterwards, if they are dissociated or extracted from their congregation by a follow-up operation dedicated to that purpose. In the seeing, they are absolute" (82). Insisting on this absoluteness, Massumi nevertheless writes as if the elements of the confound have a prior existence as separate entities. Humberto Maturana expunges this vestige of realism when, in developing the theory of autopoiesis, he makes the point that someone experiencing a hallucination would be unable to distinguish the hallucination from reality.[23] In Maturana's view, what we perceive *is* reality for us.

In the "Einstein's Brain" project, Canadian artists Alan Dunning and Paul Woodrow stage what we might think of as Maturana's claim (although they come to this view via their own independent paths and not necessarily through Maturana). They are keenly conscious of the ironic overtones of their chosen title, referencing Roland Barthes' essay by a similar title. Commenting on Einstein's brain as a fetishized object, they write, "His brain has passed into the world of myth, cut up and minutely examined but revealing little."[24] The title points up the fact that the brain as fetishized physical object, considered in isolation from the world, cannot possibly account for the richness of human experience. In his meditation on the subject, Barthes related the duality of physical brain and prodigious mind to a split between Einstein as the researcher and Einstein the knower of the world's innermost secrets.[25] Rooted in the physical brain, Einstein's mind nevertheless seemed to have nearly occult powers of insight, at least in the popular imagination. This oscillation between ordinary physical reality and occult power translates in the "Einstein's Brain" project into a desire to use advanced technology to reveal the constructedness of our everyday world.

The "Einstein's Brain" project has been in process for five years and has taken a variety of forms in different installations, but a common idea unites all the instantiations.[26] The artists (like Maturana) are committed to the realization that the world of consensual reality does not in any sense exist "out there" in the forms in which we perceive it. Rather, the world we know is an active and dynamic construction that emerges from our interactions with the flux.

> We think of the body as separate from the world—our skin as the limit of ourselves. This is the ego boundary—the point at which here is not there. Yet, the body is pierced with myriad openings. Each opening admits the world—stardust gathers in our lungs, gases exchange, viruses move through our blood vessels. We are continually linked to the world and other bodies by these strings of matter. We project our bodies into the world—we speak, we breathe, we write, we leave a trail of cells and absorb the trails of others. The body enfolds the world and the world enfolds the body—the notion of the skin as the boundary to the body falls apart.[27]

They self-consciously position their work in opposition to military and corporate uses of virtual reality, which continually aim for greater and greater realism. They point out that when virtual reality illusions are engineered with the goal of seamlessly reproducing the "real" world, the effect (wittingly or not) is to reinforce existing structures of authority and domination–structures that in their desire to preserve the status quo find it in their interest to foreclose alternative constructions of reality and moreover to keep them from coming to mind as possibilities. By contrast, Dunning and Woodrow conceive of the simulation technologies as deliberately imperfect, so as to make clear their construction as "reality engines connected not to the generation of a reality but as a means of attending to a consciousness that in turn fashions a reality" (7).

To resist the domination in VR of a "realism rid of expression, symbol or metaphor" that is "sustained by the authorities of homogeneity and seamlessness" (1), Dunning and Woodrow's create a "cranial landscape" merging symbolic and semiotic markers with the landscape of experience (5). Their work often has a somewhat idiosyncratic range of reference overlaying the consistency of their vision, rather as if a magpie had collected shiny bits from here and there because they attracted her attention and then wove them into a nest of breathtaking coherence and careful design. So the inspiration for the cranial landscape comes partly from the *Carte du tender*, Madeleine de Scudèry's 1654 romantic narrative map inscribing onto a landscape names indicating the predictable heating up and cooling off of a love affair. Appropriating the name of one of these sites, the "Forest of Vowels," Dunning and Woodrow create a semiotically marked landscape that exists for the user as a negotiable surface and also as a changing landmass tied in with the user's responses as they are registered through reading her brain waves and other physiological indicators. Another source of inspiration is the *derive* of the Situationists, a random walk through the city governed by the principle that every turn and meander should be taken at random rather than with the intent to arrive somewhere. Guy Debord's

1957 map of Paris showing arrows marking the course of a *derive* suggested to Dunning and Woodrow that even something as apparently static and durable as city architecture might be re-imagined as emerging in complex interplay with human enactments. So their plan for "Forest of Vowels" calls for the association of external events in the real world with the landscape of the virtual world. These associations include feeding in the moon's changing gravitational forces so they alter the form of the virtual world's landmasses, tying fluctuations in the stock market to the growth patterns of trees and plants, and connecting the daily attendance figures at Graceland to the changing cultural paradigms of the virtual world (6).

Another idiosyncratic influence is "The Stone Tapes," a story produced by Nigel Kneale on BBC Television on December 25, 1972. The story takes the form of a mystery centered on an apparently haunted building, which has somehow recorded in its inorganic stone traumatic events that happened there; the building has the capacity to play these recordings back by transferring them directly to the brains of people who come inside the building, so that it appears to the people as if they are spectators of the original events. The story appeals to Dunning and Woodrow on multiple levels. The building displays qualities that make it appear as if it can operate as a subject, thus blurring the boundary between an exterior static architecture and dynamic interior world of human emotion. Moreover, the humans haunted by the building must confront the apparent reality of an illusion generated from inside their own brains, blurring the boundary between their perceptions of the real world and the illusions activated for them by the stone building.

The incongruities between a virtual reality experienced by those who are "haunted" and the consensual reality experienced by onlookers are dramatically staged in Dunning and Woodrow's installation "The Madhouse." In this installation, participants wearing VR goggles engage in behaviors that can only appear strange and bizarre from the viewpoint of those who do not see or hear the simulations. In a similar mode is their plan for an art work that would force participants to question consensual reality by creating deliberately unstable and deficient renderings of the virtual world. In this projected work vision is blurred, detail is shifting and inconstant, slower or faster frame rates suggest a rendering engine behind the scenes, left- or right-hand sides of stereoscopic vision blink out, depth perception is lost, objects only appear when one is in motion, and the edges of the worlds visibly reinvent themselves. They write of their motivation for these strategies:

As Western artists, we developed from a world where we learned to objectify our bodies, to separate our minds from our bodies and viscera, where we learned to distinguish matter from mind and where the construction and placement of objects was the focus and culmination of our intentions and desires. Developments in cultural and social theory and in technology have suggested that we and other artists shift their attention away from a graspable, predominately corporeal world to one which is increasingly slippery, elusive, and immaterial. Mind and matter, combining in the cognitive body, are interdependent. The world we inhabit is in flux, comprised of increasingly complex connections and interactions. In this world there are no fixed objects, no unchanging contexts. There are only coexistent, nested multiplicities (8).

In the "unbroken field of transformations" that for them constitute the emergent dynamic we call reality, virtual reality art can play a vital role in shaking the belief that our bodies and the world exist independent of relation. They intend their art to enact engagements that make vividly real the fact that everything in our world, including (or rather especially) the human mindbody, emerges from our relation with the on-going flux.

These ideas and strategies come together in the installation exhibited at the TechnOboro Gallery in Montreal in September 2001 and demonstrated in prototype at the Digital Arts Conference at Brown University, Rhode Island, in April 2001, which is where I saw it. The centerpiece of the installation is the ALIBI, the "anatomically lifelike interactive biological interface," an anatomically correct life-size model of the human body stuffed with a wide variety of sensors, including theramin proximity sensors, touch sensors, aroma sniffers, pressure sensors, sound sensors, and carbon dioxide sensors. Participants wear goggles that can be arranged to show only the simulated world or (by removing the blinders from the lens area) convert the scene to a "mixed reality" in which both the simulation and the real room are visible. They are thus able to see simultaneously the virtual reality projection and the artifactual body, which lies on a light table in the center of a room. Made from a cast of a male model, the body is painted with thermochromic paint that appears as a lovely dark blue when cool but turns white when warmed by someone's touch, fading again to blue as the area cools to the ambient temperature. Participants can interact with the body by touching it on the thigh, abdomen, legs, and so on, by whispering in its ear, even breathing into its mouth. These interactions, when sensed by the system, activate and change the simulated worlds being imaged in the goggles. The blue body thus acts as a navigational interface, opening portals to a variety of simulated world as the appropriate body areas are touched, massaged, and otherwise manipulated. One user wears a helmet capable of sensing her electroencephalic activity, including alpha, beta, theta and delta brain waves. These data are fed into the simulation, along with other biological data collected from the user such as blood pressure, pulse rate, and galvanic skin response. The data trigger the performance of simulated images, with sunbursts, polygons, and flashes of light appearing in response to the user's reactions. Moreover, the amplitude and frequency of the participant's brain waves are converted to MIDI files and used to create a soundscape for the simulation, which serves as an acoustic transform of her ongoing physiological responses.

Two other components complete this complex work. On the back wall are projected images of revolutionary historical events, including authentic footage of statues being pulled down during the Russian Revolution and speeches by Mao. This corresponds to the "Stone Tapes" motif, establishing a visual connection between past and present and "haunting" participants with events that have changed the course of history and that continue to be remembered as dramatic instantiations of the revolutionary impulses. Further complexity is added by the presence of a "viewing room," in which other visitors can see the artifactual body being manipulated by users who engage in actions and behaviors that remain inexplicable to those who cannot see the virtual worlds. The effect, once again, is to call into question consensual reality by fragmenting the space so that different versions of "reality," virtual and actual, compete and conflict in their representational stimuli.

The most striking part of the installation from my point of view are the feedback loops between the user's responses, her interactions with the artifactual body, and the production of the simulated world. Imagine the scene. You are in some initial brain state that generates images and sounds in the simulated world you see. While watching these displays, you begin to touch the body in its sensitive areas, opening portals to other simulated worlds, which trigger new responses in you, which feed back into the simulation to change it, which makes you want to touch the body in new ways, which further changes the simulated images and sounds, which in turn generate yet more responses from you. The loop is endless, and endlessly fascinating, forming, as the authors say, a "single intelligent symbiotic system."[28] Commenting on a different art work, Dunning and Woodrow explain the relation of the body to the VR world in ways that apply with special force to this installation. "These worlds are not external to the body, but, are properly thought of as being inside the body. This accounts for the apparent invisibility of the body in a virtual space. The body disappears because it is turned in on itself. The ego-boundary is no longer the point at which the body begins and ends in relation to an external environment, but is, rather, . . . the very limit of the world."[29] Using a different metaphor in "The Stone Tape, the Derive, the Madhouse," they comment, "It is as if we are inside ourselves, like a three-dimensional eye which constructs itself as it moves through internal haptic space."[30] Relationality here is not beside the point; it *is* the point of a mindbody that realizes itself through its playful and intense interactions with evolving virtual worlds, which in the view of these artists include our perceptions of both real and simulated worlds. In this sense all human experience is a "mixed reality," emerging from another kind of brightness confound in which technology, the world, and human embodiment all play a role.

III. Relation as Enculturation

A notable characteristic of "NØtime" is its collaborative nature, as is the case with "Traces" and "Einstein's Brain" as well. The human collaborators, listed by name and sometimes by affiliation, indicate the range of expertise necessary to construct the installations, including such skilled contributors as computer scientist and cultural critic Phoebe Sengers for the "Traces" project, software designer and CSCW engineer Hideaki Kuzuoka for the "Einstein's Brain" project, and software designer Gerald Jong for the NØtime project. For convenience I have referred to works using the names of the artists who had the initial idea as creator, but they more than anyone else realize how deeply their collaborators have shaped the projects and how essential their contributions have been. Less prominently featured and usually identified by model name and technical capacity are the silicon collaborators, the intelligent machines and software packages without which these works would have been impossible to create. It is not merely whimsical to refer to the machines as collaborators, for their capabilities and limitations are as important to the project's shape as the capabilities and limitations of the human designers. These silicon collaborators include computers, software programs, sensing systems, music synthesizers, tracking systems, motion detectors, and a host of other processors, interfaces and actuators. The complexity of the collaborations between many different humans and intelligent machines indicates that in a deep sense all of these projects are distributed cognitive systems. Moreover, cognition takes place not only in the minds of the artists and the logic gates of the machines but also in the participants who interact with the art works. As "Einstein's Brain" in particular makes clear, the user's interactions in the installations are not merely passive viewing of preexisting works but active components in the work's construction. Among these three art works, NØtime insists most visibly and interactively on the distributed cognitive collaborations that construct it and especially on the role played by the global community of intelligent machines we call the Internet. It also locates the arena of relationality with which it is concerned in the broadest geographic terms. Whereas "Traces" focused on the immediate proximity of the body and "Einstein's Brain" on the room-sized spaces where the artifacts were placed and the simulations projected, the reach of "NØtime's" enactments is global, although it simultaneously insists as well on the importance of local interactions and proximity.

Victoria Vesna originally conceived of NØtime as a response to the common postmodern condition of having no time, living amidst the multiple conflicting commitments and stresses that people negotiating complex urban environments find to be an inevitability of contemporary life. Her playfully paradoxical idea was to create avatars that could take over portions of our lives and live them for us while we were busy doing other things. As the project evolved, the idea of collaborative interactions that together create a "person" or "life" took a somewhat different turn, focusing on a nested series of relations between the local and the remote, the individual and the collective, the proximate and the distributed, the immediate and the long-term. As with "Traces" and "Einstein's Brain," the effect is to create a space of intense interaction and feedback in which the subject experiences herself as emerging from relational dynamics rather than existing as a pre-given and static self.

The art work consists of a distributed cognitive system that includes a physical installation located in a gallery space and a remote component played out over the Internet. The physical installation consists of a gridwork of five supporting legs and four trusses that, draped with cream-colored spandex fabric, forms a beautifully translucent 3D spiral thirteen feet high, which the visitor enters with dawning delight of a snail discovering a shell of palatial dimensions. When the visitor positions herself at the center of the installation, the translucent sheeting functions like a borderland between inside and outside, for it creates a sense of enclosure at the same time it allows shapes and sounds to be discernible through the fabric.[31] On the wall is a projection flashing the names of participants who have previously created "bodies" in NØtime. When the visitor sees a name she recognizes or likes, she steps forward and the "body" corresponding to the name is displayed on another wall.

Like Penny, Dunning, and Woodrow, Vesna is critical of the tendency in military and corporate VR to move toward greater realism. Rather than participate in this tendency by creating an anthropomorphic avatar, Vesna prefers to break with realistic representation and visualize the information/energetic "body" as a tetrahedron consisting initially of the six lines and four apices required to outline the tetrahedral shape. The tetrahedron, messages flashing on the walls explain, is privileged because it alone of all polygons has the greatest resistance to an applied load. When the load exceeds the critical tolerance, a tetrahedron will not dimple or bend like other polyhedral structures. Rather the tetrahedron turns inside out, thus making it "unique in being its own dual." These characteristics were why Buckminster Fuller chose the tetrahedron as the basic unit of construction for geodesic domes. They also relate to the tetrahedral shape of carbon stereochemistry, which makes the tetrahedron the essential shape for all carbon-based life on earth. The six edges of the tetrahedron Vesna calls "intervals" and associates them with the components essential to human life as identified by the Indian chakra system, adding color coding so that the family interval appears red, the finances interval orange, the creativity interval yellow, the love interval green, the communication interval blue, and the spirituality interval purple.

The apices are also named, but here the naming scheme focuses on the cultural constructions that Richard Dawkins called "memes," ideas, bits of song, and other concepts that propagate rapidly through the culture, acting as ideational viruses that use humans as their conceptual replication system, much as Dawkins envisioned the "selfish gene" as doing through the physical body.[32] When a user creates a body as tetrahedron, she chooses the length of the intervals, which reflects the relative importance she gives to the six components and determines the shape of that particular tetrahedron. In addition, this initial choice affects the overall shape as other tetrahedrons are added onto the virtual "body." To complete the body, the user names the four apices with words or short phrases representing the memes she wants to circulate through the virtual space of her "body." Once the body is complete, it is correlated with a 3D soundscape that the on-site visitor can navigate by changing position within the installation. Gerald Jong's custom software, entitled "fluidiom," coordinates the visitor's position as registered by motion sensors with this soundscape, creating an acoustic experience unique to the interactions between a specific virtual body and a user's unique movements within the space.

Once the basic body is constructed, it can grow only through collaborators who are willing to spend time in the physical space. The longer an on-site visitor stays gazing at someone's tetrahedron, the more intervals are added to the figure. The body's owner can then add more memes, allow friends to add them, or distribute cards that enable visitors to add them at an on-site Internet connection. In keeping with the installation's theme, however, growth cannot continue indefinitely. Enacting the finitude that makes time, space and life span limited commodities for all humans, a body deconstructs when it reaches a size of 150 intervals. The event is announced in advance at the web site, and people are invited to witness the virtual collapse. At that point the overgrown body is archived in a file accessible only to the owner, who has the option to start the growth process again with the same basic tetrahedron or to craft another one.

Through its distributed architecture, collaborative procedures and sculpturally striking on-site installation, "NØtime" enacts the human body as an emergent phenomenon coming into existence through multiple agencies, including the owner's desires, the cultural formations within which identities can be enacted and performed, and the social interactions that circulate through the global networks of the World Wide Web. The phenomenon of "no time" is thus transformed from an indicator of a declining quality of life into a site for creative play and collaborative interaction. But only, of course, if we make the time to visit the installation, participate in the Web site, and extend the bodies of our fellow humans by physically committing ourselves to relational interactions that last longer than the thirty seconds usually accorded a gallery installation. Relationality requires care, attention, and dynamic interaction, all of which take the time that "NØtime" paradoxically insists we have after all.

IV. Relation as the Posthuman

In *How We Became Posthuman*, I argued that a range of developments in such fields as cognitive science, artificial life, evolutionary psychology and robotics were bringing about a shift in what it means to be human that differs so significantly from the liberal humanist subject it could appropriately be called posthuman. Among the qualities of the liberal humanist subject displaced by technoscientific articulations of the posthuman are autonomy, free will, rationality, individual agency and the identification of consciousness as the seat of identity. The posthuman, whether understood as a biological organism or a cyborg seamlessly joined with intelligent machines, is seen as a construction that participates in distributed cognition dispersed throughout the body and the environment. Agency still exists, but for the posthuman it becomes a distributed function. Consciousness for the posthuman ceases to be seen as the seat of identity and becomes instead an epiphenomenon, a late evolutionary add-on whose principal function is to narrate just-so stories that often have little to do with what is actually happening. In the crises precipitated by the deconstruction of the liberal humanist subject, one kind of response is represented by attempts to reinstate the lost qualities through mastery of increasingly powerful computational and informational technologies. If consciousness is reduced to an epiphenomenon, perhaps its sovereign role can be reinstated by losing the body and resituating the mind within a computer. If agency is distributed, perhaps it can be regained by creating more powerful prostheses, more extensive implants, more smart weapons. These responses share a reluctance to accept human finitude; they remain intent on imposing the will of the individual onto the world seen as an object to dominate. In these constructions, the subject remains inviolate even while losing the body, and the boundaries of the subject continue to be clearly delineated from an objective world. In an important sense, these responses carry on the worst aspects of the liberal humanist subject even as they turn toward the posthuman.

Another kind of response is enacted by the virtual reality art works discussed above. Here the posthuman is embraced as the occasion to rethink the mind/body split and the premise that mind and body, like the rest of the world, preexist our experiences of them. As we have seen, the relational stance enacted by these works puts the emphasis instead on dynamic interactive processes from which both mindbody and world emerge together. The significance of these works in this posthuman moment is profound, for they operate with a performative intensity that makes us realize the importance of emergent relationality in mind and body, transforming these two "elements" into the mindbody that in turn is embedded in our relations with the technoworld. Speaking to more than conscious mind, these art works provide our mindbodies with rich experiential fields that invest the relational stance with meanings that work on multiple levels, including the neocortex but reaching below and beyond it as well. They vividly demonstrate the promise of the posthuman, which despite all its problems and dangers may open us to the realization that without relation, existence (if it is conceivable at all) would be a mean and miserable thing. We do not exist in order to relate; rather, we relate in order that we may exist as fully realized human beings.

Notes

1. For a sympathetic account of the importance of emotion to human cognition, see Antonio R. Damasio, *The Feeling of What Happens: Body and Emotion in the Making of Consciousness* (New York: Harcourt, 1999).
2. I have been arguing for such a view for several years. See for example "Constrained Constructivism: Locating Scientific Inquiry in the Theater of Representation," *New Orleans Review*, 18 (1991): 76–85, reprinted in *Realism and Representation: Essays on*

the Problem of Realism in Relation to Science, Literature, and Culture, ed. George Levine (Madison: University of Wisconsin Press, 1993), 27–43.
3. Antonio R. Damasio, *Descartes' Error: Emotion, Reason, and the Human Brain* (New York: Putnam, 1994).
4. Mark Hansen, presentation at UCLA, May 2001.
5. See, for example, Thomas Laqueur, *Making Sex: Body and Gender from the Greeks to Freud* (Cambridge, MA: Harvard University Press, 1992).
6. Andy Clark, *Natural Born Cyborgs (Why Minds and Technologies Are Made to Merge)* (New York and London: Oxford University Press, forthcoming 2003).
7. Oliver Sacks in *The Man Who Mistook His Wife for a Hat* (New York: Harper Perennial, 1990) recounts the case of Christine, who had lost her proprioceptive sense through neurological disorder and consequently felt that she operated her body as if it were a puppet.
8. Simon Penny, "Representation, Enaction and the Ethics of Simulation," in *First Person Plural*, ed. Pat Harrington and Noah Wardrip-Fruin (Cambridge, MA: MIT Press, forthcoming 2003).
9. Sid Perkins, "Lamprey cyborg sees the light and responds," *Science News*, 158, no. 20 (November 11, 2000): 309. See also the conference paper delivered at Artificial Life VII (Portland OR: August 2000): www.smpp.nwu.edu/~smpp pub/RegerEtAlv_2000.pdf.
10. Kevin Warwick, "Cyborg 1.0," *Wired* 8.02 (Feb. 2000), available on-line at www.wired.com/wired/archive/8.02/warwick.html.
11. Kevin Warwick describes this device and its function in *I, Cyborg* (London: Garnder's UK, 2003). The operation is also described on his website, www.kevinwarwick.org as "Project Cyborg 2.0."
12. Mark Vernon, "Telematics will transform the driving experience," *Financial Times* (Wednesday, June 6, 2001, FT-IT Review), 7–12.
13. Andy Clark, *Being There: Putting Brain, Body, and World Together Again* (Cambridge, MA: MIT Press, 1998).
14. Andy Clark, "Natural Born Cyborgs?" hosted at The Third Culture Web site maintained by John Brockman, www.edge.org/3rd_culture/clark/clark_index.html. This article is excerpted from his forthcoming book, *Natural Born Cyborgs*.
15. Edwin Hutchins, *Cognition in the Wild* (Cambridge, MA: MIT Press, 1996).
16. Terrence W. Deacon, *The Symbolic Species: The Co-evolution of Language and the Brain* (New York: W. W. Norton, 1997), 111.
17. Of course there were also other factors, including faster processor speed, greater memory storage, and most importantly, the advent of the Internet and the World Wide Web.
18. Don Idhe, *Technology and the Lifeworld: From Garden to Eden* (Bloomington: Indiana University Press, 1990), 106 and passim.
19. Francisco Varela, Evan Thompson and Eleanor Rosch, *The Embodied Mind: Cognitive Science and Human Experience* (Cambridge, MA: MIT Press, 1991), 198.
20. Simon Penny, "Virtual Reality as the Completion of the Enlightenment Project," *Culture on the Brink: Ideologies of Technology*, eds. Gretchen Bender and Timothy Druckrey (Seattle: Bay Press, 1994), 231–263, especially 238.
21. Simon Penny, Jeffrey Smith, Phoebe Sengers, Andre Bernhardt, and Jamieson Schulte, "Traces: Embodied Immersive Interaction with Semi-Autonomous Avatars," unpublished essay, 2. I am grateful to Simon Penny for giving me permission to quote from this essay prior to its publication.
22. Brian Massumi, "The Brightness Confound," *Body Mécanique: Artistic Explorations of Digital Realms*, ed. Sarah J. Rogers (Columbus, OH: Wexner Center for the Arts, 1998), 81–94, especially 81.
23. In my view this claim needs to be modified. Some people who experience migraine headaches see visual auras but are quite conscious these auras are not reality. Other visions, such as those inscribed by Hildegard in *Scivias*, are taken by the perceiver as real but are distinguished from ordinary reality, in this case by being identified with the divine. A similar case is presented by the auditory hallucinations experienced by the science fiction writer Philip K. Dick near the end of his life, which he finally decided were communications from an extraterrestrial intelligence. Oliver Sacks in *The Man Who Mistook His Wife for a Hat* (New York: Harper Perennial, 1990) reports auditory and visual hallucinations that seemed very real to the patient but were also understood

by the patient as coming from somewhere other than consensual reality, in one case as replayed memories (132–55).

24. Alan Dunning and Paul Woodrow, "Einstein's Brain," at wwwl.acs.ucalgary.ca/~einbrain/EBessay.htm, 2.
25. Roland Barthes, "The Brain of Einstein," *Mythologies* (New York: Hill and Wang, 1971), 68–70.
26. Collaborators vary from project to project but include, among others, Martin Raff of the MRC Laboratory for Cell Biology, University College, London; Pauline van Mourik Broekman, *Mute* magazine, London; Hideaki Kuzuoka, Department of Engineering, University of Tzukuba, Japan; Nick Dalton, Bartlett School of Architecture, University College, London; and Arthur Clark, Department of Neurology Health Sciences, University of Calgary, Calgary AB Canada.
27. Dunning and Woodrow, "Einstein's Brain," 7.
28. Alan Dunning and Paul Woodrow, "The Stone Tape, the Derive, the Madhouse," presented at the New Media Institute at the Banff Centre, September 2000, 6.
29. Dunning and Woodrow, "Einstein's Brain," 5.
30. Dunning and Woodrow, "The Stone Tape," 3.
31. Don Idhe in *Technology and the Lifeworld*, 72 ff. remarks that many people want the advantages of technology without having it intrude on their lives, a contradictory desire made manifest in the wish that powerful technologies exist but that they also be transparent. The translucent enclosure seems to acknowledge this wish and also resist it by evoking transparency and simultaneously denying it.
32. Richard Dawkins, *The Selfish Gene* (New York: Oxford University Press, 1990).

CHAPTER **12**

Gene(sis)

STEVE TOMASULA

In the Beginning, God *said,* "Let us make man in our image, after our likeness," and He formed man of clay and breathed into his nostrils the breath of life, punning *adam,* Hebrew for "man," with *admah,* "earth." Soon afterwards, Adam, in God's image, created language—Man's first creation—his every utterance the birth of another word as he cried out names for the other animals in Eden. Some 7,000 generations after Adam (according to DNA theory), Eduardo Kac creates the transgenic artwork *Genesis,* reenacting these primal conflations of language and earth and thereby reanimating the myth that is most central to the West's conception of humankind, nature, and progress.

Entering the exhibition space of *Genesis* (see Figure 12.1), the viewer stands before a large projected image: a circular field suspended in blackness and reminiscent of astronomical photographs—a sky filled with galaxies, each composed of millions of suns—circled by how many Edens? As in those photographs, however, scale belies creation. For the God's-eye view afforded by Kac's *Genesis* comes from a micro-videocamera not a telescope, and the "galaxies" are actually bacteria in a Petri dish. Each bacterial body is written in the same genetic language as our bodies, as are all bodies, even if some of them carry a gene unlike the genes of any body. That is, in Kac's eden, some of the animals carry a synthetic gene he fashioned, not from mud, but by arranging genetic material into an order that did not exist in Eden, and today does not exist in nature.

Specifically, Kac's genesis begins with the genetic alphabet: the chemical bases, adenine, guanine, cytosine and thymine, abbreviated as A, G, C, T. By being chained together, these chemical bases make up the rungs of the DNA molecule, the double helix whose sequences of letters—genes—serve as both blueprint and material for the creation of life. Just as the dot-dot-dot | dash-dash-dash | dot-dot-dot of Morse Code can form a message, here an S-O-S, sequences of three genetic bases, for example, AGC | GCT | ACC, form particular amino acids. Particular strings of amino acids form particular proteins, while particular proteins form the particular cells of particular organisms, be they a serpent, an apple, or the rib of a man. Thus each DNA molecule is both material and message, both the book and its content: a book that is its message embodied. Alter this sequence and the new message will produce a different book: a mutation, for example, that brings into existence the larynx that allows human speech, or

Fig. 12.1 Eduardo Kac, *Genesis* (1999). Transgenic installation linked to the internet. Artifacts from installation, dimensions variable. Courtesy the artist.

a Frankenfruit, as environmentalists refer to genetically engineered fruits and vegetables. Or the cells that make up the bacteria in *Genesis.*

Although the sequences of letters that make up the "artist gene" in *Genesis* are artificial, they are not arbitrary. Significantly, they embody a sentence from the Biblical Genesis: "Let man have dominion over the fish of the sea, and over the fowl of the air, and over every living thing that moves upon the earth." To translate this human language into the language of the cell, the AGCTs of DNA, Kac used Morse Code as an algorithm. The dots and dashes of Morse Code easily translate into the ones and zeros used by a digital computer to represent the alphabet—information in a form that can easily be sent around the globe or across the microscopic distances within an integrated circuit. Similarly, in *Genesis,* information is given its physical corollary: after translating the biblical passage into the dots and dashes of Morse Code, the dots were replaced by the genetic base cytosine (C); dashes were substituted with thymine (T); word spaces were replaced by adenine (A); and letter spaces were replaced by guanine (G). This unique string of AGCTs constitutes a gene that does not exist in nature, an "art gene."

The "art gene" carrying the coded biblical passage was then combined with another gene that produces a protein that glows cyan when illuminated by ultraviolet light. Both genes were inserted into a species of *E. coli* similar to that found in the human intestinal tract but which is unable to live outside of the medium in the Petri dish. Art and science are thus collapsed into one another through two characteristics of *E. coli:* its ability to carry DNA from unrelated organisms and its facility for self-replication. Together, they make *E. coli* useful as a living factory for genetically engineered products, such as insulin; they also allow it to function as a microscopic "scribe" copying out the narrative carried within the "artist gene." These genetically engineered bacteria were then placed in a Petri dish along with a strain of *E. coli* that will glow yellow under an ultraviolet lamp but that do not carry the Genesis gene.

Like one of the seventy scholars who first translated Genesis from Hebrew into Greek, then, Kac has translated Genesis into a new language, and like them, embodied it in a "book" that is both a product and reflection of his times. Consider the illuminated manuscript, and how its body expressed medieval culture. Its materials were all natural, its text linked to the earth by inks and pigments extracted from minerals, berries, or flowers, and scratched onto sheepskin with quills from a goose. Writing the text was an act of physical as well as mental labor. The words themselves were written with no separation just as creation was thought to be a single parchment, God's book, an uninterrupted Great Chain of Being from the lowest dregs to the celestial spheres where, as Augustine put it, the angelic and blessed pass their nontime reading a "language without any syllables," a text that is unequivocal and eternal because it is the face of the Word itself.[1] In Eden, it was believed, God, man, and animals all spoke the same language in which words and things had the direct one-to-one correspondence Adam gave them. Or as Emerson later put it, "Every word was once an animal."[2] In this way, written words were natural objects: visible traces of God's mind, as was the rest of the world, shapes that could be read for meaning just as a later age taught itself to read the history of weather in the rings of trees. Letters, words, sentences, and pages merged into sacred books of mysteries serene as the *primum mobile* in their gilt capitals and painted illustrations, their ornaments and imposing page layouts, displayed on high altars for the adoration of the faithful.

Few of the materials of Kac's "book" are natural; even its biological materials are highly mediated by technology. Yet this fact is barely noticeable, seeing as it has become "natural" for us to spend most of our time in artificial light, artificial heat, eating and sleeping not when we are hungry or tired but when the clock says it is time. In the dim templelike atmosphere of a gallery, viewers are drawn closer by the beauty of *Genesis*, its projection of the Petri dish, round as a rose window and luminous as stained glass. A diffuse blue light reflects off lettering on walls that complete what can be thought of as a triptych: on the right-hand panel are the words extracted from the biblical text, "Let man have dominion over . . . every living thing." The left-hand panel displays its genetic translation—the string of AGCTs used to encode the biblical passage in the bacteria, printed out in a computer's block letters without separation just as genes are found before mapping unveils the mystery of their identity and function. The gallery space is thus transformed into a polyglot in which the same passage is presented in three languages: a natural human language, a language of chemicals, and Morse Code, that first electronic language whose first transmitted words—"What hath God wrought?"—ushered in an age of global communication. Reading this polyglot, we begin to understand how to a contemporary sensibility all the world is a text—even unto the lowest dregs commonly found in the colon—and how, like that world, Kac's book is densely coded. Standing at a pulpit that presents the Petri dish as if it were an open book, viewers/readers realize that what they have been admiring in Kac's staging is the beauty of bacteria, the beauty of the flower in the crannied wall, that if understood, could reveal all in all.

Yet the artistry and significance of *Genesis* is not in Kac's creation of aesthetic objects. Instead, its meaning unfolds as its viewers participate in the social situation he has orchestrated. Visiting *Genesis* at home via the Internet, or by using a computer in the gallery that is likewise networked through the Internet, viewers constitute a worldwide community able to write upon Kac's text. By clicking their mouses, they control an ultraviolet light trained on the Petri dish. When they do, the "rose window" flashes blue as if animated by a primordial spark, the bacteria glow. Those carrying the text of Genesis as part of their bodies give off cyan light; those without it give off yellow. More importantly, as viewers activate the ultraviolet light they become Kac's coauthors by accelerating the natural mutation rate of the bacteria. Some descendants retain their original color, others exchange plasmids with one another and give off color combinations, such as green, while still more lose their color. Operating the light to observe this evolution within Kac's microcosm, the viewer realizes how impossible it is to walk in the Garden without altering it. Looking down on this microcosm, finger on the button, it is hard to not want to alter the bacterial garden if for no other reason than to see what will happen. Understanding that changing the bodies of the bacteria also changes the message they carry, we realize that the seduction of *Genesis* is also the seduction of science, of word and body, of art and world—all intimately linked.

No one knows the origins of Genesis, the biblical text Kac incorporates into his microcosm. For centuries it circulated in various forms along with other creation myths until it was written down, sometime in the eighth century B.C.E. Thus, as is said of the *Odyssey* and other scribal texts, the "author" was the aggregate of all the people who wove and rewove oral teachings, reworking, corrupting, and embellishing the stories to fit their circumstances. This is why the inconsistencies we find in Genesis today, including two contradictory stories of creation, were of so little consequence to those first "users" that they could all be taken up and passed on together. As biblical historian Karen Armstrong writes, believers of all three monotheist religions regarded the creation of a myth in the best sense of the word: as a symbolic account that helped people to orient themselves to ontological, and theological questions as well as their present circumstances. It was only long after Genesis was written down that it began to ossify into an official doctrine believed to be factually true.[3]

Indeed, contemporary scholars such as Gerald Bruns distinguish between the open text of scribal cultures, and the closed text of print cultures: that is, between the text that is continually turning into new versions of itself, and the text that has reached its final form and is thus closed to revision. In the Middle Ages it was common for readers to add their comments to a manuscript by writing between its lines, or in its margins, altering a text as they saw fit and passing it on as though the alterations were part of the original book.[4] Since "original" was thought to be "that which was there from The Origin," writing was an act of proliferation, not the "creation" of a unique utterance.[5] Conversely, reading was the act of eliciting from a text that which had remained hidden, or unspoken. In this sense, every text was ripe with more than it said, with myths being the most open of texts, the most incomplete in that myths held the most *potential* meaning. The authority of a text resided in its ability to remain fecund, to be the first word, not the last word. Midrash, the Jewish practice of scriptural explication, was (and to this day still is) the practice of incorporating all of the previous commentary into the text. The text itself was conceived of as always being in need of refiguring to present circumstance. That is, the point of Midrash was not literal interpretation, but to guide people through the complexities and contradictions of their own lives, their own moment in history. The text in this sense was always being made new. And since making it new was figured as a way of life, it was obvious who had the authority to say what the text meant. It was obvious who had the responsibility to understand what it meant: Everyone.[6]

Similarly, Kac's *Genesis* opens itself up as a myth for our times in the sense of John Dryden's description of translations as "transfusion," that is, the "transfusion of new life into an old text."[7] The thousands of people who transmitted the Biblical Genesis as oral teachings, its coauthors, find their corollary in Kac's coauthors: the thousands of engineers, scientists, and technicians on whom *Genesis*'s existence depends. Their labor offers up a vocabulary of "gene splicing," and "interactivity," and "nucleotide polymorphisms" without which Kac's *Genesis* could not be written. Incorporating the traces of this labor as layers in his own palimpsest, Kac creates an allegory of Origins, of Nature, and man's relation to them. By enabling ordinary readers all over the globe to join in the rewriting of this text, he stresses the communal nature of allegory—how authorship itself has become communal in an age when physical diaspora is mitigated by global communication, a development anticipated by Morse Code. Indeed, at the turn of our century, the increased speed and interaction of global communication has accelerated an evolution of reading as the practice of reading between the lines, to reveal all that is unsaid. *Grande Histoire* has become *petite histoires* in which the body has been the only closed book—a naturally impermeable text that could be reread, but not rewritten.

But biotechnology has opened up ever wider spaces for new authors to write between the lines, just as biotechnology revealed how the structure of *E. coli* bacteria would allow Kac to copy in the text of Genesis. With the sequencing of the gene, the practice of rewriting "the fish of the sea, fowl of the air and every other living thing" is becoming so common as to precipitate a shift in our conception of nature analogous to the shift in the conception of earth at the advent of the telescope. The critics of Copernicus who refused to accept that the earth revolved around the sun, Thomas Kuhn wrote, were not entirely wrong. To them, "earth" meant "fixed, immovable position." Looking through Galileo's telescope and seeing evidence for the earth's orbit and rotation thus entailed a semantic leap as well as a shift in perspective.[8] The world could only change, after Galileo, to the degree that language changed. Similarly, it is becoming easy to think of animals not as fixed "objects" in nature but as rearrangeable packets of DNA. Over the past decade, the list of patents issued worldwide for bioengineered products is long and varied and includes a combination of cow embryos with human genes in attempts to grow human replacement parts and tomatoes with the genes of a cod fish to make them less susceptible to freezing. Chickens carry salmon genes while sheep receive tobacco genes, and worms, after Methuselah, have been engineered to increase their life span to the equivalent of 600 human years.[9] Similarly, the plants of the earth increasingly include strawberries engineered for size, potatoes engineered to produce vaccines, and basic crops like wheat and corn that have been genetically altered to kill insects.

As our garden becomes populated with more, and more extreme, varieties of transgenic plants and animals, as these techniques are increasingly applied to humans, can the Adamic conception of the self remain any more constant? Dramatic advances such as the cloning of our primate cousins receive the most attention, but it is perhaps the thousands of small steps that coalesce, like myths, into habits of mind that have the most profound effects: calls for genetic national identity cards; the permission we give on the back of our driver's license for our bodies to become recyclable material—permission that allowed Matthew Scott to receive the hand of a cadaver by transplant, the hand that John Doe, its previous "owner" had used to write his name, to clasp in prayer, now taking up a new name, new prayers. Artificial skin, artificial bone, in Petri dishes like the one used in *Genesis*, researchers at the University of Massachusetts Medical School have been able to grow cartilaginous ears and noses.[10] Other labs claim to have discovered a genetic explanation for a growing list of human phenomena from shyness to rape to altruism; first steps to practical applications soon follow, such as those taken by researchers at Yale University who by manipulating a gene identified as important to memory have created a strain of supersmart mice. Once the genetic tree of knowledge is completely sequenced, won't we begin in earnest to rewrite genes to increase longevity, manipulate skin color, personality, indeed, all the traits that make us us?—to completely throw off the original sin and destiny of biology? Considering how conceptions of the self have had profound consequences for laws, for customs—for how people order society and conduct themselves and behave toward others—can we do without springboards to meditation such as Kac's *Genesis?*

When the prospect of "personal evolution" (i.e., the prospect of individuals altering the genes of their descendants) became a reality, the U.S. National Bioethics Advisory Commission turned to religious traditions as one factor in formulating its recommendations on how public policy should respond to these new technologies.[11] In so doing, they cited the centuries people have used these traditions to guide their own behavior in the face of a changing world. By putting a global audience in collective control of his *Genesis*, by making their actions impinge on an excerpt from the Biblical Genesis, Kac puts his audience in a position to consider tradition—or its erasure—as one factor in their response to the biological course we are just beginning to navigate. The evolution in a Petri dish we communally alter underscores how the use of technology is not always planned, its consequences not always foreseen, nor benign. Standing in the box formed by the walls of *Genesis*, it is easy for viewers to reverse the scale and think of themselves in the position of the bacteria with ultraviolet light streaming down (possibly through a hole in the ozone layer?). We're invited to contemplate consequences of interfering with evolution when Kac translates, at the end of the exhibit, the DNA code of his original message back into English:

LET AAN HAVE DOMINION OVER THE FISH OF THE SEA AND OVER THE FOWL OF THE AIR AND OVER EERY LIVING THING THAT IOVES UA EON THE EARTH

The now-corrupted sentence calls to mind other literatures of constraint: those texts, such as Raymond Queneau's *One Hundred Million Million Poems*, that have been generated out of a self-imposed rule. In Queneau's work, a traditional fourteen-line sonnet is combined with ten other fourteen-line sonnets in such a way that any one line can be combined with the thirteen lines of any of the other sonnets. Thus, the poem as a whole allows the meaning held as a potential within the dull mass of language to emerge: a potential of 10^{14} sonnets, a quantity of text, as François Le Lionnais notes, "far greater than everything man has written since the invention of writing, including popular novels, business letters, diplomatic correspondence, private mail, rough drafts thrown into the wastebasket, and graffiti."[12] Conversely, Kac's corruption also calls to mind literatures of nonconstraint, such as the hypothetical copy of *Hamlet* that typing monkeys would eventually produce by chance, given enough monkeys, given enough time. With over 3 billion genetic letters in the book that is the human, *Genesis* asks us to consider the ramifications of typos—and their transmission to future generations. Unbridled, typos cumulate into gibberish quickly, for as Alice learned in Wonderland, even a sentence of only ten words has 3,628,800 combinations, only one or two of which will make sense. Mutating any three letter word, say APE, into another three letter word, say MAN, by randomly switching one letter at a time takes thousands of generations to hit the right combination. But if the changes are governed by the constraint that each step must make sense, then the mutation can be made in only eight steps:

APE
ARE
ARM
AIM

RIM
RAM
RAN
MAN

Thus can be seen the apparent paradox of how the application of a constraint directs rather than stifles creation: the application of a constraint allows the process to ignore all the other constraints that would take it into other directions. Before man's intervention, "survival of the fittest" was the dominant constraint under which changes were made to the book of each organism, including humans. Although gene management has resulted in hairless Chihuahuas, seedless watermelons, indeed every strain of plant and animal not seen in Eden, it is only with the advent of bioengineering that changes could be made that skip intervening steps.

As Kac's *Genesis* illustrates, which potential literature will be offered up from among the thousands of potentials dormant in the mud of genetic language will depend on the constraints under which change operates. So it is instructive to note how much of both the Biblical and the artist's *Genesis* is concerned with lineage. Indeed, the Hebrew innovation with regard to the creation myths that circulated among the Israelites was to use them to shape their identity as a people—an identity traced through their bodies in a direct line of descendancy to Adam and Eve who were fashioned in the likeness of God. Thus, the mother of this people was named Eve, *hawwa* in Hebrew, related to *hay* "living," the mother of all the living to follow. Reconstruction of genetic trees estimate that this woman—not the first woman, but the last woman from whom every person now alive on earth is descended—lived 143,000 years ago.[13] For 5,700 generations, then, or 120 million years if we count our ancestry back to the original cells, our biological identity has been shaped one letter at a time. In Kac's *Genesis*, however, we see an icon for our newfound ability to rewrite ourselves—instantly, and in ways whose ramifications might not become apparent for generations. In an age when people are increasingly looking to chromosome stains to explain the difference between Cain and Abel—as well as differences in sexual orientation, intelligence, personality, and hundreds of other human traits—Kac's *Genesis* reminds viewers of the wisdom in tempering change with reflection.

That is, Kac's *Genesis* calls us to consider which identity we are fashioning for ourselves, for our species, for nature, by the constraints we do or do not place on the potential literature of our bodies. Will the constraint of survival be replaced by economic gain? It was not until 1967 that the U.S. Federal Trade Commission ruled that blood could be bought and sold.[14] Up until then, blood with all of its metaphorical richness was considered a gift that could be given, like life, but was too sacred to be bought and sold. Today, the world market for blood is a $19 billion-a-year business and constitutes only a segment of a biotrade that includes on-line auctions for human eggs and sperm (see www.ronsangels.com) among other human "components," from whole corpses to fetal "products."

Will the only constraint placed on these new potential literatures of the body be technological progress? Can constraints not be political? Does the ability to manipulate a gene, say for one of the 5,000 diseases now known to be inherited, carry with it the responsibility to do so? Who has the authority to alter the germ line of future generations? Who has the authority to determine the fate of the tens of thousands of embryos accumulating in storage tanks, the leftovers of reproduction technologies that allow couples to select the most genetically viable embryos while abandoning the rest? Will the constraints of biotechnology be social and/or aesthetic?—preferences for skin color or hair texture? Will they be legal?—such as the legal fights over who can copyright a person's genetic information? Kac's *Genesis* asks us to consider these issues by having us revisit the language of "dominion over every living thing." By making us his coauthors, he emphasizes how the name we give ourselves can be in the spirit of "masters" or "caretakers" of our garden, how our collective actions will be our Midrash.

Notes

1. St. Augustine, *The Confessions of Saint Augustine*, trans. Edward B. Pusey (New York: Random House, 1999), 312–3.
2. Quoted in Ben Marcus, *The Age of Wire and String* (New York: Alfred A. Knopf, 1995), xi.
3. Karen Armstrong, *A History of God: The 4,000 Year Quest of Judaism, Christianity and Islam* (New York: Alfred A. Knopf, 1993), 12–39, 100–1.
4. Gerald L. Bruns, *Inventions: Writing, Textuality, and Understanding in Literary History* (New Haven: Yale University Press, 1982), 47–8.
5. Paul Ricoeur, *Time and Narrative*, trans. Kathleen McLaughlin and David Pellauer, Volume 2 (Chicago: University of Chicago Press, 1985), 163.
6. Gerald L. Bruns, *Hermeneutics Ancient and Modern* (New Haven: Yale University Press, 1992), 105–6.
7. Cited in Bruns, *Inventions*, 49.
8. Thomas S. Kuhn, *The Structure of Scientific Revolutions*, 2nd ed. (Chicago: University of Chicago Press, 1970), 115–6.
9. Hopkin in *Your Bionic Future*, eds. Carol Ezzell and Glenn Zorpette, special issue of *Scientific American Presents*, 10, no. 3 (1999), 34.
10. Ezell and Zorpette, *Your Bionic Future*, 13.
11. Gina Kolata, "Little-Known Panel Challenged to Make Quick Cloning Study," *New York Times* (18 March 1997), late ed., C1.
12. Raymond Queneau, *Oulipo: A Primer of Potential Literature*, ed. and trans. Warren F. Motte Jr., (Normal: Dalkey Archive Press, 1998), 3.
13. Luigi Luca Cavalli-Sforza, *Genes, Peoples, and Languages* (New York: North Point Press, 2000), 79.
14. Douglas Starr, *Blood: An Epic History of Medicine and Commerce* (New York: Alfred A. Knopf, 1998), 195.

CHAPTER 13
Transgenic Art Online

EDUARDO KAC

"Genesis" (1998/99) is a transgenic artwork that explores the intricate relationship between biology, belief systems, information technology, dialogical interaction, ethics, and the Internet. The key element of the work is an "artist's gene," that is, a synthetic gene that I invented and that does not exist in nature. The Genesis gene was created by translating a sentence from the biblical book of Genesis into Morse Code, and converting the Morse Code into DNA base pairs according to a conversion principle specially developed for this work. The sentence reads: "Let man have dominion over the fish of the sea, and over the fowl of the air, and over every living thing that moves upon the earth." This sentence was chosen for its implications regarding the dubious notion of (divinely sanctioned) humanity's supremacy over nature. Morse Code was chosen because, as first employed in radiotelegraphy, it represents the dawn of the information age—the genesis of global communications.

I selected the King James English version (KJV), instead of the Hebrew original text, as a means of highlighting the multiple mutations of the Old Testament and its interpretations, and also to illustrate the ideological implications of an alleged "authoritative" translation. King James tried to establish a final text by commissioning several scholars (a total of 47 worked on the project) to produce this translation, meant to be univocal. Instead, this collaborative effort represents the result of several "voices" at work simultaneously. Most of the Old Testament books were written in Hebrew, while parts of the books of Daniel and Ezra were written in Aramaic. The King James Bible was translated in 1611 after consultation of previous translations to multiple languages, that is, it is a translation of many translations. In the preface of the authorized version, the translators wrote: "Neither did we think much to consult the Translators or Commentators, Chaldee, Hebrew, Syrian, Greek or Latin, no nor the Spanish, French, Italian, or Dutch."

Following centuries of oral traditions, the Bible was written over a long time span by many authors. It is unclear exactly when the Bible was written down. However, it is believed that the text was fixed in scrolls during the period from 1400 B.C. to 100 A.D. Because the first versions of the text had no connection between letters, no spaces between words and sentences, no periods or commas, and no chapters, the material encouraged multiple interpretations. Subsequent translations and editions attempted to simplify and organize the text—that is, to arrest its continuous transmutation—only to generate more versions. The division of the Bible into chapters was carried out by Stephen Langton (d. 1227), who later became the Archbishop of Canterbury. Father Santes Pagninus, a Dominican priest, divided the Old Testament chapters into verses in 1528. With the advent of moveable-type printing in 1450, yet newer versions proliferated, all different in their own way, with both deliberate and accidental changes.

The biblical passage from KJV employed in my transgenic work "Genesis" is emblematic, as it speaks of dominion. King James is the founding monarch of the United States. Under his reign, the first successful colonies were established. In his own words, King James sought to propagate "Christian religion to such people as yet live in darkness." The colonizers brought his authorized translation. The genesis of the New World was built on dominion "over every living thing that moves on the earth."[1]

I employed Morse code not out of a technical need, but as a symbolic gesture both meant to expose the continuity of ideology and technology and to reveal important aspects of the rhetorical strategies of molecular biology. Samuel Morse embraced the radical Protestant movement of the 1830s known as Nativism. The nativist platform was racist, anti-immigrant, anti-Catholic, and anti-Semitic. All his life Morse hated and feared American Catholics, supported denying citizenship to the foreign born, and wrote pamphlets against the abolishment of slavery. In my work "Genesis," the translation of the KJV Genesis passage to Morse represents the continuity from fierce British colonialism to the bigotry of nativist ideology. The industrialization of North America, in tandem with technological hegemony, was based on the gargantuan profits amassed from the slave trade in the eighteenth century. In 1844, Morse sent the first telegraphic message, from Baltimore to Washington, DC: "What hath God wrought!" The translation from KJV/Morse to a gene is meant to reveal the continuity between imperialist ideology and the reductionistic view of genetics, both focused on suppressing the complexity of historic, political, economic, and environmental forces that make up social life.[2]

In addition, Morse code is a central metaphor in molecular biology. In his influential essay "What Is Life?" (1943), physicist Erwin Schrödinger promoted an atomistic view of biology and predicted key characteristics of genetic material more than a decade before the structure of DNA was understood. He wrote: "It has often been asked how this tiny speck of material, the nucleus of the fertilized egg, could contain an elaborate code-script involving all the future development of the organism.... For illustration, think of the Morse code. The two different signs of dot and dash in well-ordered groups of not more than four allow of thirty different specifications."[3] The metaphor of the "code-script" proposed by Schrödinger took centerstage in molecular biology and became an epistemological instrument in this field. This begs the question, that I seek to ask with "Genesis," of how meaning is constructed in science. How do we go from the metaphor of "genes as code" to the "fact" that "genes are code"? Is it by the progressive erasure of the initial conditions of enunciation of a metaphor?

In the Genesis net installation, the gallery display enables local as well as remote (Web) participants to monitor the evolution of the work. This display consists of a Petri dish with the bacteria,[4] a flexible microvideo camera, an ultraviolet (UV) light box, and a microscope illuminator. This set is connected to a video projector and two networked computers. One computer works as a Web server (streaming live video and audio) and handles remote requests for UV activation. The other computer is responsible for DNA music synthesis. The original music, which employs the genesis gene, was composed by Peter Gena.[5] The local video projection shows a larger-than-life image of the bacterial division and interaction seen through the microvideo camera. Remote participants on the Web interfere with the process by turning the UV light on. The fluorescent protein in the bacteria responds to the UV light by emitting visible light (cyan and yellow). The energy impact of the UV light on the bacteria is such that it disrupts the DNA sequence in the plasmid, accelerating the mutation rate. The left and right walls contain large-scale texts applied directly on the wall: the sentence extracted from the book of Genesis (right) and the Genesis gene (left). The back wall contains the Morse translation.

In the context of the work, the ability to change the sentence is a symbolic gesture: it means that we do not accept its meaning in the form we inherited it, and that new meanings emerge as we seek to change it. Employing the smallest gesture of the online world—the click—participants can modify the genetic makeup of an organism located in a remote gallery. This unique circumstance makes evident, on the one hand, the impending ease with which genetic engineering trickles down into the most ordinary level of experience. On the other, it highlights the paradoxical condition of the nonexpert in the age of biotechnology. To click or not to click is not only an ethical decision, but also a symbolic one. If the participant does not click, he allows the Biblical sentence to remain intact, preserving its meaning of dominion. If he clicks, he changes the sentence and its meaning, but does not know what new versions might emerge. In either case, the participant is implicated in the process.

In the nineteenth century, the comparison made by Champollion based on the three languages of the Rosetta Stone (Greek, demotic script, hieroglyphs) was the key to understanding the past. Today, the triple system of Genesis (natural language, DNA code, binary logic) is the key to understanding the future. "Genesis" explores the notion that biological processes are now writerly and programmable, as well as capable of storing and processing data in ways not unlike digital computers. Further investigating this notion, at the end of the first showing of "Genesis," at Ars Electronica '99, the altered biblical sentence was decoded and read back in plain English, offering insights into the process of transgenic interbacterial communication. The mutated sentence read: "LET AAN HAVE DOMINION OVER THE FISH OF THE SEA AND OVER THE

FOWL OF THE AIR AND OVER EVERY LIVING THING THAT IOVES UA EON THE EARTH." The boundaries between carbon-based life and digital data are becoming as fragile as a cell membrane.

Notes

1. See Kenneth L. Barker, ed., *The NIV: The Making of a Contemporary Translation* (Grand Rapids: Zondervan, 1986); Eugene H. Glassman, *The Translation Debate* (Downers Grove, IL: InterVarsity Press, 1981); D. A. Carson, *The King James Version Debate* (Grand Rapids, MI: Baker, 1979).
2. See Samuel Irenaeus Prime, *Life of Samuel F. B. Morse* (New York: Appleton, 1875); Jeffrey L. Kieve, *The Electric Telegraph: A Social and Economic History* (Newton Abbot: David and Charles, 1973); Paul J. Staiti, *Samuel F. B. Morse* (Cambridge, UK: Cambridge University Press, 1989).
3. Erwin Schrödinger, *What Is Life?: The Physical Aspect of the Living Cell With Mind and Matter and Autobiographical Sketches* (Cambridge, UK: Cambridge University Press, 1992), 61. See also Richard Doyle, *On Beyond Living: Rhetorical Transformations of the Life Sciences* (Stanford: Stanford University Press, 1997), 25–38.
4. This work was carried out with the assistance of Dr. Charles Strom, formerly Director of Medical Genetics, Illinois Masonic Medical Center, Chicago. Dr. Strom is now Medical Director, Biochemical and Molecular Genetics Laboratories Nichols Institute/Quest Diagnostics, San Juan Capistrano, CA.
5. See Peter Gena and Charles Strom, "A Physiological Approach to DNA Music," in *Digital Creativity: Crossing the Border. The Proceedings of CADE 2001: The 4th Computers in Art and Design Education Conference,* eds. Robin Shaw and John McKay (Glasgow: The Glasgow School of Art Press), 129–34.

CHAPTER **14**

Gene(sis): Contemporary Art Explores Human Genomics

ROBIN HELD

On June 26, 2000, The Human Genome Project, a public consortium, and Celera Genomics, a private company, jointly announced the completion of a "working draft" of the human genome. Supporters hail this event as a scientific revolution, anticipating benefits in improved diagnosis of disease, early detection of genetic predisposition, gene therapy, and personalized drug prescription, among others. Detractors raise the specter of eugenics, warning that the use of genetic testing as a basis for reproductive decision making can be compared to the practice of selective breeding. As anticipation and anxiety mount, the debate surrounding the potential impact of recent genomic developments is rapidly becoming a defining issue of our contemporary culture. Social and ethical issues of human genetics—including the misuse of genetic information and potential threat to personal privacy, the potential discrimination against individuals carrying altered genes, free will versus genetic determinism—which have been debated throughout the twentieth century, are now intensified with the completion of the rough draft.

It is important to note that scientists face years of work in understanding the various levels of the genome's operation. Among the insights that have thus far emerged from the publication of the two groups' research are these three. There are fewer human genes than were estimated—approximately 30,000 total, rather than the 100,000 that had been previously been supposed.[1] Almost half of the human genome is comprised of transposable elements—parasitic "genes" unrelated to the needs of human existence. For example, more than 200 human genes actually come from bacteria. The genetic differences between the human genome and the genomes of other species—such as the mouse, the roundworm, the fruit fly, and yeast—is less than had been supposed. However, at this point, the impact of these insights is, in large part, speculation.

I. Genomics, Art, and Culture

Recently, geneticist Eric Lander noted the reflexive relationship between genetics and culture, remarking that the ultimate meaning of the human genome would be decided not by scientists alone, but would be fought out in the arenas of art and culture.[2] Like other arenas of culture, contemporary art is deeply implicated in the determination of the ultimate meaning of the human genome. Indeed, a number of artists have engaged genomic themes, materials and processes from many perspectives. Their work provides several ways to explore the potential social, emotional, and ethical implications of these complex issues.

Gene(sis): Contemporary Art Explores Human Genomics, a major exhibition organized by the Henry Art Gallery, University of Washington (April–August 2002, traveling), presents some of the most powerful new work created in response to recent developments in human genomics. It also demonstrates the changes genetics has had on notions of artistic practice and the artist-subject, providing artists with new tools, new materials, and new issues for critical exploration. The exhibition features three new projects developed by artists in collaboration with scientists and commissioned especially for the exhibition, alongside major works previously exhibited elsewhere. Interweaving humorous commentary, multimedia installations, documentary images and pseudo- (or actual) scientific laboratory situations, the exhibition elucidates technical advances for a lay audience. More importantly, it exploits the power of contemporary art to provoke, to question, and to articulate new paradigms, providing conditions necessary for a deeper understanding and a fuller discussion of genomic issues.

Gene(sis) suggests several ways to divide and examine the onslaught of genetic information and genomic hysteria in public discourse. It is organized into four general themes (SEQUENCE, SPECIMEN, BOUNDARY, SUBJECT) that, as curator of the exhibition, I developed in consultation with geneticists, bioethicists, historians of science, and artists. Neither the exhibition as a whole, nor the specific subjects it addresses, are intended as position statements on recent genomics issues. Instead, each theme organizes a series of issues and urgent questions that recur in public discourse about genomics, from the scholarly to the sensationalist.

II. SEQUENCE: Language and Media Representations of Genomics

The image of the human genome as a "Book of Life"—an image infused with biblical references—has remained a guiding metaphor for the theories and practices of molecular biologists since its emergence in the 1960s.[3] Human genome projects have been characterized as enormous efforts of information gathering and word processing, with a mission of "reading" and "editing." The DNA sequence is frequently conceived as the "word," the human genome as a linguistic text written in DNA code.[4] This problematic view permeates current language about genomics. "Breaking the code," "reading the book of life," "mapping our genes," "drafting the human blueprint"—such phrases imply the success of reading, managing and controlling this (textual) genomic information. Many historians of science warn that "information," "language," "code," "message" and "text," while incredibly productive as analogies, present the danger that such rhetorical phrases can become confused with factual definition. Several artists are working with this problematic language in compelling ways: engaging the metaphors it inspires, exploring its semiotic arbitrariness, or investigating the linguistic ways in which its truth claims are advanced. Others are proposing alternative models for our cultural understanding of genomics. What follows are a sample of artworks included in a section entitled "SEQUENCE."

In his installation entitled *Genesis* (1999, see Figure 12.1)—which inspired the title of the Henry exhibition, Chicago-based artist Eduardo Kac begins with a biblical quotation that describes the supposed domination of humans over nature. He translates this passage into Morse code (the first new language of the telecommunications age). He then converts the Morse code directly into genetic language and then into actual genetic material, producing the poetic equivalent of "junk" DNA. This DNA mutates in response to viewer attention in both the museum and over the Internet, activity that is projected on one wall of the gallery. Online viewers can raise and lower the levels of the ultraviolet lights focused on the cells, thus changing the rate of mutation. *Genesis* exemplifies how our understanding of the world, including the ways we make meaning out of recent genomic developments, is entwined with changes in language and technology.

In their performance *Cult of the New Eve* (2000; see Figure 14.1), the artists' collective Critical Art Ensemble uses the apocalyptic language of an imaginary cult to explore the rhetoric surrounding the announcement of recent genomic developments. *Cult of the New Eve* also explores the fears and anxieties we project on these developments. Critical Art Ensemble proposes that the theological

Fig. 14.1 Critical Art Ensemble, *Cult of the New Eve* (2000). Video documentation and performance artifacts, courtesy the artists. Photo Credit: Critical Art Ensemble.

language surrounding the completion of the "rough draft" of the human genome is one way to avoid discussion of the eugenic history on which such advances are founded. In response, they have taken theological language to an extreme in this performance, foregrounding the effects that language has on our understanding of current events and their social implications. In this performance, "cult" members wear special uniforms and proselytize faith in the New Eve (a woman from Buffalo, New York) offering a sacrament of beer, into which her DNA has been brewed, and special cookies, into which her DNA has been baked. This performance is followed by public dialogue on the potential effects of genomic developments on our daily lives. Critical Art Ensemble developed *Cult of the New Eve* in consultation with University of Washington geneticist Mary-Claire King.

No-Die (2000: see Figure 14.2) is an imaginary advertisement for a future in which genomic advances allow us to double or triple our life span. An alliance of biotechnology and the "new economy" makes this future available to medical consumers as a patented product called *No-Die* that "you can ask for by name." Maira Kalman, a New York-based artist, has created *No-Die* as a cartoon

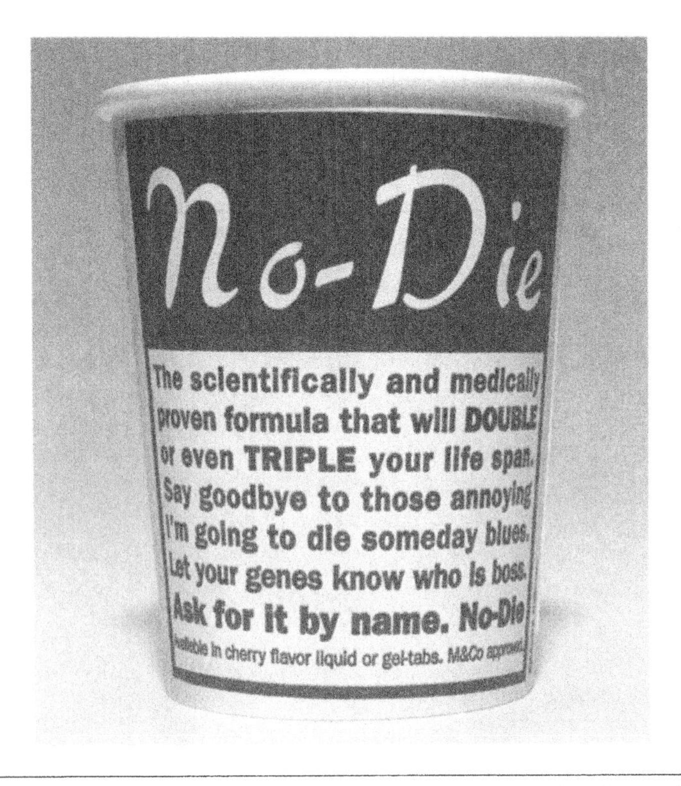

Fig. 14.2 Maira Kalman, *No-Die* (2000). Paper coffee cup, 4 × 3 1/4 inches. Courtesy Creative Time, New York. Photo Credit: Robert Glasgow.

printed on paper coffee cups to be distributed at coffee shops, bodegas, and delis. Kalman's cup is one of four such artist-designed coffee cups distributed by Creative Time, a nonprofit public art organization based in New York.

No-Die, and the other coffee cups in this series, addresses serious issues of genomics' societal implications in the form of DNA jokes. These cups are offered at public sites as potential catalysts for public dialogue about genomic issues. As components of an art exhibition exploring human genomics, these cups bring this art and the important issues it raises outside the confines of the museum to address with humor a broader, more diverse audience than that which is usually addressed in a museum exhibition.

III. SPECIMEN: Genetic Information, DNA Ownership, and Personal Privacy

The first practical impacts of human genomics are anticipated in healthcare. Genomics is widely expected to revolutionize medicine, providing knowledge that will enhance control over specific human diseases and certain aspects of behavior.[5] Ethical review will be demanded of genetic testing and screening, presymptomatic diagnosis and the development of new therapies and technologies in genetic engineering. The appetite for genetic information and the potential dangers of misuse and threats to personal privacy cannot be underestimated. The impact of human genomics on legal and political debates about privacy, ownership, control of access, consumer information, and choice could alter key societal values.

Several artists are exploring issues of DNA ownership, personal privacy, and ethics. Some are drawing on their personal encounters as medical consumers and employees; others are looking to the history of these issues and their intersection with race and sexual difference. Questions being addressed include: as biotechnology provides tools to extend life, how does it affect healthcare choices? How might genetic information be used in discriminatory ways? Who makes up the rules about use and abuse of new technologies? How much do you want to know about your future? What follows are selected artworks from a section entitled "SPECIMEN."

This installation by Glasgow artist Christine Borland explores moments from the history of genetics. Key components are *HeLa* (2000: see Figure 14.3) and *Spirit Collection* (1999). According to Borland's explanatory text panels, HeLa cells are the cells of an African-American woman who died of cervical cancer in an U. S. hospital in the 1950s. These cells subsequently became important in medical research in the prevention of polio and the virus that causes AIDS, and are still used as a teaching tool today. As was customary at the time, permission was never acquired from the woman to use her cells nor is she given attribution for the role her cells have played in the development of drugs that have saved innumerable lives. In Borland's installation, live HeLa cells in culture are presented under a microscope. On a monitor, viewers can watch the genetic

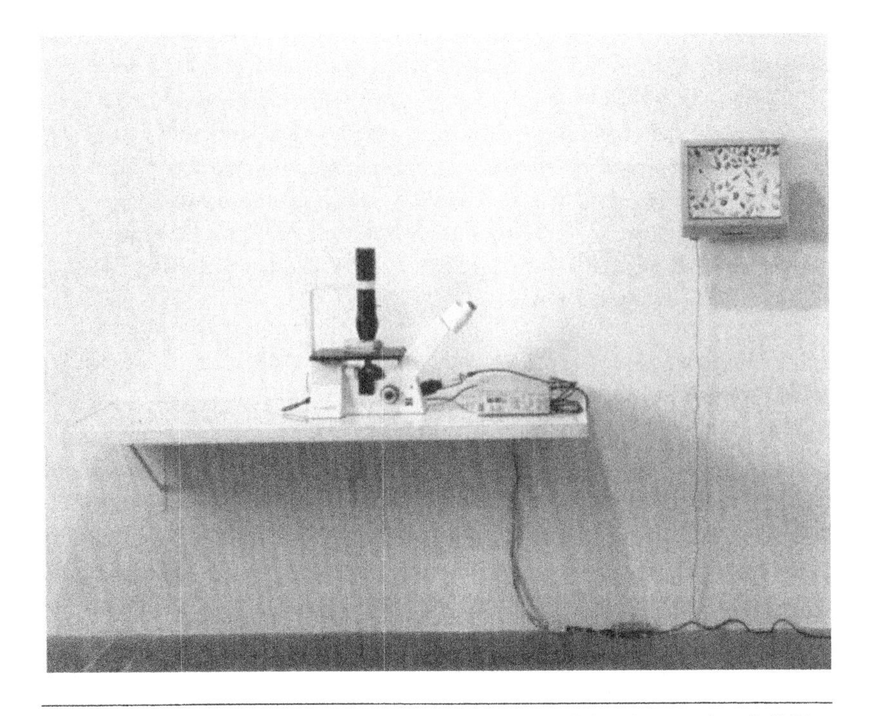

Fig. 14.3 Christine Borland, *HeLa* (2000). Video, microscope, cells in culture, monitor, shelf, 21 × 64 × 22 inches. Courtesy the artist and Sean Kelly Gallery, New York.

material mutate in a context in which Borland has highlighted issues of race, class, personal privacy, and DNA ownership.

Spirit Collection, a companion piece to *HeLa,* includes 100 hand-blown glass orbs suspended from the ceiling. In each orb is suspended a leaf skeleton from a tree on the Glasgow campus at which Borland consults with geneticists. This tree is purported to be a clone from the ancient tree under which Hippocrates, the "father of medicine" taught his students. Positioned near *HeLa, Spirit Collection* positions *HeLa*'s explorations of issues of privacy, DNA ownership, and consumer information in the context of the long and canonic history of western medicine.

−86 Degrees Freezer (1995: see Figure 14.4) is one panel of a 12-panel, black-and-white photographic series of "still lives" created by San Francisco artist Catherine Wagner. In this series, Wagner documents the mundane, everyday activities of the scientific laboratory, especially the activity of managing and properly storing genetic materials. In *−86 Degrees Freezer,* the artist makes issues of DNA ethics, privacy, and ownership tangible and accessible. Wagner does not document the historic benchmarks of science—its great men and women, its victories and defeats—instead, she documents the ways in which genomics

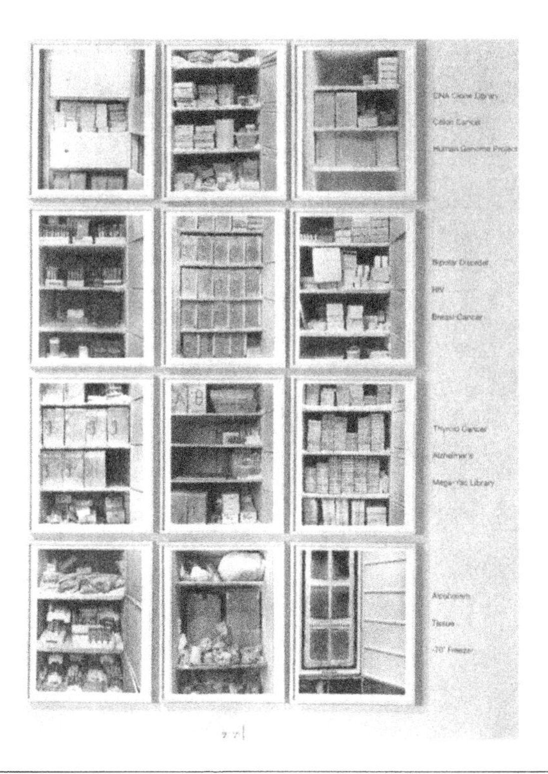

Fig. 14.4 Catherine Wagner, *–86 Degrees Freezer (Twelve Areas of Concern)* (1995). 12-panel typology, gelatin silver prints, 20 × 24 inches each; 6 × 5 feet installation. Collection of the Museum of Fine Arts, Houston.

produces knowledge through the labor of archiving, labeling, managing, controlling, and maintaining genetic materials and information.

Like many artist included in *Gene(sis),* Seattle-based Susan Robb seeks to dismantle the opposition of art and science. Her studio is full of the trappings and accessories of science, including pipettes, Bunsen burners, Petri dishes, scalpels, and scientific textbooks. But Robb takes liberties with scientific protocols, crafting sculptures from unscientific materials such as Play-Doh, lint, moss, cake crumbs, and glass cleaner, as well as saliva and other bodily fluids. She might begin her experiment with a question like, "What does it look like chemically when one falls in love?" Thus inspired, she creates a sculpture of this imagined chemical process, cellular activity, or genetic sample.

To create her photographic series *Macro-Fauxology* (2000, Figure 14.5), Robb used this art-and-science process to create sculptures representing her idea of specific organisms and activities. After photographing the sculptures with a macro lens, she destroys them, leaving only the photographic trace of their

Fig. 14.5 Susan Robb, *Macro-Fauxology: sucrosefattyacetominophen* (2000). Cibachrome print, 10 × 10 inches. Courtesy the artist.

existence—colorful and fantastical images reminiscent of both scientific imagery and science fiction.

IV. BOUNDARY: Permeability of Species Boundaries

Scientists once believed that species boundaries were for the most part impenetrable, although myths of hybridity haunt our most ancient cultural imaginings. In Greek and Roman mythology, both the Chimera and the Centaur are changelings: half animal, half human. Human–animal hybrids are also prevalent in Chinese and Buddhist mythology where they play a key role in the spiritual transition of souls.

Recent genomic developments make possible the production of hybrid creatures that only a few years ago were thought scientifically impossible. We have even become accustomed to the notion of genetically engineered sheep, pigs, goats, mice, and cattle. In early January 2001, the first transgenic primate, a rhesus monkey, was developed as a tool for studying human diseases.

Several artists are exploring transgenics—the formation of new combinations of genes by isolating one or more genes from one organism and introducing them into another. In the context of rapid developments in transgenics, when boundaries between species and between living and nonliving forms are blurred, what are the ethical questions we should be asking? Whose values will guide such developments? What follows is a selection of a few artworks from a section entitled "BOUNDARY."

In her photographic series, *Transgenic Mice,* New York-based artist Catherine Chalmers highlights the production of genetically engineered mice, an industry that has experienced enormous growth almost wholly attributable to genetics research. *Pigmented Nude* (2000: see Figure 14.6) and *Rhino* are programmed from birth to develop tumors, inherit glaucoma, or reproduce milder pathologies. "Pigmented nude" is a mouse strain developed for research into human skin and hair–texture defects. "Rhino" is a mouse strain developed for research into human immunology and inflammation problems.

These genetically engineered mice are produced at a rate of approximately 50 million a year. They are cheaper to breed than primates and, until recently, a loophole in the USDA's 1966 Animal Welfare Act left these mice unprotected and

Fig. 14.6 Catherine Chalmers, *Transgenic Mice: Pigmented Nude* (2000). C-print, 30 × 40 inches. Courtesy the artist and RARE, New York.

Fig. 14.7 Eduardo Kac, *GFP Bunny* (1999-present). Photographic documentation of living artwork, dimensions variable. Courtesy of Julia Friedman Gallery, Chicago. Photo Credit: Chystelle Fontaine.

their standard of care unmonitored. A change in their status could cut deeply into the skyrocketing profits ($200 million per year) of private "mouse ranches" nationwide. Chalmers' series *Transgenic Mice* highlights this genomics growth industry that has until recently escaped public scrutiny or calls for accountability.

Eduardo Kac's *GFP Bunny* (2000: see Figure 14.7), or "Alba" as she is affectionately known, is a transgenic animal, created by splicing the DNA of a Pacific Northwest jellyfish with that of an albino rabbit. The effect of this combination is that under ultraviolet light of a particular intensity, the albino rabbit glows fluorescent green. "Alba," was created by the artist, in collaboration with French geneticists, as part of a much larger performative artwork that includes interdisciplinary dialogue on the cultural and ethical implications of genetic engineering, and the integration into society of a transgenic creature. This project also advocates expanding the present boundaries of artmaking to include the invention of life, an ability that has only recently become part of an artist's toolkit.

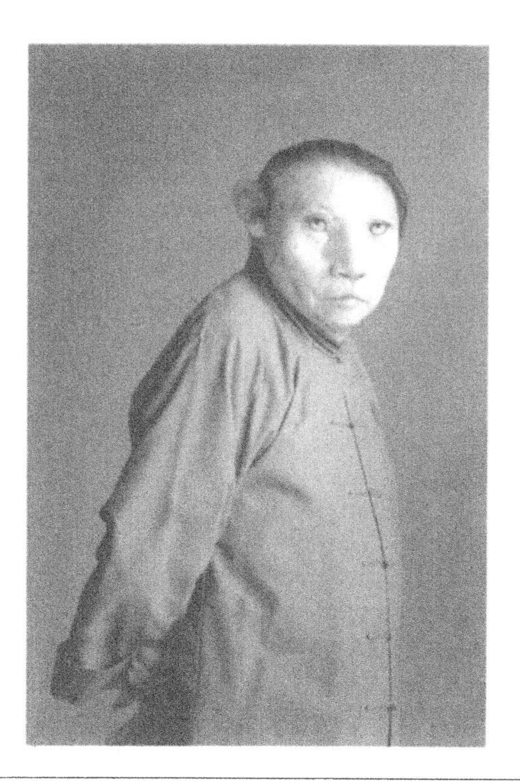

Fig. 14.8 Daniel Lee, *Judgment* (1994): *Juror No. 6 (Leopard Spirit)*. Digital C-Print, 25 × 37 inches. Courtesy the artist and O.K. Harris Works of Art, New York.

New York-based artist Daniel Lee has imagined transgenics from a very different angle than Kac. In his large-scale, digitally altered photographic series entitled *Judgment*, Lee has created a cast of underworld mythological creatures based on Chinese mythology—a jury of mythological chimeras who are half-animal, half-human hybrids. According to legend, there are 108 different creatures, including the human being, in the Chinese Circle of Reincarnation. Each of these creatures is judged after its death in a mythological court under the earth. The judge and jurors of this court try the dead souls to determine their destiny in the afterlife. Illustrated here is *Leopard Spirit* (1994: see Figure 14.8).

V. SUBJECT: Self, Family, and Human "Nature"

In the years since the cloning of Dolly the sheep, public discussions have turned away from the specter of human replicants to less frightening possibilities, such as the production of genetically identical tissue grown for people with Parkinson's and other diseases. Infertile couples wanting a biologically related child or grieving parents longing to replace a child they have lost now openly express the desire to clone.[6] Procloning bioethicists have embraced access to cloning as a "reproductive right," as just one among an array of assisted reproduction technologies. Currently, the cloning of mammals is an inefficient process that can require hundreds of attempts to create an embryo and implant it successfully. However, this does not preclude the possibility that human cloning is taking place somewhere in the world—other than the United States or in most of Europe and Japan, where legislative bans have been placed on such research. In the wake of such recent genomic possibility, notions of self must be reimagined.

Genomic developments have also thrown the idea of difference into relief. Population geneticists have demonstrated that there is no specific gene for race and that, in fact, human beings are genetically quite similar. Meanwhile, minor genetic differences between humans—such as race and ethnicity—remain among the most lethal and intractable causes of human conflict. What impact might increased knowledge of genetic human variation and the development of a precise scientific description of the biology of race have on human societies?

Kinship and other relations are also being defined anew. Kinship is a complicated cultural construction, one that is not necessarily based on biological relationships.[7] New reproductive practices are generating a variety of meanings, identities, and relationships with which we are only now beginning to grapple, but many are aligned with a constricted essentialist notion of biological kinship. What impact will these choices have on our notions of self and family? How might the increased emphasis on genetic kinship play against the many other forms of family we recognize in contemporary society?

Several artists have recently begun exploring the ways in which the new genomics allow us to redefine or re-imagine key aspects of human identity, including race, sex, sexuality and the "essential" aspects of human nature. The following are artworks included in a section entitled "SUBJECT."

Amsterdam-based artist Margi Geerlinks creates large-scale, digitally altered photographs that focus on the anxieties and anticipation surrounding genomics. In *Twins* (1998–1999: see Figure 14.9), an elderly woman appears to be wiping age from the face of her mirror image, her twin, herself. In this act she performs one of the perceived promises of the scientific revolution of genomics—the potential to exercise some control over the aging process and the degeneration of cells, even increasing the human life span.

Houston artist Dario Robleto has created an installation entitled *If We Do Ever Get Any Closer At Cloning Ourselves Please Tell My Scientist-Doctor To Use Motown Records As My Connecting Parts b/w The Polar Soul* (1999–2000: see Figures 14.10 and 14.11). This installation conflates our fantasies about human cloning and a high point in the history of American pop music—Motown. Robleto's installation is an imaginary cloning chamber in which the disc jockey samples your favorite Motown hits.

Many of the specimens in his laboratory were created from the artist's mother's record collection. His mother lovingly played these albums for years.

Fig. 14.9 Margi Geerlinks, *Twins* (2000). Fujichrome on Perspex and dibond, 50 × 67 inches. Courtesy Stefan Stux Gallery, New York and TORCH Gallery, Amsterdam.

Fig. 14.10 Dario Robleto, *If We Do Ever Get Any Closer At Cloning Ourselves Please Tell My Scientist-Doctor To Use Motown Records As My Connecting Parts b/w The Polar Soul* (1999–2000). Installation, 8 × 17 1/2 × 10 feet. Originally commissioned by ArtPace, A Foundation for Contemporary Art, San Antonio. Photo Credit: Seale Studios.

Fig. 14.11 Dario Robleto, *If We Do Ever Get Any Closer At Cloning Ourselves Please Tell My Scientist-Doctor To Use Motown Records As My Connecting Parts b/w The Polar Soul* (1999–2001). Detail of "Cloning Chamber."

The record grooves contain her skin cells and hair—minute biological residue making up a DNA library. Robleto melted these records, ground them to a fine powder, and then cast them into various shapes. In the cloning chamber, a video monitor shows disc jockey Justin Boyd playing an original, commissioned remix of the songs from the melted-down records. The artist's recuperation of his mother's genetic material, his infatuation with popular music culture, and his commentary on our fantasies about cloning offer a witty take on the genomic disc jockey revolution. Like a disc jockey or a geneticist, Robleto's practice is recombinant, taking samples from science and pop culture and splicing them together in new ways.

The Garden of Delights (1998: see Figure 14.12), is a group portrait created by Chicago artist Iñigo Manglano-Ovalle that engages our current complex understanding of the notion of family. It is composed of forty-eight individual DNA portraits, arranged in sixteen family groupings. Although the individual portraits are genetic, the members of these families are not necessarily bound by these genetics. Manglano-Ovalle has included families composed of a father,

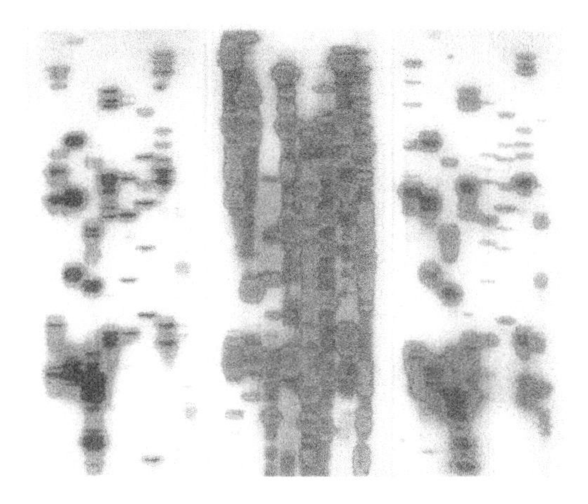

Fig. 14.12 Iñigo Manglano-Ovalle, *The Garden of Delights: Lu, Jack and Carrie* (1998). Cibachrome DNA print, 60 × 23 inches. Courtesy Max Protetch Gallery, New York. Photo Credit: Dennis Cowley.

a mother, and their biological offspring; gay partners with a child from a former marriage; a group of friends; a man and his two aliases. Thus, *The Garden of Delights* does not limit the notion of family to one based on genetic essentialism; instead, it embraces the wider array of families that society recognizes.

Manglano-Ovalle, who is of Spanish and Colombian heritage, selected with purpose the number 16 for his triptychs. In the eighteenth century, during a colonial period in which Spain was obsessed with maintaining racial purity, "scientific" caste paintings were produced showing the sixteen possible child-products of the mixed-race unions of indigenous peoples, African slaves, and "pure"-blooded Spaniards. Race and the role of science in the production of the colonial body, along with shifting notions of family are explored in this series of paintings.

VI. In Closing

That is just a sampling of the almost 100 artworks included in *Gene(sis)*. An extensive Website and CD-ROM exhibition catalogue document the exhibition, public programs, and dialogue activities in each city to which the exhibition travels. An educational "tool box" containing exhibition curriculum guides, guidelines for establishing public dialogue, and other program materials are also available to participating venues.

Gene(sis) promises to be a particularly effective and provocative exhibition for several reasons. It provides a unique cross-section of compelling new artwork created in response to recent genomic advances and debates. At the same time, it addresses concrete audience needs: finding a means of access to comprehending highly technical material with potentially profound social implications; fostering public dialogue among populations, legislators and specialists; and enabling high-level collaboration between the fine arts and hard sciences.

Notes

1. Celera Genomics puts the number at 26,558; The Human Genome Project consortium puts the number at about 31,000. See *The Economist* (Feb 17, 2001), 1.
2. Dr. Eric Lander, Whitehead Institute, Center for Genomic Research, quoted in *The New York Times* (September 12, 2000).
3. Lily E. Kay, *Who Wrote the Book of Life? A History of the Genetic Code* (Stanford: Stanford University Press, 2000), xv.
4. See Kay, *Who Wrote the Book of Life?*
5. "Medicine and the New Genetics," *Human Genome Project Information*. www.ornl.gov/hgmis/medicine/medicine.html.
6. See Margaret Talbot, "A Desire to Duplicate," *The New York Times Magazine* (February 4, 2001), 49–68. See also Brian Alexander, "(You)2," *Wired* (February 2001), 122–35.
7. See the discussion on genetics and kinship in Dorothy Nelkin, *The DNA Mystique: The Gene as Cultural Icon* (New York: W. H. Freeman and Company, 1995), 58–78.

The Editors and Contributors

Editors

Robert Mitchell is Assistant Professor of English at Duke University. His research focuses on eighteenth- and nineteenth-century theories of sympathy, as well as contemporary intersections between information technologies, genetics, and commerce. His essays have appeared in the edited collection *The Politics of Community* and the journals *1650–1850: Ideas, Aesthetics, and Inquires in the Early Modern Era* and *Romanticism on the Net*. He is coeditor, with Phillip Thurtle, of *Semiotic Flesh: Information and the Human Body* (Seattle: University of Washington Press, 2002).

Phillip Thurtle is Assistant Professor of Sociology and Anthropology at Carleton University in Ottawa. His research focuses on identity and biology in the American eugenics movement, the use of new media in the representation of popular science, and the material culture of information processing. His work has appeared in the *Stanford Humanities Review*, the *Journal of Immunology*, and the *Journal of the History of Biology* as well as the edited collection *Shifting Ground: Transformed Views of the American Landscape*.

Contributors

Richard Doyle is Associate Professor in English at Penn State University. He is the author of *On Beyond Living: Rhetorical Transformations of the Life Sciences* (Stanford CA: Stanford University Press, 1997) and *Wetwares! Experiments in Postvital Living* (Minneapolis, MN: University of Minnesota Press, 2003).

Mary Flanagan is a digital artist and cybercultural critic teaching in the Department of Film and Media Studies at Hunter College in New York. Her essays on digital art, technoculture, and gaming have appeared in periodicals such as *Art Journal*, *Wide Angle*, *Convergence*, and *Culture Machine*, as well as several books, and her coedited collection *Reload: Rethinking Women + Cyberculture* was published by MIT Press in 2002. She is also the creator of *The Adventures of Josie True*, the first Web-based adventure game for girls. Flanagan's computer-based artwork has shown internationally at venues including the Whitney Museum of American Art's Artport, New York Hall of Science, and the Moving Image Gallery, Auckland, NZ.

N. Katherine Hayles is Professor of English and Design/Media Arts at the University of California, Los Angeles, and teaches and writes on the relations of literature, science, and technology in the twentieth and twenty-first centuries. Her recent book, *How We Became Posthuman: Virtual Bodies in Cybernetics, Literature, and Informatics,* won several prizes, including the Rene Wellek Prize for the Best Book in Literary Theory, 1998–1999. Her most recent book is *Writing Machines* (MIT Press, 2002). She is currently at work on *Coding the Signifier: Rethinking Semiosis from the Telegraph to the Computer,* scheduled for completion this year.

Robin Held is Assistant Curator at the Henry Art Gallery at the University of Washington. She recently exhibited the show "*Gene(sis): Contemporary Art Explores Human Genetics.*"

Eduardo Kac is an Associate Professor of Art and Technology at the School of the Art Institute of Chicago as well as research fellow at the Centre for Advanced Inquiry in Interactive Arts (CAiiA) at the University of Wales, Newport, UK. Professor Kac is an international artist, acclaimed for his pioneering use of transgenics, telepresence, and biotelematics. Kac's work has been exhibited internationally at venues such as Exit Art and NY Media Arts Center, New York; OK Contemporary Art Center, Linz, Austria; InterCommunication Center (ICC), Tokyo; Chicago Art Fair; and the Julia Friedman Gallery, Chicago. He also has published his writings in numerous journals on art and media.

Elisabeth LeGuin is Assistant Professor of Music at the University of California at Los Angeles. One of the foremost Baroque cellists in the United States, Professor LeGuin is currently finishing her book *Bocherini: An Essay in Carnal Musicology.*

Timothy Lenoir is Professor of History and chair of the Program in History and Philosophy of Science at Stanford University. Lenoir is the author of *The Strategy of Life: Teleology and Mechanics in Nineteenth Century German Biology* (D. Reidel, 1982, paperback edition by the University of Chicago Press, 1989), *Politik im Tempel der Wissenschaft: Forschung und Machtausübung im deutschen Kaiserreich* (Campus Verlag, 1992), and *Instituting Science: The Cultural Production of Scientific Disciplines* (Stanford University Press, 1997). He is also editor of *Inscribing Science: Scientific Texts and the Materiality of Communication* (Stanford University Press, 1998).

Mark Poster is Professor of History at the University of California at Irvine. He is the author of numerous books, such as *The Second Media Age* (Blackwell, 1995), *The Mode of Information* (University of Chicago Press, 1990), *Cultural History and Postmodernity* (Columbia University Press, 1997). His most recent text is entitled *What's the Matter with the Internet?* (University of Minnesota Press,

2001). A collection of pieces old and new with a critical introduction by Stanley Aronowitz is published as *The Information Subject* (G & B Arts International, 2001).

Steve Tomasula is Assistant Professional Specialist in English at the University of Notre Dame. A renowned writer of art criticism and fiction, his work has been published in *Fiction International, Leonardo, Circa, Kunstforum*, and the *New Art Examiner.*

Anne C. Vila is Associate Professor in the Department of French and Italian, University of Wisconsin-Madison. Professor Vila is the author of *Enlightenment and Pathology: Sensibility in the Literature and Medicine of Eighteenth-Century France* (Johns Hopkins University Press, 1998) and articles on Diderot, Rousseau, Sade, Balzac, literature and medicine in eighteenth- and nineteenth-century France, the Encyclopédie, and cultural studies. She is currently preparing a book entitled *Virile Minds: Ambiguous Bodies: Thinkers in French Literature and Medicine, 1700–1840.*

Bernadette Wegenstein is Visiting Professor and Lecturer in the Department of Comparative Literature, SUNY Buffalo. She has published *Die Darstellung von AIDS in den Medien: Semio-linguistische Analyse und Interpretation* (Wiener Universitätsverlag, 1998) as well as articles for the *Journal of Semiotics* and for the edited collection *Reload: Rethinking Women and Cyberculture* (eds. by Mary Flanagan and Austin Booth, MIT Press, 2002).

Kathleen Woodward is Professor of English at the University of Washington where she also directs the Walter Chapin Simpson Center for the Humanities. The editor of *The Myths of Information: Technology and Postindustrial Culture* and coeditor of *The Technological Imagination: Theories and Fictions*, she is the author of *Aging and Its Discontents: Freud and Other Fictions* and is completing a book on the cultural politics of the emotions entitled *Statistical Panic and Other New Feelings.* Her essays on the broad crossdisciplinary topics of technology and culture, aging, and the emotions have appeared in *New American History, Discourse, differences*, and *Cultural Critique.*

Index

Printed in Great Britain
by Amazon